AGRICULTURE ISSUES AND POLICIES

SUGARCANE

PRODUCTION, CONSUMPTION AND AGRICULTURAL MANAGEMENT SYSTEMS

AGRICULTURE ISSUES AND POLICIES

Additional books in this series can be found on Nova's website under the Series tab.

Additional e-books in this series can be found on Nova's website under the e-book tab.

AGRICULTURE ISSUES AND POLICIES

SUGARCANE

PRODUCTION, CONSUMPTION AND AGRICULTURAL MANAGEMENT SYSTEMS

ELEANORE WEBB
EDITOR

nova publishers
New York

For permission to use material from this book please contact us:
Telephone 631-231-7269; Fax 631-231-8175
Web Site: http://www.novapublishers.com

Library of Congress Cataloging-in-Publication Data

Sugarcane : production, consumption and agricultural management systems / editor: Eleanore Webb.
pages cm. -- (Agriculture issues and policies)
ISBN: 978-1-61122-356-9 (hardcover)
1. Sugarcane. I. Webb, Eleanore. II. Series: Agriculture issues and policies series.
SB231.S85 2014
633.6'1--dc23 2014029140

Published by Nova Science Publishers, Inc. † New York

CONTENTS

PREFACE

Sugarcane is a C4, perennial, sucrose-storing grass belonging to the genus *Saccharum* (Arceneaux, 1965) that originated in Asia, and it is a cultivated crop in tropical and subtropical countries throughout the world. Among the countries cultivating sugarcane, Brazil is the largest producer. Sugarcane has been harvested for human and animal consumption for centuries, and in recent decades, it has been used for fuel production by juice fermentation (first-generation ethanol). The primary sugarcane by-products are molasses, used as ruminant feed and as a sugar substitute, and bagasse, a source of fibers for animal diets and bioelectricity. This book discusses the production, consumption and agricultural management systems of sugarcane.

Chapter 1 - Sugarcane has been harvested for human and animal consumption for centuries, and in recent decades, it has been used for fuel production by juice fermentation (first-generation ethanol). The primary sugarcane by-products are the molasses, used as ruminant feed and as a sugar substitute, and the bagasse, a source of fibers for animal diets and bioelectricity. Sugarcane bagasse is one of the largest agro-industrial lignocellulosic resources generated as a by-product of the sugar and ethanol industry. Such biomass is a raw material intended to be used for the industrial production of second-generation (2G) bioethanol. The 2G ethanol from sugarcane has aroused great interest due to the need to reduce pollutant emissions as well as achieve fossil-fuel independence and energy security. Moreover, the ethanol production comes from a renewable resource and has a socioeconomic impact by promoting increased job opportunities. Sugarcane bagasse is essentially composed of cellulose, hemicellulose, and lignin. These biocomposites are highly organized in the plant cell wall as a rigid structure that is critical for cell growth and morphogenesis. Lignocellulosic biomass has long been recognized as a recalcitrant material due to its high natural resistance to degradation and biological conversion. This property makes the use of untreated material economically infeasible. To overcome the biomass recalcitrance, pretreatment strategies must be applied to deconstruct the cellulose-lignin-hemicellulose organization and properties for it to be viable to break down the cellulose into fermentable sugars by cellulolytic enzymes. The recalcitrance of a lignocellulosic biomass is determined by its inherent properties such as the heterogeneity of the physicochemical and structural complexity, which collectively are responsible for the organization of the plant biomass at the macro-, micro-, and nanoscales. The limiting factors related to the biomass recalcitrance at the macroscale include the plant anatomy, tissue organization, and cell diversity. At the microscale, the chemical composition, lignin concentration and localization, cell wall thickness, and lignin sealing of the cellulose and hemicellulose sheathing contribute to the recalcitrance. At the nanoscale, the length and crystallinity of the cellulose microfibrils and the cell wall matrix porosity impair biomass biodegradability. This review provides an overview of the physicochemical and structural features of sugarcane to understand their contribution to its recalcitrance, highlighting their intrinsic heterogeneity and properties.

Chapter 2 - Sugarcane (*Saccharum sp.*) is a C4 grass cropped in more than 70 countries. As a result of its photosynthetic cycle, this plant is highly efficient in turning solar radiation into biomass. In Brazil, sugarcane is cropped in about 10 Mha to obtain mainly sugar and ethanol, the latter of which is considered by international agencies an "advanced biofuel". However, as a monoculture with a crop cycle of six years, sugarcane is grown with agricultural practices that can potentially affect SOM dynamics and consequently interfere with the carbon balance of sugarcane ethanol. In this chapter the authors reviewed available data on management practices in sugarcane production, focusing on soil organic carbon (SOC) dynamics and greenhouse gas (GHG) emission impacts. Over the past 20 years there were significant improvements introduced in sugarcane agro-systems. One of most important improvements was the modification from the pre-harvest "burning management" to "green harvest" with maintenance of the dry leaves and tops in the field. This practice has potential benefits to the agricultural system and to SOM dynamics. Recent studies have verified a soil C accumulation rate of 1.5 Mg ha^{-1} $year^{-1}$ and a potential reduction in N fertilization by 36% - 40% within 30 and 45 years after implementing a green harvest system. In this way, from a GHG perspective, the "green harvest" of sugarcane could reduce GHG emissions from N fertilizers whose emission factor has been estimated at

1.11% in plant cane and 0.76% in ratoon cane cultivated in Brazil. Tillage operation can also affect key biogeochemical processes associated with soil C and N cycling. Despite the fact that soil tillage is an agricultural activity only performed every 5 to 6 years in sugarcane, these operations increase the mineralization of soil organic carbon (SOC) and the emission of CO_2. Conventional tillage implemented during the reformation of the field can cause losses equivalent to 80% of the C potentially accumulated in the soil during one year of "green harvesting". Meanwhile, reduced and minimum tillage practices have smaller effects on CO_2 emissions, accounting for losses of 12% and 2% of that rate of accumulation, respectively. In order to meet the increasing demand for ethanol, Brazil should increase the planting area of sugarcane in upcoming years. Thus it is crucial to implement sustainable management practices in this agro-system that supports carbon accumulation, improves soil quality and minimizes GHG emissions from soils, thereby reducing the carbon footprint of the ethanol and increasing the environmental benefits from fossil fuels replacement with sugarcane ethanol.

Chapter 3 - This chapter approaches the economic valuation of environmental impacts related to soil erosion and silting-up of water streams, designed to allow the transfer of recovery costs to a policy of payment for ecosystem services. The aim of this study is to evaluate the contribution of silting-up mitigation to funding the environmental recovery of riparian areas found in sugarcane farms. The city selected for study is Arealva, located in the Central-West region of São Paulo State, Southeast of Brazil. Spatial dynamic models were conceived to simulate past land cover/land use changes (2005-2010) and future landscape scenarios (2010-2020) in the study area. The main observed changes that took place from 2005 to 2010 were: sugarcane expansion (6,012.71ha (49.68%)), mostly extending over grazing lands, and deforestation (3,107.16ha (22.33%)), predominantly converted into pastures. Three sets of scenarios were defined: i) stationary scenarios, in which the transition rates observed in former years were held constant (business as usual); ii) non-stationary scenarios with a partial recovery of environmentally protected areas along riversides (70% by 2015), and; iii) non-stationary scenarios with a full recovery of environmentally protected areas along riversides (100% by 2015). The regarded impacts are dependent on the estimated amount of lost soil, assessed by means of the Universal Soil Loss Equation (USLE). The authors also estimated the sediment accumulation rate in order to calculate siltation. The envisaged scenarios for environmental recovery can reduce environmental impacts up to 16% (US$41,479.29 to US$56,789.40) yearly. The riparian and alike vulnerable areas (prone to erosion and silting-up) can be recovered through a financing mechanism, relying either on water use charging or even on a taxation strategy implicitly considering the payment for ecosystem services. The silting mitigation would approximately contribute with US$13.83 to US$18.94 $ha^{-1}.year^{-1}$. In this way, sugarcane farms would have a financial incentive to restore and maintain the environmentally protected areas within their domain, reducing the environmental impacts related to silting-up processes.

Chapter 4 - The ethanol industry is of great importance to the Brazilian economy since the sugarcane is one of the most important monocultures in the country. Although its activities are regulated by numerous rules in order to minimize the environmental impacts, the sector is worried about the amount of waste that results from the sugar and alcohol production process. Among the waste from this process, the bagasse, filter cake and vinasse can be highlighted. The environmental impacts of using these waste/byproducts in the agriculture are still not completely elucidated and have been emphasized, mainly due to the adverse effects on the aquatic and soil environments. Thus, the present chapter will gather the available data regarding the use of bagasse, filter cake and vinasse, highlighting their chemical characteristics, the effects on the soil and aquatic environments due to its use in agriculture.

Chapter 5 - The sugarcane industry generates huge amount of sugarcane bagasse ash waste worldwide. The management of this solid waste material has resulted in increased economic, social and environmental concerns in the world. Over the years, the sugarcane ashes have been mainly disposed as soil fertilizer. More recently, the recycling of such solid waste into clay-based ceramics appear to be a viable economic and environmental option. The prospective benefits of using sugarcane bagasse ash waste to produce clay-based ceramics include conservation of natural resources and use of costless raw materials. This chapter presents a brief review on the valorization and reuse of sugarcane bagasse ash from the sugarcane industry in the production of sustainable clay-based ceramics for civil construction.

Chapter 6 - This paper proposes a method for the extraction of sugarcane region contour from a LANDSAT - TM image using image enhancement, recursive splitting technique by quadtree structure, region merging and non-linear anisotropic diffusion via Partial Differential Equation. In this context, the proposed methodology comprises preprocessing steps: Initially is produced a LANDSAT TM false-color image. After this, is applied an enhancement technique. This technique is based on the spatial domain. This enhancement operation use an adaptive average of the pixel value, based on a specifically function which adjusts the intensity of each pixel based on its relative magnitude with respect to the neighboring pixels. The recursive splitting technique using the quad tree structure consists of splitting the image into four homogeneous subregions of identical size. Each subregion is checked for homogeneity using a predefined threshold based on prior knowledge of objects presented in the scene. The splitting process proceeds recursively until no regions can be subdivided. In the end, the result is the input image organized according to the quad tree structure, where all homogeneous regions are explicitly represented. In order to meet the goal, the resulting regions are firstly structured by using the neighborhood structure. Next, the resulting regions are classified using similarity criteria, in this case regions presenting high probability of similarity are merged. The algorithm for contour filling is applied to the regions. The sugarcane contours are segmented using techniques such as, anisotropic diffusion detector that is used to previously focus the edge structure due to its notable characteristic in selectively smoothing the image, leaving the homogeneous regions strongly smoothed and mainly preserving the physical edges, i.e., those that are really related to objects presented on the image. The resulting regions are extracted by using techniques well-known, such as, vectorization, and polygonization. TM-LANDSAT images from 2008, bands 3, 4 and 5 were used. The results showed that the proposed methodology is promising for application involving extraction of cultures, because it has made possible the extraction of regions usually related to sugarcane culture.

Chapter 7 - Intensive efforts are being made to use the renewable lignocelluloses biomass for the production of energy and high value-added chemicals due to the global challenge of the face of depleting fossil carbon resources and growing concerns about environmental issues. Sugarcane bagasse is a by-product of the cane sugar industry, which consists of mainly tree polymeric components, namely cellulose (40-45%), hemicelluloses (xylan, 28-30%), and lignin (19-21%). The carbohydrate of sugarcane bagasse can be used to generate valuable products of commercial interest. In the period of the last few decades, carbohydrate resource of sugarcane bagasse as a promising biomass has been explored to produce bioethanol by many technologies mainly including pretreatment, hydrolysis and fermentation. The pretreatment to selectively fractionate components of the cell wall of sugarcane bagasse has the potential application for improving cellulose hydrolysis in the bioethanol production. Physical, chemical and biological treatments have been applied and the most potential pretreatment process is brought forward. The removal of hemicelluloses and lignin during the pretreatment could be as useful raw materials for preparing high value-added products like platform chemicals or biomaterials. The hydrolysis process for carbohydrate (cellulose and hemicelluloses) by biotechnology and non-biotechnology is summarized in view of economy, efficiency and environmental issues. The different fermentation technologies are described. Biotransformation can offer fantastic opportunities for the economic utilization of sugarcane bagasse in the production of ethanol, which displays sustainable, economic, environmental, and strategic advantages.

Chapter 8 - World Sugarcane production has more than 500 years of history and integrates the agricultural activities of growing, harvesting and transportation of sugarcane with industrial production in sugar mills and distilleries. The integration of the territory is needed as an agro-industrial cluster to solve logistical problems, productivity, innovation and new productions, as productive diversification and transition from conventional sugar mill and distilleries to biorefineries, to increase competitiveness. However, it has challenges related to low agricultural productivity derived from conventional practices of crop management, climate change and other ecological and socio-economic constraints such as vulnerability (pests, diseases, drought, etc.) and environmental impacts that are a risk to food security and the conversion to biorefineries. Therefore, at the present time, sugar industry and sugarcane crops are a potential source and epicenter of renewable energy, biofuels and bio-materials, as well as a food crop, but they are becoming more widely recognized as a source of rural livelihoods in developing countries and it will require a systematic effort, innovative and multidisciplinary methodologies of analysis to determine critical points that threaten the environmental and economic sustainability to improve the profitability and productivity with a reduction in the cost of production. This paper presents an approach or a conceptual framework as a new methodology for analysis based on existing knowledge of the sugar industry and the state of the art, for evaluating diversification using the multicriteria evaluation by analytical hierarchy process (AHP) as a tool suitable for analyzing complex systems, and for the identification of alternatives to the current situation and their discussion to facilitate decision making in the use of sugarcane as raw material in biorefineries to produce sugar based value-added products and derivatives as ethanol and cogeneration, bioplastic, etc. in cleaner production.

Chapter 9 - In this chapter a system of integrated harvest for sugarcane is put forward, with the inclusion of the agricultural residue. This is based on the need of sugar manufacturing companies to obtain new renewable energy alternatives, where the agricultural residue from sugarcane (ARS) can be used as a raw material for this purpose.

However, the high costs associated with obtaining it, make refineries resort to burning practices, thereby wasting a potential energy resource.

The system model the authors developed allows the efficient flow of the operation of the whole crop to be obtained, reducing the management costs of ARS with a major improvement in productivity in sugar manufacturing: linking the activities of collection, transport, and separation of sugarcane and residues in an integrated manner, thereby generating appropriate solution to the local logistic needs in Valle del Cauca state, the major sugarcane producer of Colombia, South America. It also allowed the authors to observe the performance of technological tools to ensure compliance with the demands of the sugarcane industry while at the same time obtain the agricultural residue to generate new products.

Chapter 10 - Dirt collected with sugarcane is processed and separated from the juice in the sugar factory by filtration equipment for return to the cane fields. New technologies over the past decade have enabled performance improvements to be obtained for this key unit operation. Filter mud product still contains a reasonable amount of sugar and the transportation of high moisture mud product has considerable cost. Australia's traditional approach has been to use Rotary Vacuum Filters for processing and separating mud and other impurities from juice, but in recent years there has been interest in reducing sugar losses and transportation costs through utilisation of new technologies such as Horizontal Bed Filters, Vacuum Belt Press Filters, Membrane Press Filters and Centrifuges. Increasingly, these alternative equipment are being installed in new factories. This chapter describes the general principles of mud filtration theory and mud conditioning followed by a detailed description and review of the various filtration technologies and analysis of the relative merits associated with the equipment.

Chapter 11 - Sustainable sugarcane production and processing requires intensification of benefits and minimization of both short term and long term loses. Identification of long term loses/benefits from sugarcane production and processing is a difficult venture that entails critical scientific analysis based on collected scientific data, historical events and laboratory experiments. Moreover, most companies do not invest in research activities geared towards identifying critical long term loses or benefits. The long term effects of agronomic activities in sugarcane farming are herein discussed. Disposal of processed and unprocessed wastes from sugarcane production and processing activities are also discussed with various possible technical solutions and scientific techniques of effectively generating profits from such wastes summarized. It is hoped that the diverse green technologies of sugarcane production and processing explored herein can be of significant contribution to the management of this vital sector of the economy.

Chapter 12 - Among the disposal means for vinasse, application to agricultural land is believed to represent the most sensible economic option from both the agronomic and environmental point of view. This belief stems from the numerous studies that have been conducted in sugar producing countries to determine the impacts of the vinasse, often at high doses, on soil quality, on the sugarcane plant and on groundwater quality. The vinasse,, is very variable in chemical composition but from analyses on samples collected at regular intervals of three months during 2005 to 2008 in Mauritius contain on average 9.37 g/L of K. Its fertilizer value as found everywhere is therefore mostly as a source of potassium. However at an application rate of for instance 100 m^3/ha, vinasse can in addition represent a significant source of N (average of 122 kg N/ha) and of organic matter (average of 8-15 % dry matter).

Apart from K, organic matter and N, vinasse contains heavy metals (Cu, Zn, Ni, Mn, Pb) but their concentrations are in general negligible and most often the heavy metals are below their detection limits on the atomic absorption spectrophotometer (5 mg/kg for Cu, Zn, Ni, Pb and 10mg/kg for Mn). Analyses of soils have shown that application of vinasse may, on the other hand, initially lower soil pH, e.g. from 5.9 to 5.4. but the soil pH will invariably return to its original value a few months afterwards. At high rates of the order of 100 m^3/ha, vinasse will in addition raise the electrical conductivity of the soil, but in spite of this increase, the electrical conductivity will remain below the threshold value of 1700 μS/cm accepted for sugarcane. Despite its high K content, analyses of soils have further showed that after its application, even at 100 m^3/ha, vinasse will have no adverse bearing on the exchangeable Ca status of the soils.

Field trials have often demonstrated that vinasse gives a higher cane yield than with NPK fertilizers alone. Additionally, because of its low heavy metal concentrations, vinasse would not increase the heavy metal concentration in the sugarcane plant.

Measurements of the effects on groundwater quality of applying vinasse to soil at high rates in different soil types and rainfall regimes moreover showed that the vinasse would not enhance the leaching loss of N in the form of nitrate. The heavy metals (Cu, Ni and Zn) known to be mobile, when they were detected in drainage water, would remain well below the drinking water limits proposed by the World Health Organization (1 mg/L for Cu, 5 mg/L for Zn and 0.02 mg/L for Ni).

Indeed the studies tend to show that application of high rates of vinasse is environment friendly and will not be to the detriment of the soil quality or of the sugarcane plant. In particular vinasse will not lead to any contamination of groundwater under sugarcane fields.

In: Sugarcane
Editor: Eleanore Webb
ISBN: 978-1-61122-356-9

Chapter 1

The Influence of the Heterogeneity, Physicochemical and Structural Properties on the Recalcitrance and Conversion of Sugarcane Bagasse

Celso Sant'Anna[1,2,•], Wanderley de Souza[1,2,3] and Michel Brienzo[4]

[1]Laboratory of Microscopy Applied to Life Science, Directory of Metrology Applied to Life Science - National Institute of Metrology, Quality and Technology - INMETRO, Duque de Caxias - RJ, Brazil
[2]The National Institute of Science and Technology for Structural Biology and Bioimaging, Rio de Janeiro, RJ, Brazil
[3]Federal University of Rio de Janeiro – UFRJ, Rio de Janeiro, RJ, Brazil
[4]Laboratory of Macromolecules, Division of Metrology Applied to life Science - National Institute of Metrology, Quality and Technology - INMETRO, Duque de Caxias - RJ, Brazil

Abstract

Sugarcane has been harvested for human and animal consumption for centuries, and in recent decades, it has been used for fuel production by juice fermentation (first-generation ethanol). The primary sugarcane by-products are the molasses, used as ruminant feed and as a sugar

• Tel: +55-21-2145-3150, Fax: +55-21-679-1420, E-mail: cbfilho@inmetro.gov.br.

substitute, and the bagasse, a source of fibers for animal diets and bioelectricity. Sugarcane bagasse is one of the largest agro-industrial lignocellulosic resources generated as a by-product of the sugar and ethanol industry. Such biomass is a raw material intended to be used for the industrial production of second-generation (2G) bioethanol. The 2G ethanol from sugarcane has aroused great interest due to the need to reduce pollutant emissions as well as achieve fossil-fuel independence and energy security. Moreover, the ethanol production comes from a renewable resource and has a socioeconomic impact by promoting increased job opportunities. Sugarcane bagasse is essentially composed of cellulose, hemicellulose, and lignin. These biocomposites are highly organized in the plant cell wall as a rigid structure that is critical for cell growth and morphogenesis. Lignocellulosic biomass has long been recognized as a recalcitrant material due to its high natural resistance to degradation and biological conversion. This property makes the use of untreated material economically infeasible. To overcome the biomass recalcitrance, pretreatment strategies must be applied to deconstruct the cellulose-lignin-hemicellulose organization and properties for it to be viable to break down the cellulose into fermentable sugars by cellulolytic enzymes. The recalcitrance of a lignocellulosic biomass is determined by its inherent properties such as the heterogeneity of the physicochemical and structural complexity, which collectively are responsible for the organization of the plant biomass at the macro-, micro-, and nanoscales. The limiting factors related to the biomass recalcitrance at the macroscale include the plant anatomy, tissue organization, and cell diversity. At the microscale, the chemical composition, lignin concentration and localization, cell wall thickness, and lignin sealing of the cellulose and hemicellulose sheathing contribute to the recalcitrance. At the nanoscale, the length and crystallinity of the cellulose microfibrils and the cell wall matrix porosity impair biomass biodegradability. This review provides an overview of the physicochemical and structural features of sugarcane to understand their contribution to its recalcitrance, highlighting their intrinsic heterogeneity and properties.

Keywords: Sugarcane, Sugarcane Bagasse, Recalcitrance, Cell Wall, Cellulosic Ethanol

1. INTRODUCTION

The current expansion of energy markets as a result of new energy and environmental policies enacted in the last decade in developed, and many developing, countries is leading to a recasting of the role of agriculture. Most significant is the growing role of the sector as a supplier of raw material for the production of liquid biofuels. Modern bioenergy represents a new source of demand for agricultural products, and approximately 85% of the global economy for liquid biofuel production is in the form of ethanol. Most of the bioethanol produced in the world is from sugarcane in Brazil and corn in the United States, which together are responsible for approximately 60% of the total production. The production of energy from ethanol derived from sugarcane is more efficient than that from corn or others energy crops such as sugar beets and palm/vegetable oils. (Goldemberg, 2008).

Sugarcane is a C4, perennial, sucrose-storing grass belonging to the genus *Saccharum* (Arceneaux, 1965) that originated in Asia, and it is a cultivated crop in tropical and subtropical countries throughout the world. Among the countries cultivating sugarcane, Brazil is the largest producer (ca. 40% of worldwide production), utilizing it to produce sugar, renewable energy resources such as ethanol, and electricity. Brazil has integrated industrial plants based on sugarcane agribusiness that have been developed over several decades. Currently, the Brazilian sugarcane industries combine sugar, ethanol, and electricity production by the integral use of the sugarcane by-products, the sucrose juice and bagasse. Sugarcane bagasse, the waste remaining after the sugar juice is extracted by crushing, is a fiber-rich material that can be used as a fuel in boilers to generate steam and electricity, so-called bioelectricity. Sugarcane also generates straw composed of the tops and leaves of the sugarcane stalks, which is currently left in the field for soil fertilization; however, it is intended to be used for energy generation (Goldenberg, 2008).

A unique feature of this plant is the accumulation of high concentrations of sucrose, approximately 0.7 M (Moore, 1995). Sugarcane is generally used for the production of sugar, generating approximately two-thirds of its world production (Lakshmanan et al., 2005). In addition to producing sugar, sugarcane has received considerable attention, particularly in Brazil, because the ethanol derived from it has low production costs, low pollutant emissions, and represents an important source for a renewable biofuel (Lakshmanan et al., 2005; Goldemberg et al., 2008). Thus, the possibility has emerged of sugarcane becoming a global trade commodity and an important energy resource.

In addition to the bioenergetic factors, the sugarcane agribusiness produces other by-products for end-use and intermediate feedstocks such as food for animal feed, brown sugar and brandy, vinasse for use as a fertilizer, plastic packaging, and bagasse for use in electricity generation in biorefineries, among others.

Important advantages are observed in sugarcane culturing: reduced water abstraction and effluent discharge due to the rational use and deployment of recycling techniques; less use of agrochemicals due to increased nutrient recycling and biological pest control; and direct social implications due to income generation and the large number of workers formally involved. From the point of view of resource availability for food production, the countries in the tropical region of the world have land available for both sustainable food production and bioenergy. It is currently estimated that approximately 1% of the world's arable land is used for biofuel production, with expectations of that increasing to 4% by 2030. One great advantage of using sugarcane as a bioenergy resource is that it does not directly compete with food production. Moreover, the Brazilian model for biofuel production from sugarcane is very competitive with petroleum-based fuels, which meets the socioeconomic and environmental concerns (Goldemberg, 2008).

The production of ethanol from lignocellulosic material involves five major steps: (1) preparation, (2) pre-treatment, (3) enzymatic hydrolysis, (4) fermentation, and (5) distillation (Mosier, 2005). For ethanol production, pretreatment step is essential to disassemble the lignocellulosic complex, resulting in the biomass becoming amenable to further chemical and biological pretreatments that make the cellulose more accessible for efficient enzymatic hydrolysis (Agbor et al., 2011). The enzymatic digestibility of native samples without a pretreatment produces a very low yield, less than 20% (Yang and Wyman, 2008).

The high degree of compaction and complexity of the lignocellulosic biomass structure makes its transformation into fermentable sugars much more difficult (Lynd et al., 2008), and the cost of ethanol production from lignocellulose remains high and is still prohibitive at the industrial scale. Nevertheless, technological advances have reduced the projected costs of bioethanol obtained by the enzymatic hydrolysis of cellulose. An effective pretreatment method is one that increases the accessibility of the cellulose and the complete solubilization of the sugar monomers without forming degradation products that inhibit fermentation. The choice of pretreatment has a direct effect on the cost and efficiency of the subsequent hydrolysis and fermentation processes. A single pretreatment may not be suitable due to the high diversity of the natural biomass in terms of its chemical, physical, and structural complexity. Thus, physical, chemical, and biological pretreatments, or their combinations, have been developed for an efficient hydrolysis process (Sun and Cheng, 2002; Mosier, 2005). The yield is related to a combination of the intrinsic characteristics and properties of the biomass source and the pretreatment method chosen (Agbor et al., 2011). Technological research efforts have been directed toward searching for the optimal methods to identify, evaluate, develop, and demonstrate the efficiency of processing biomass by enzymatic hydrolysis after pretreating it to overcome its recalcitrance.

The recalcitrance of the cell wall is determined by multiple factors of the plant tissue organization at the macro-, micro-, and nanoscales. Limiting factors that affect the enzymatic hydrolysis of biomass at the macroscale include the plant anatomy and cell types within the plant tissue. At the microscale, factors contributing to the recalcitrance include the chemical composition, concentration and location of the lignin, cell wall thickness, and the crosslinks between the cellulose, hemicellulose, and lignin. At the nanoscale, the limited porosity of the cell wall matrix and the degree of the cellulose microfibril crystallinity hinder the penetration of cellulase and the accessibility of the cellulose, thereby contributing to the recalcitrance of the biomass (Grethlein, 1985; Himmel and Picataggio, 2008; Donohoe et al., 2009). While the macroscale traits in sugarcane are well known, understanding the integrity of the cell wall constituents at the micro- and nanoscopic levels has been the subject of few studies.

Focusing on the biomass recalcitrance, the following discussions will review the physicochemical and structural features of sugarcane at the macro-, micro-, and nanoscale levels to understand their contributions to biomass recalcitrance and highlight their intrinsic properties and heterogeneity.

2. Macro Scale Recalcitrant Traits: Plant Anatomy, Tissue Organization and Cell Type Diversity

The sugarcane plant is formed by a root system, stalk, and leaves, which external part can be identified in Figure 1. The sugarcane stalk consists of a uniform set of structures with the typical features of monocot anatomy (Esau, 1977; Evert, 2006). The stalk is sectioned by internodes that are intercalated by nodes. The internode is a long cylinder where the sucrose is stored, and the node is the connection point between the internodes. The length and diameter of the stalk vary based on the plant variety and growing conditions. The internode surface is covered by wax, and each structure behaves as an independent unit during stalk growth with the leaves attached at the nodes.

Figure 1. General view of sugarcane plant.

The internode reaches its maximum elongation, cell wall thickening, and sucrose content before a leaf dies. The internodes ripen from the bottom up, with lower internodes ripening while the upper part of the stalk is still under development (Bull, 2000).

Moving from the periphery to the center, a cross-sectional view of a sugarcane stalk displays the general structures of the (i) epidermis, (ii) cortex, and (iii) ground tissue with embedded vascular bundles. The epidermis is the outermost, water-impermeable stalk structure responsible for protecting the internal stalk, and it is composed of a strong, dense, single layer of intercalated long and short cells that have a cuticular membrane and epicuticular waxes in the outer periclinal cell walls (Moore, 1987). The epidermis has been considered the most recalcitrant and is the larger fraction of sugarcane bagasse (Figure 2). The morphological, compositional, and structural characteristics of the epidermis make this fraction more resistant to acidic, thermal, and enzymatic treatment than the node and internode fractions (Brienzo et al., 2014). The internal cortex and ground tissue beneath the epidermis is primarily composed of sucrose-storing parenchyma cells and vascular bundles.

Figure 2. Sugarcane bagasse from sugarcane crush at ethanol/sugar industrial condition showing the higher concentration of epidermis fraction (arrows).

The cortex is comprised of two or three layers of thick-walled, highly lignified hypodermic cells, which contribute to strengthen the stalk. Between the cortex and ground tissue is a layer formed by small thin-walled cells. The ground tissue contains thin-walled parenchyma cells (0.6–1.2 μm wide) that form a storage tissue in which the vascular bundles are embedded. The vascular bundles are secondary helical thickenings (ca. 2.5 mm wide) surrounded by a thick layer of highly lignified and mechanically supportive sclerenchyma cells that form a protective structure (Sant'Anna et al., 2013) (Figure 3).

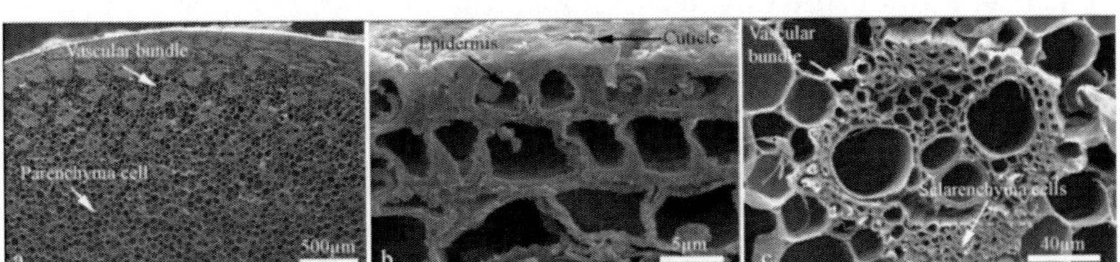

After Sant'Anna et al., 2013.

Figure 3. SEM images of sugarcane anatomical structure. a: Low magnification image showing the epidermis, parenchyma cells and vascular bundles. b: cuticular membrane and epicuticular waxes in the outer periclinal epidermal cell walls is observed. c: Image of vascular bundles in the fundamental parenchyma.

Relative to the internode (Figure 4a), the node region (Figure 4b) has a more dense shell of sclerenchyma cells, with the number of sclerenchyma cells elevated to a remarkable degree.

Due to the significant heterogeneity of the cell wall morphology and the high degree of lignification, the vascular bundles are the most recalcitrant plant structure. These differences in the vascular bundle morphology in the node and internode regions may significantly contribute to the higher recalcitrance of the node relative to the internode (Brienzo et al., 2014). From the standpoint of burning bagasse for energy production, this heterogeneity is not negative.

However, for ethanol generation from the bagasse, whether bioethanol or cellulosic ethanol, the heterogeneity becomes a negative factor affecting the pretreatment and processing design.

There are distinct differences in the sugarcane stalk node and internode recalcitrance that have been demonstrated by acidic pretreatments, enzymatic hydrolysis, and thermal degradation, and they aggravate the conversion process of the fermentable sugars into bioethanol, especially when the epidermis is considered (Brienzo et al., 2014).

3. Micro Scale Recalcitrant Traits: Chemical Composition, Lignin Localization, Cell Wall Structure, and the Lignin Seal Protecting the Cellulose and Hemicellulose Sheathing

3.1. Chemical Composition and the Lignin Seal Protecting the Cellulose and Hemicellulose Sheathing

Lignocellulosic biomass, including that of sugarcane, is organized in different specialized tissues that can have different cell wall structural organizations and compositions.

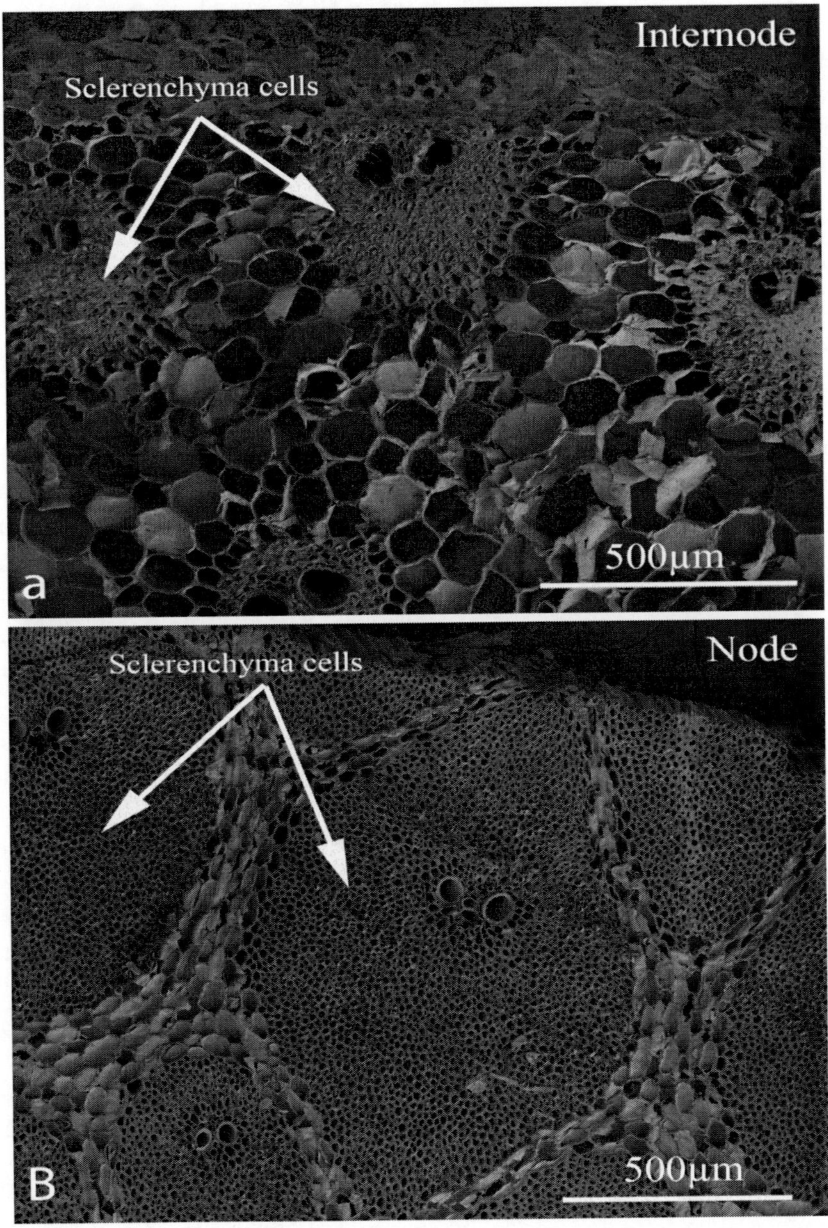

Figure 4. SEM images of internode (a) and node (b) regions showing the differential of vascular bundles. Note the larger number of sclerenchyma cells of node vascular bundles.

The plant cell wall encompasses the entire cell as a physical barrier that protects the interior contents. The cell wall also provides mechanical support and rigidity as a result of the highly organized layers and arrangements of the microfibril components embedded in the amorphous matrix. Lignocellulosic materials include cellulose, xylan, galactan, arabinan, mannan, pectin, lignin, and ash, and their chemical compositions vary based on the species, location, and harvest method, as well as the botanical fraction. The major cell wall component is cellulose, a homopolysaccharide formed by β-(1,4)-linked D-glucoses, which corresponds to 50–60% of the cell wall biomass (Perez and Mazeau, 2004). Cellulose is a water insoluble, high molecular weight, tertiary structure organized as nanofibrils that collectively are responsible for the rigidity and support of the plant cells. The molecule is organized with a non-rigid amorphous region and a highly ordered crystalline region. Due to its insolubility in water, crystallinity, and association with hemicellulose and lignin, cellulose is not an easily degraded compound. Therefore, for the glucose hydrolysis process of degradation, the synergistic action of three key enzymes is necessary: endo-1,4-β-D-glucanase (EC 3.2.1.4), exo-1,4-β-D-glucanase or cellobiohydrolase (EC 3.2.1.91), and β-glucosidase (EC 3.2.1.21) (Nidetzky et al., 1994). Hemicellulose is a low molecular weight, water soluble, amorphous heteropolysaccharide composed of a mixture of monosaccharides such as glucose, mannose, xylose, and arabinose (Saha, 2003). Pectin is an acidic, branched polysaccharide that retains a large amount of water and is the main constituent of the middle lamella. Lignin is an amorphous polymer associated with cellulose in the cell walls that has the functions of stiffening and waterproofing, as well as providing mechanical strength to the vegetable tissue (Donaldson, 2003). Lignin is the second most abundant polymer in most plants, and it represents the primary barrier to the enzymatic degradation processes that convert the biomass into sugar. For this reason, several pretreatment strategies have been developed for lignin removal to improve the saccharification process.

Sugarcane bagasse is composed primarily of cellulose, xylan, arabinan, lignin, and ash. The botanical fractions such as the leaves and stalk have different xyloglucan and arabinoxylan contents (De Souza et al., 2013), with the epidermis having less arabinoxylan and more cellulose and lignin than the internode and sucrose-free node of the stalk (Brienzo et al., 2014). The sugarcane bagasse, which has a heterogeneous stalk structure (node and internode covered by epidermis), has a wide range of average structural polysaccharide and lignin contents. Structural carbohydrate content ranges from 65% to 77% and lignin content ranges from 14% to 24%, depending on the variety of classical and precision breeding (Benjamin et al., 2013; Masarin et al., 2011). In general, the precision breeding varieties have more arabinoxylan and less lignin than the classical breeding varieties (Benjamin et al., 2013).

The chemical composition is changed by pretreatment with acidic/basic reagents, organic solvents, and physical processes and can be affected by the catalyst concentration, temperature, and reaction time (severity of the process). The more common assay for determining the composition precisely has been a two-step acid hydrolysis (72% and 4% H_2SO_4, sequentially), where the sugars are determined by liquid or gas chromatography and the lignin, gravimetrically. Fourier transform infrared (FT-IR) spectroscopy has been used as an analytical tool to qualitatively determine the chemical changes in the lignocellulosic material after pretreatment and to characterize the constituents of the plant biomass such as the lignin, extractives, hemicellulose, methoxyls, and aromatic hydroxyl groups (Faix, 1991). FT-IR spectroscopy can also be used to predict changes in the chemical composition of the lignocellulosic material. A relatively rapid technique, diffuse reflectance infrared Fourier transform spectroscopy, has been proposed as a faster and cheaper analysis method (Kelley et al., 2004) to predict the glucose, xylose, and lignin content with more than 90% accuracy (Meder et al., 1999). The system calibration needs a large number of samples with different compositions to create a set of large result ranges for correlation. Interpreting the spectral data is not trivial due to peak overlapping and broadening, and it requires a multivariate calibration to establish an association between the chemical data matrices and to calibrate the selected frequency values in relation to a chosen variable, such as a functional group. Principal components analysis has been used for FT-IR and near-infrared data to evaluate the chemical composition (Liu et al., 2010) and changes in the lignocellulose composition during biodegradation (Brienzo et al., 2007).

The aim of pretreatments is to reduce the lignin content, as it is known to negatively affect enzymatic hydrolysis. Monitoring the lignin content is a typical analysis performed to evaluate pretreatment effects. The more severe the pretreatment (e.g., acidic), the more product can be formed from the degradation of the sugars, such as furfural and hydroxymethylfurfural from xylose and glucose, respectively. One component that negatively affects enzymatic hydrolysis is the acetyl groups, which are side chains on the xylan backbone. Removing the acetyl group from the biomass enhances enzymatic hydrolysis (Chang and Holtzapple, 2000), but the acetyl groups present in the liquid during hydrolysis can also inhibit enzymatic activity. While furfural and hydroxymethylfurfural do not provoke enzymatic inhibition (Hodge et al., 2008), they can inhibit the fermentation step. In addition, high concentrations of phenol-derived products such as vanillin, syringaldehyde, trans-cinnamic acid, and hydroxybenzoic acid can inhibit the cellulose activity (Ximenes et al., 2010). Chromatography assays are the main tools utilized for quantifying the components of the lignocellulosic material and evaluating the pretreatment, enzymatic hydrolysis, and fermentation products. As a typical grass lignin, the lignin in sugarcane bagasse is composed of syringyl, guaiacyl, and small amounts of *p*-hydroxyphenyl. The lignocellulosic lignin is esterified with a small amount of *p*-coumaric alcohol, while ferulic acid has more etherified linkages than esterified (Sun et al., 2003). The basic compounds in lignin are composed of the same phenylpropanoid skeleton with different degrees of substitution on the phenyl ring. The *p*-hydroxyphenyl has no methoxy group, while guaiacyl has one, and syringyl has two. It is widely accepted that the higher the degree of methoxy-group substitutions, the less recalcitrant is the lignin. In fact, sugarcane bagasse submitted to a sequential pretreatment (alkali or alkaline peroxide) first released lignins rich in syringyl units, which contain more non-condensed ether structures, while the remaining lignins in the residual pretreated bagasse were rich in guaiacyl units that have more condensed structures (Sun et al., 2003). These results suggested that most of the easier lignin fraction was removed, while the much more recalcitrant lignin structure remained in the pretreated material.

3.2. Cell Wall Structure and Lignin Localization

To enable ethanol production from lignocellulosic material, deconstructing the plant cell wall is required. Moreover, it is necessary to study and develop methods for understanding the cell wall characteristics at different structural scales. The morphology of the cell wall and the lignin distribution are two morphological traits that contribute to the cell wall recalcitrance (Chundawat et al., 2011). The chemical composition of the sugarcane cell wall, which is well-understood (De Souza et al., 2012), does not describe the structural diversity. The structural information for the cell wall architecture needs to be defined in detail at the micro- and nanoscale levels. Although there are several microscopy studies describing fine details of the arrangement of the cell wall in other plants (e.g., Chundawat et al., 2011; Ding et al., 2012; Xu et al., 2006; Abdul Khalil et al., 2008; Abdul Khalil et al., 2010, Ma et al., 2011), surprisingly few publications are available describing the cell wall organization in sugarcane (De Souza and Sant'Anna, 2012; Sant'Anna et al., 2013, Abud et al., 2013). Consequently, the morphological traits affecting the recalcitrance of the cell wall relative to the micro- and nanoscopic levels needs to be investigated in detail.

Some examinations of the sugarcane cell wall organization by approaches such as light, electron, and atomic force microscopy have recently gained attention (Sant'Anna and De Souza, 2012; Sant'Anna et al., 2013, Abud et al., 2013). Microscopy has also been used to investigate the cell wall deconstruction after pretreatment and enzymatic hydrolysis (De Souza and Sant'Anna, 2012). The structural localization of the cellulose and lignin in the sugarcane tissues was determined by light microscopy using safranine, a fluorescent dye that can simultaneously reveal the localization of both molecules (Figure 5). Use of this strategy revealed a high concentration of cellulose in the secondary cell walls of the sugarcane tissue, while the lignin was shown to be densely concentrated in the cell corners and middle lamella (Sant'Anna et al., 2013). Coletta et al. (2013) investigated the lignin distribution in sugarcane bagasse and the delignification process by acid and alkali pretreatment using confocal and fluorescence lifetime imaging microscopy. According to these authors, the lignin is heterogeneously distributed in the biomass and is arranged by weak interactions in the cross-linked lignin. In addition, the lignin redistribution and redeposition on the cell wall's outermost external regions was found to be an effect of the biomass acid pretreatment.

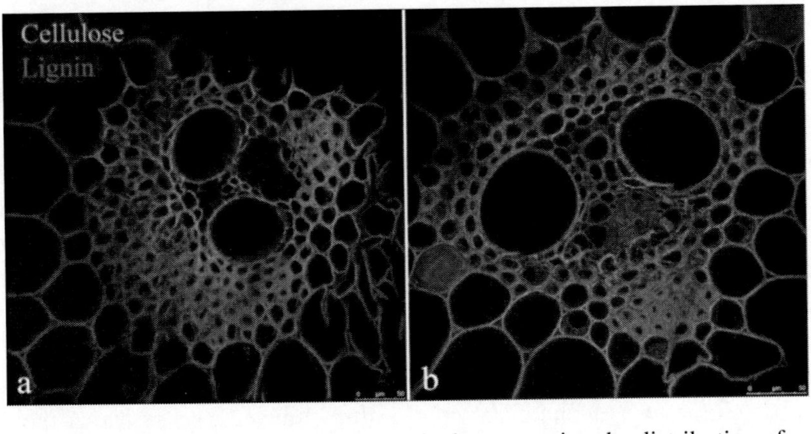

Figure 5. Safranin stained sugarcane internode demonstrating the distribution of cellulose and lignin in two vascular bundles (a, b). Red fluorescent signal showing the distribution of lignin and green fluorescent signal showing the distribution of cellulose.

According to these authors, the lignin is heterogeneously distributed in the biomass and is arranged by weak interactions in the cross-linked lignin. In addition, the lignin redistribution and redeposition on the cell wall's outermost external regions was found to be an effect of the biomass acid pretreatment. Moreover, there was a heterogeneous response by the biomass to the pretreatment as a function of the differential lignin distribution.". The figure and the figure legend should be inserted just such sentence and just before: "Chimenez et al. (2014) used the autofluorescence pattern of sugarcane bagasse to study the orientation of the lignin fraction and the effects of a bleaching treatment. Using this strategy, it was shown that lignin is longitudinally oriented in the bagasse fibers, and the pretreatment both decomposes the lignin components and partially modifies the lignin orientation.

Cellulose accessibility is a key factor in the efficient bioconversion of lignocellulosic biomass to fermentable sugars. The structural diversity of the cell walls in plant tissues, which is related to the specialized functions of the cell walls, influences the accessibility of the cellulose to cellulolytic enzymes (Donohoe et al., 2009). Consequently, the plant cell wall ultrastructure needs to be understood in order to correlate the structural organization (e.g., cell wall morphology and thickness) with the recalcitrance. Transmission electron microscopy (TEM) images reveal that the cell wall in sugarcane is divided into 3 layers that are strongly bonded together, giving rise to the typical dense architecture of cell walls: (1) a thinner, non-lignified primary cell wall; (2) a thicker, highly lignified secondary cell wall, which is subdivided into the S1, S2, and S3 layers; and (3) the middle lamella that provides support to the adjacent cells.

TEM images of ultra-thin sections of the sugarcane cell wall distinguished the secondary cell wall sublayers, and measurements showed that S2 is the thickest layer, with a thickness varying between 500 nm and 600 nm. The S3 sublayer varies from 300 nm to 400 nm, while the S1 sublayer is the thinnest, with a thickness of 200 to 300 nm (Sant'Anna et al., 2013). Scanning and atomic force microscopy images showed several lignin-rich particles on the surface of the sugarcane cell walls with diameters varying from 30 nm to 60 nm (mean diameter = 43 nm) as the result of a diluted acid treatment (De Souza and Sant'Anna, 2012; Abud et al., 2013). Ultrastructural cytochemistry using potassium permanganate staining in ultra-thin sections provided the clue that the lignin was extruding in droplets from the cell wall after the thermo-chemical treatment (Abud et al., 2013).

Regarding the physicochemical characteristics, little attention has been given to determining the structural aspects of the different fractions/regions of the sugarcane biomass or to the response of these fractions to different pretreatments based on their specific recalcitrance. For the first time, a comparison study of sugarcane fractions was recently conducted by scanning electron microscopy to investigate the morphological changes of the epidermis, node, and internode after an acid pretreatment (Brienzo et al., 2014). The morphological data showed that the epidermis was the least damaged fraction, followed by the node, and then the internode. In conjunction with the enzymatic hydrolysis yield analysis, it revealed that the internode was more susceptible to enzymatic activity than the node, while the epidermis was the most resistant.

4. Nanoscale Recalcitrant Traits: Physicochemical Properties, Cellulose Microfibril Arrangement, and Cell Wall Matrix Porosity

The challenge of biomass conversion has been attributed to its natural resistance, defined as recalcitrance. The natural cellulose structure is resistant to enzymatic hydrolysis primarily as a result of the highly ordered crystalline organization and the degree of polymerization, and the consequent characteristic of water-insolubility. It is secondarily resistant due to hemicellulose surrounding the microfibrils and covalently linking them to the lignin, thus reinforcing the structure and making the macromolecules a compact and rigid matrix. Beyond this complex organization, other factors exist that also contribute to the biomass recalcitrance, including physicochemical properties such as the cellulose accessibility and reactivity, which are influenced by the protective effects of the lignin and hemicellulose. The following sections will discuss some of the physicochemical properties of the biomass and cellulose microfibrils. These properties are evaluated for their impact on the biomass recalcitrance and conversion. Although physicochemical properties act at the nanoscale structure of the cell wall, their effects contribute to the characteristics of the material as a whole.

4.1. Accessible Surface Area

The accessibility of the material can be related to the internal/external surface area and the pores that exist in the cellulose material, which can be increased by the pretreatment effects of removing the hemicelluloses and lignin. The features of a porous surface are characterized by the pore shape, pore volume, and pore-size distribution, which can enhance the physical contact between the enzymes and substrate. Studies have tried to connect particle size and fiber porosity to specific surface area, and both are related to the enzymatic access to the substrate (Nahzad et al., 1995). The accessibility of the cellulose has been recognized as an important factor for the digestibility of lignocellulose substrates (Arantes and Saddler, 2011).The specific surface area is defined as the accessible area and its relation to the mass of material, and it is expressed as adsorbed substance or area per gram of fiber (g/g fiber or m^2/g fiber, respectively). The specific surface area can be measured by techniques such as dye interactions (Chandra et al., 2008); solute exclusion, determined with a series of differently sized molecular probes that do not interact with the fibers (Stone and Scallan, 1968); gas permeability (Carey et al., 1973); scanning electron microscopy (Chinga et al., 2002); light scattering (Springer et al., 2000); mercury intrusion (Rigby et al., 2002); nuclear magnetic resonance (^{1}H NMR) thermoporometry (Ishizawa et al., 2007); and fiber scanning colorimetry, which can be performed on wet samples (Park et al., 2006). The cellulose surface area can be estimated using nitrogen adsorption (Brunauer-Emmett-Teller method) with specific adsorption to the cellulose. However, the nitrogen also adsorbs to regions of the cellulose where the enzymes cannot reach. The Simon stain technique has been shown to be efficient in comparing the relative accessibility of pretreated samples, allowing a semi-quantitative estimate of the porosity (Chandra et al., 2008). Of the dyes used, Direct Blue has a smaller molecular size and weaker affinity to cellulose than Direct Orange, which preferentially adsorbs on the cellulose surface. However, due to its molecular size, the Direct Blue can penetrate areas that are inaccessible to Direct Orange (Yu and Atalla, 1995).

The accessible surface area of a lignocellulosic material is affected by the removal of the lignin and hemicellulose. The cellulose surface area can be exposed by removing the protective components, which are directly related to the enzymatic digestibility. The accessible surface area can be enhanced by removing the lignin and hemicellulose, and probably by decreasing the crystallinity, and thus they are correlated with the cellulose digestibility. Many researchers try to explain biomass recalcitrance by identifying the most influential factor of several, ignoring their relationships to one another. For example, an enzyme (liquid phase) needs physical contact with the cellulose molecule (solid phase) before a reaction can occur by the following steps: (i) adsorption of the enzymes on the cellulose surface; (ii) catalytic reaction releasing products such as oligomers and cellobiose; and (iii) desorption of the cellulase to the liquid phase (Sun and Cheng, 2002). Clearly, the higher the degree of cellulose exposure to the enzymatic activity, the better will be the enzymatic hydrolysis yield. In fact, pretreatment disrupts or disorganizes the lignocellulosic structure to promote pores and voids as the hemicellulose and lignin are removed or modified, thereby contributing to more cellulose exposure. The cellulose accessibility in lignocellulosic material can be determined by the internal/external surface area. The external surface area of the cellulose is related to its exposure, which is influenced by chemical and physical barriers such as the amount of lignin and hemicellulose present and the particle size. The internal surface area depends on the capillary structure of the fibers and pores produced by the pretreatment. The internal surface area is normally less than the external surface area, but it is dependent on the pretreatment effects. The fibers can swell during pretreatment with polar solvents and thereby increase the internal surface area. However, drying the fibers can reduce the internal surface area as a consequence of the collapse of the previously formed capillaries and voids. The pores or capillaries in the fibers are an important avenue for the chemicals and enzymes to penetrate the biomass, and their collapse decreases the physical exposure of the cellulose with a consequent reduction in the enzymatic digestibility (Luo and Zhu, 2011).

The internal/external surface areas are influenced by the pretreatment conditions. Bagasse from different varieties of sugarcane that were pretreated with a dilute acid adsorbed a range of 70–170 milligrams of dye per gram of fiber. The acid-pretreated internode, node, and epidermis fractions adsorbed 310, 105, and 80 mg/g, respectively (Brienzo et al., 2014), which explains the different responses to pretreatment by the different sugarcane bagasse varieties and their fractions, and demonstrates the effects of biomass heterogeneity.

4.2. Arrangement of the Cellulose Microfibrils

The arrangement of the cellulose microfibrils is a biomass trait related to the cell wall recalcitrance at the nanoscale. This arrangement has been well studied in others crops, including corn stover, by atomic force microscopy

(AFM) (Yarbrough et al., 2011). AFM measurements performed on corn provided evidence to suggest that a given pretreatment should focus on the lignin removal, while retaining the native microfibrillar structure of the cell wall (Ding et al., 2012). Despite the importance of the nanoscale aspects related to its resistance to degradation, only recently has the microfibrillar architecture of the sugarcane cell wall been elucidated (Abud et al., 2013). In that study, internodal parenchyma cell wall fragments were obtained for the first time by subcellular fractionation and were imaged by AFM (Figure 6a, b). The nanostructural pattern of the sugarcane cell wall was determined to be overlapping layers of randomly orientated elementary microfibrils with diameters varying from 15–21 nm, which is similar to values found for cellulose microfibrils in other crops (McCann et al., 1990; Kirby et al., 1996; Morris et al., 1997). The response of the microfibrillar arrangements to acid pretreatment was also investigated on the surface of parenchyma cell walls (Abud et al., 2013). The AFM image revealed a heterogeneous effect on the cell wall arrangement (Figure 6c-e). In some regions, the general structure of the cell wall appeared unchanged (Figure 6c), while there was a loss of cellulose microfibrils in other regions, accompanied by the removal of cell wall matrix in the form of globular structures (Figure 6d). The formation of the globular structures as an effect of the pretreatment may lead to pore formation, and the enhancement of matrix porosity improves the ability of the cell wall to be digested by cellulases (Chundawat et al., 2011; Corrales et al., 2012). Other regions had a complete loss of the cell wall microfibrillar array (Figure 6e). These findings suggest there are cell wall regions with more or less recalcitrance, which may be related to the local cellulose microfibril orientation and lignin concentration.

However, it is important to mention that the AFM analysis was performed on a sugarcane cell type with low lignification. To the best of our knowledge, the thick, highly lignified cell walls of sugarcane sclerenchyma cells, for example, were not investigated for their cellulose microfibril organization or their response to pretreatment. In addition, a comparison between the cell wall nanostructure from the node and internode regions was not done. There is a vast field remaining to be scientifically explored by the bioethanol industry.

4.3. Cellulose Crystallinity

Cellulose microfibrils contain regions with highly and lowly organized structure, defined as crystalline and amorphous regions, respectively. In the organization of cellulose microfibrils, it is well known that hydrogen bonds link the cellulose molecules (intra- and inter-chain), promoting the supramolecular structure with a crystalline region. Native cellulose is a polymorphic structure defined as cellulose I, which treatments can then convert to polymorphs II, III, and IV. Native cellulose is found in the form of two crystalline phases known as I_{α} and I_{β} (Atalla and Vander Hart, 1984).

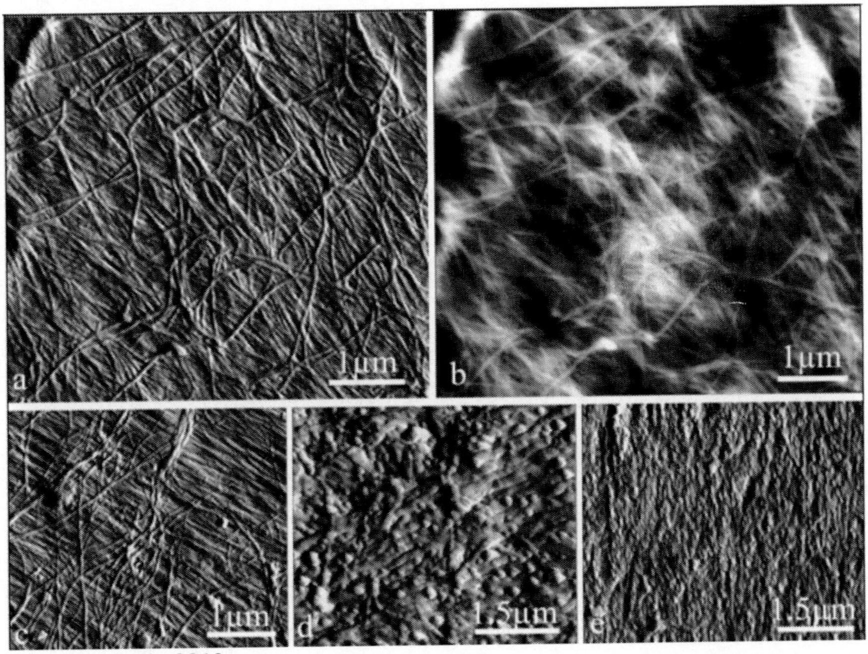

After Abud et al., 2013.

Figure 6. Topographic (a) and height (b) AFM images of the cell wall fragments showinh the arrangement of cellulose microfibril in the different layers. (c) AFM image of sulfuric acid pretreated sugarcane cell wall demonstrating undisturbed cell wall region. (d) Cell wall showing filament loss and the formation of globular structures. (e) Region of the cell wall that was strictly affected sulfuric acid pretreatment. Note the absence of filaments.

The crystallinity of a lignocellulosic material depends on factors such as the nature of the crystal lattice, proportion of crystalline region, and the size and orientation of the crystals (Ward, 1950). The crystalline and amorphous cellulose materials have different reactivities that negatively influence the enzymatic digestibility, particularly the initial hydrolysis rate (Chundawat et al., 2011).

Moreover, a lower enzymatic hydrolysis rate can be predicted for later stages after the amorphous fraction solubilization, when the material is rich in the crystalline fraction (Fan et al., 1980). The crystalline cellulose is less accessible for enzymatic digestion than the amorphous cellulose, because the crystallinity affects the efficiency of the enzyme's contact with the cellulose. The crystallinity has a direct effect on the enzymatic hydrolysis of lignocellulose, but the effect is negative (Yoshida et al., 2008; Sinitsyn et al., 1991; Thompson and Chen, 1992). Enzymes can degrade amorphous cellulose five to ten times faster than crystalline cellulose (Gama et al., 1994), as indicated by most of the amorphous component being removed during the first stage of hydrolysis and the initial rate of enzymatic hydrolysis being affected by decreased cellulose crystallinity (Laureano-Perez et al., 2005). The crystallinity and specific surface area are closely associated, and the swollen cellulose can have an increased surface area, suggesting an increased adsorptive capacity, as well. The adsorption of endoglucanase on crystalline and amorphous cellulose reportedly differs, supporting the notion that the two substrates have different reactivities (Klyosov et al., 1986). Hall et al. (2010) observed the degradation of the cellulose structure in two fractions (crystalline and amorphous) and suggested the decrease in the hydrolysis rate could be explained by changes in other factors such as the crystallinity. The crystallinity index of a lignocellulosic material is frequently measured by X-ray diffraction, and different methods for its calculation have been used such as peak height (Segal et al., 1959), subtraction of the amorphous spectrum as background (Chung and Scott, 1973), and area under the peak (Hermans and Weidinger, 1948) (Figure 7). An alternative is ^{13}C NMR, but this requires an extensive acquisition time to resolve the peaks and is not applicable to low levels of crystallinity (Bansal et al., 2010).

The crystallinity index measurements can also be influenced by variation in the sample drying conditions (Weimer et al., 1995) and the presence of residual proteins or cells (Converse, 1993).

Studies using pure cellulose have shown that some pretreatments can decrease the cellulose crystallinity. However, the cellulose crystallinity cannot be properly measured if the cellulose is not isolated or pure. With a lignocellulosic material, an X-ray measure of the crystallinity is of all the constituents of the lignocellulosic material, including the lignin and hemicelluloses that contribute as the amorphous fraction. Thus, there are contradictions in the literature regarding the correlation of the enzymatic hydrolysis yields with the crystallinity index, which differs between pure cellulose (negative effect) and lignocellulosic material (positive effect). Some pretreatments increase the crystallinity of the biomass by removing the amorphous components. However, the pretreatment also effectively enhances the hydrolysis by increasing the accessibility of the enzyme to the cellulose by removing the lignin and hemicelluloses that formed a physical barrier to the cellulolitic enzymes. Pretreatments such as diluted acid and steam explosion can improve the enzymatic hydrolysis, but the biomass crystallinity increases as a function of the severity employed (Saddler et al., 1982).

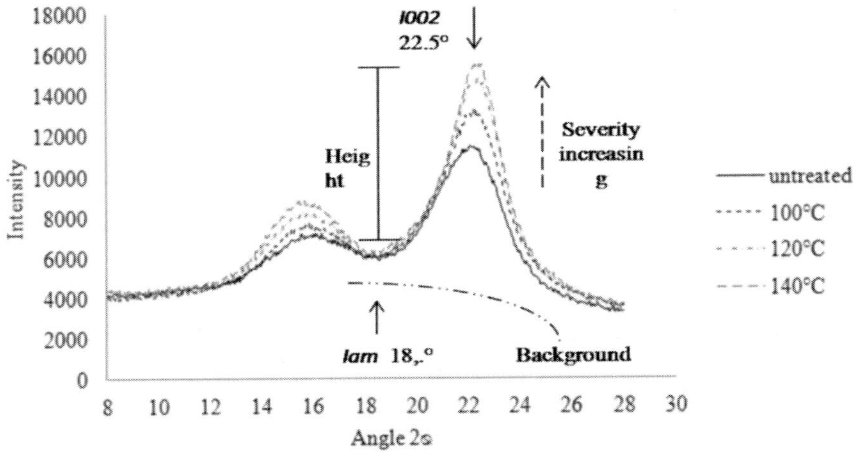

Figure 7. X-ray diffraction of acid-pretreated sugarcane bagasse. *I002* shows the maximum intensity at 2Θ = 22.5°. *Iam* represents the minimum intensity (2Θ = 18.7°) used to calculate the crystallinity index based on the peak height. The dotted line represents the background.

The sugarcane bagasse crystallinity index has been reported in the literature as ranging from 35% to 50%, with sensible increase variations based on the pretreatment type and its severity. The crystallinity indexes of the untreated sugarcane epidermis, internode, and node fractions have been reported as 51.12%, 45.82%, and 42.78%, respectively (Brienzo et al., 2014). Some sugarcane bagasse pretreatments and their typical crystallinity index values are steam explosion with CO_2 and SO_2 at 56.4% and 65.5%, respectively (Corrales et al., 2012); peracetic acid at 62% (Zhao et al., 2008); and sodium hydroxide at 66% (Rezende et al., 2011). Briefly, the crystallinity index of a biomass can be reduced with ball milling and ionic liquids, and it can be increased with acidic, alkaline, and oxidative auto-hydrolysis or steam explosion.

4.4. Degree of Cellulose Polymerization

The number of glucose units in a cellulose molecule defines its degree of polymerization (DP). Cellulose is a linear homopolymer of anhydroglucose units linked by β-1,4 linkages with a DP up to 20,000. Native or extracted (isolated) cellulose is polydispersive, containing molecules with different numbers of glucose units, and thus different DPs and molar masses. The cellulose polysaccharide is insoluble at a high DP, but short cello-oligomers (cellodextrins) with DPs between 2 and 6 are soluble, while those between 6 and 12 are slightly soluble (Klemm et al., 1998). Cellulases are multicomponent enzyme systems that act synergistically to hydrolyze cellulose. Endoglucanase and cellobiohydrolase act directly on the cellulose fibers, releasing short cello-oligomers and cellobiose, respectively. Cellobiohydrolase is most efficient on crystalline structures and releases cellobiose from the cello-oligomer hydrolysis, while β-glucosidase hydrolyses cellobiose to glucose (Gan et al., 2003).

Reducing the cellulose DP causes an increase in the number of reducing ends, and more accessible chains are available to the actions of the enzymes, specifically exoglucanase. The exoglucanase acts on the chain ends and does not cause a significant decrease in the DP. Endoglucanase acts on the interior part of the cellulose chain and can decrease the DP (Pala et al., 2007). Similar to crystallinity, the DP is not an independent factor of the biomass recalcitrance, since altering the DP by a pretreatment can influence the chemical composition or other physicochemical properties. The relationship between the physical and chemical factors reveals the complexity of the lignocellulosic structure and requires a lot of structural information to discern. In this context, the DP correlates with the information regarding what happens to the cellulose after pretreatment. The decreases in the rate of hydrolysis have been attributed to the increase in recalcitrance that can occur from structural changes to the substrate. The primary changes are the result of the pore size and crystallinity, as well as the DP (Sinitsyn et al., 1991). However, it is not easy to predict a substrate's digestibility prior to hydrolysis for samples with differences in the crystallinity index and cellulose DP (Ramos et al., 1993). The prediction of an enzyme's hydrolysis must be made considering several physicochemical parameters, because isolated parameters such as the crystallinity or other factors are insufficient (Mosier et al., 2005). The DP of cellulose can be determined using high-performance size-exclusion (gel permeation) chromatography (Kennedy et al., 1995), laser light scattering detection (Pang and Rudin, 1992), and chemical methods based on measuring the reducing power of the insoluble fibers, where the DP is determined by the relationship between the total cellulose and the insoluble reducing power (Pala et al., 2007). Although each of the methods can be used to provide similar results for the DP, more sensitive assays, such as those for the enzyme mode of action or depolymerization, require gel permeation chromatography to provide more details. Before the analysis, the cellulose is dissolved without any modification to the chain length in solvents such as metal complex solutions (cupriethylenediamine), organic solvents, inorganic acids (nitric acid), and ionic solutions (*N*,*N*-dimethylacetamide/LiCl). Zhang and Lynd (2005) evaluated the chemical method for DP measurements and found a phosphoric acid treatment to be necessary for swollen filter paper cellulose, while it was unnecessary for substrates such as avicel (microcrystalline cellulose). Although the method is simple, it has the disadvantages of suffering from interference if protein is present on the medium and incomplete accessibility to the chain ends of the insoluble cellulose (Kongruang et al., 2004).

The common method to determine the DP of cellulose is a viscosity measurement that uses cupriethylenediamine as the solvent in a capillary viscometer. The viscosities determined as centipoises are converted to the DP of the cellulose with the formula $DP^{0.905} = 0.75\cdot[954\cdot\log(X) - 325]$, where X is viscosity in centipoises. For the cellulose DP measurement of a lignocellulosic biomass, a prior delignification should be done, such as with acetic acid (10%) and sodium chlorite (5%) (Mazumder et al., 2000). The viscosity correlates with the molecular weight and depends on the concentration and temperature. A common solvent for the DP measurement dissolves the cellulose in organic solvents or acids.

The cellulose molecule is generally partially cleaved during extraction, with a consequent increase in the number of reducing ends and a reduction in the DP. The DP values of pretreated sugarcane bagasse can vary significantly according to the process severity (i.e., temperature, reaction time, and catalyst concentration).

Some pretreatments of sugarcane bagasse and the resulting cellulose DP are listed: peracetic acid, 1030 to 1550 (Zhao et al., 2008); and supercritical CO_2, alkaline, and ozone steam explosion, 600 to 1100 (Puri, 1984).

The challenge of converting sugarcane bagasse is attributed to the morphology and structural features, chemical composition, and the physicochemical properties. These properties that correlate to the recalcitrance can be minimized by generating a bagasse with less heterogeneity by dissecting the epidermis from the sugarcane to provide a less recalcitrant stalk (Figure 8).

Figure 9 shows an alternative route for reducing the bagasse heterogeneity and recalcitrance by removing the epidermis previous to the crush. The removed epidermis could then be pretreated using the proper conditions or used in energy cogeneration.

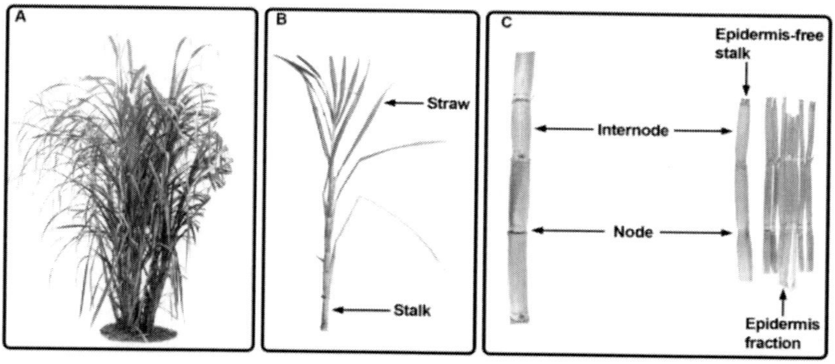

Figure 8. Sugarcane stalk and leaf system showing the heterogeneity of the plant tissue. A) Sugarcane clump of several stalks; B) Stalk and straw (leaves); C) Node and internode stalk fractions, epidermis and epidermis-free stalk.

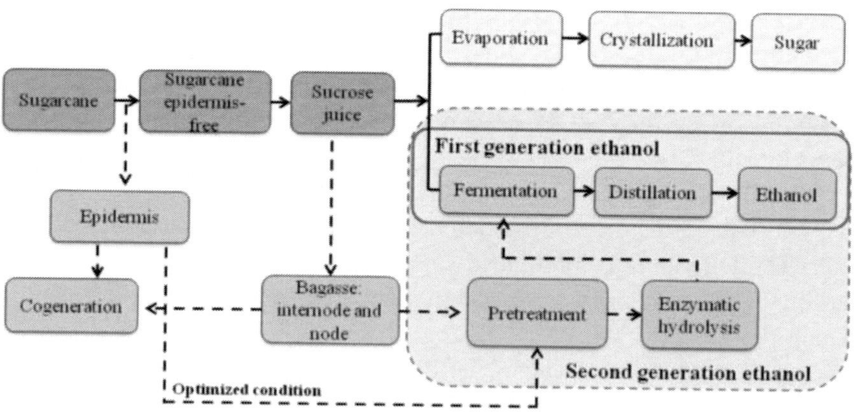

Figure 9. Flow diagram of ethanol (first and second) generation and proposed preliminary epidermis removal to decrease the bagasse heterogeneity and recalcitrance. The epidermis integrates into the process by energy cogeneration or an optimized pretreatment prior to 2G ethanol conversion. Dashed arrows indicate the alternative route optimization to produce 2G ethanol and cogeneration.

CONCLUSION

Sugarcane is a monocot plant organized into different systems that have different tissues, cell types, organizations, and functions. The organization of a sugarcane tissue is defined by the cells types and their composition. The sugarcane stalk has a highly developed strategy to store sucrose in the parenchyma cells of the inner stalk region, while the sclerenchyma fibers and vascular bundles provide the structural support and rigidity of the plant. The distribution of the cells differs from one tissue type to another, thus explaining the biomass heterogeneity. Sugarcane is a heterogeneous material formed by specialized structures such as the epidermis, node, and internode, and each sugarcane fraction has a specific organization and function in the plant. Moreover, these fractions have specific morphologies and physicochemical properties that were developed by the plant during its evolution, and to overcome these natural barriers and achieve their disassembly and deconstruction, specific chemical or physical pretreatments are necessary. High energy inputs are needed in the process to make the biomass conversion possible, and sometimes this is cost prohibitive for industrial applications. The pretreatment to break down the lignocellulosic structure must deal with the different fractions (epidermis, node, and internode) of the sugarcane bagasse, because each fraction has a different response to the pretreatment. Given the different degrees of recalcitrance of the fractions, the epidermis will be less damaged by the pretreatments than the node and internode. The diverse responses to a given pretreatment make the conversion process of fermentable sugar into ethanol ineffective.

Moreover, it can be hypothesized that the less recalcitrant fractions deconstructed in the early stages, such as the sugar degradation, have the potential to inhibit compound formation. Given this scenario, strategies to reduce the bagasse heterogeneity and consequent recalcitrance are necessary to produce 2G ethanol. Dissecting the epidermis from the sugarcane before extracting the juice to render a less recalcitrant stalk appears to be a good strategy to optimize the production of 2G ethanol from sugarcane. The resulting material fractions (epidermis, node, and internode) can be subjected to individual pretreatments according to their intrinsic recalcitrance. Furthermore, the high heating value of the epidermis gives this fraction great value as a fuel for energy cogeneration (bioelectricity).

ACKNOWLEDGMENT

The authors thank the Brazilian Innovation Agency (FINEP), the Brazilian Council for Research and Development (CNPq) and National Institute of Metrology, Quality and Technology (Prometro grant number 550105/2012-8) for financial support.

REFERENCES

Abdul Khalil, H. P. S., Siti Alwani, M., Ridzuan, R., Kamarudin, H., Khairul, A., 2008. Chemical composition, morphological characteristics, and cell wall structure of Malaysian oil palm fibers. *Polym. Plast. Technol. Eng.*, 47, 273–280.

Abdul Khalil, H. P. S, Ireana Yusra, A. F., Bhat, A. H., Jawaid, M., 2010. Cell wall ultrastructure, anatomy, lignin distribution, and chemical composition of Malaysian cultivated kenaf fiber. *Ind. Crop. Prod.,* 31, 113–121.

Agbor, V. B., Cicek, N., Sparling, R., Berlin, A., Levin, D. B. 2011. Biomass pretreatment: Fundamentals toward application. *Biotechnol. Adv.*, 29:675–685.

Arantes, V., Saddler, J. N., 2011. Cellulose accessibility limits the effectiveness of minimum cellulase loading on the efficient hydrolysis of pretreated lignocellulosic substrates. *Biotechnol. Biofuels*, 4, 3.

Arceneaux, G. 1967. Cultivated sugarcanes of the world and their botanical derivation. *Proc. Int. Soc. Sugar Cane Technol.*, 12, 844–854.

Atalla, R. H., Vander Hart, D. L., 1984. Native cellulose: A composite of two distinct crystalline forms. *Science,* 223, 283-285.

Bansal, P., Hall, M., Realff, M. J., Lee, J. H., Bommarius, A. S., 2009. Modeling cellulase kinetics on lignocellulosic substrates. *Biotechnol. Adv.,* 27, 833-848.

Benjamin, Y., Cheng, H., Görgens, J. F., 2013. Evaluation of bagasse from different varieties of sugarcane by dilute acid pretreatment and enzymatic hydrolysis. *Industrial Crops and Products*, 51, 7-18.

Brienzo, M., Silva, E. M., Milagres, A. M. F., 2007. Degradation of eucalypt waste components by *Lentinula edodes* strains detected by chemical and near-infrared spectroscopy methods. *Appl. Biochem. Biotechnol.*, 141, 37-50.

Brienzo, M., Ferreira, S., Vicentim, M. P., Souza, W., Sant'anna, C., 2014. Comparison study on the biomass recalcitrance of different tissue fractions of sugarcane culm. Bioenergy Research. In press.

Bull, T., 2000. The sugarcane plant. In: Manual of cane growing, Hogarth, M., Allsopp, P., eds. Bureau of sugar experimental stations, Indooroopilly, Australia.

Carey, C. L., Leekley, R. M., Hultman, J. D., Nagel, S. C., 1973. Determination of pore size distribution of pigment coatings. *Tappi J.*, 56, 134-138.

Chandra, R., Ewanick, S., Hsieh, C., Saddler, J. N., 2008. The characterization of pretreated lignocellulosic substrates prior to enzymatic hydrolysis, part 1: A modified Simons' staining technique. *Biotechnol. Prog.*, 5, 1178-1185.

Chang, V. S., Holtzapple, M. T., 2000. Fundamental factors affecting biomass enzymatic reactivity. *Appl. Biochem. Biotechnol.*, 84-86, 5-37.

Chinga, G., Helle, T., Forseth, T., 2002. Quantification of structure details of LWC paper coating layers. *Nordic Pulp Paper Res.*, J. 17, 313-318.

Chimenez, T. A, Gehlen, M. H., Marabezi, K., Curvelo, A. A. S. Cellulose. 2014. *Characterization of sugarcane bagasse by autofluorescence microscopy,* 21, 653-664.

Chundawat, S. P. S., Bellesia, G., Uppugundla, N., Sousa, L., Gao, D., Cheh, A., 2011. Restructuring the crystalline cellulose hydrogen bond network enhances its depolymerization rate. *J. Am. Chem. Soc.*, 133, 11163-11174.

Chung, F. H., Scott, R. W., 1973. A new approach to the determination of crystallinity of polymers by X-ray diffraction. *J. Appl. Cryst.,* 6, 225-230.

Coletta, V. C., Rezende, C. A. Conceição, F. R, Polikarpov, I., Guimarães, F.E.G. 2013. Mapping the lignin distribution in pretreated sugarcane bagasse by confocal and fluorescence lifetime imaging microscopy. *Biotechnol. Biofuels*, 6, 43.

Converse, A. O., 1993. Substrate factors limiting enzymatic hydrolysis. In: Saddler J. N. Bioconversion of forest and agricultural plant residues. Wallingford: CAB International.

Corrales, R. C. N. R., Mendes, F. M. T., Perrone, C. C., Sant'Anna, C., Souza, W., Abud, Y., Bom, E. P. S., Ferreira-Leitão, V., 2012. Structural evaluation of sugar cane bagasse steam pretreated in the presence of CO_2 and SO_2. *Biotechnol. Biofuels*, 5, 36.

De Souza, A. P., Leite, D. C. C., Pattathil, S., Hahn, M. G., Buckeridge, M. S., 2013. Composition and structure of sugarcane cell wall polysaccharides:

Implications for second-generation bioethanol production. *Bioenerg. Res.*, 6, 564-579.

De Souza, W., Sant'Anna, C., 2012. Microscopy as a tool to follow deconstruction of lignocellulosic biomass. In: Current microscopy contributions to advances in science and technology. Antonio Méndez-Vilas (ed), Espanha, Formatex Research Center, pp 639 – 645.

Ding, S. Y., Liu, Y. S., Zeng, Y., Himmel, M. E., Baker, J. O., Bayer, E. A., 2012. How Does Plant Cell Wall Nanoscale Architecture Correlate with Enzymatic Digestibility? *Science,* 338, 1055-1060.

Donaldson, L. A. 2001. Lignification and lignin topochemistry — an ultrastructural view. *Phytochemistry,* 57: 859 - 873.

Donohoe, B. S., Selig, M. J., Viamajala, S., Vinzant, T. B., Adney, W. S., Esau K. 1977. Anatomy of seed plants, 2nd ed. New York: Wiley.

Evert R. F. 2006. Esau's plant anatomy: Meristems, cells and tissues of the plant body. Their structure, function and development, 3rd ed. New Jersey: Willey, Hoboken.

Faix, O., 1991. Classification of lignins from different botanical origins by FT-IR spectroscopy. *Holzforschung*, 1991, 45, 21-27.

Fan, L. T., Lee, Y., Beardmore, D. H., 1980. Mechanism of the enzymatic hydrolysis of cellulose: Effects of major structural features of cellulose on enzymatic hydrolysis. *Biotechnol. Bioeng.*, 22, 177-199.

Gama, F. M., Teixeira, J. A., Mota, M.. 1994. Cellulose morphology and enzymatic reactivity: a modified solute exclusion technique. *Biotechnol. Bioeng.*, 43, 381-387.

Gan, Q., Allen, S. J., Taylor, G. 2003. Kinetic dynamics in heterogeneous enzymatic hydrolysis of cellulose: an overview, an experimental study and mathematical modelling. *Process Biochem.*, 38, 1003-1018.

Goldemberg, J. The Brazilian biofuels industry. 2008. *J. Biotechnology for Biofuels,* 1, 6.

Grethlein, H. E. 1985. The effect of pore size distribution on the rate of enzymatic hydrolysis of cellulosic substrates. *BioTechnology,* 3, 155-160.

Hall, M., Bansal, P., Lee, J. H., Realff, M. J., Bommarius, A. S., 2010. Cellulose crystallinity – a key predictor of the enzymatic hydrolysis rate. *FEBS Journal*, 6, 1571-1582.

Hermans, P., Weidinger, A., 1948. Quantitative X-ray investigations on the crystallinity of cellulose fibers. A background analysis. *Journal of Applied Physics,* 19, 491-506.

Himmel, E. M., Picataggio, S. K. 2008. In: Himmel, E. M. editor. Biomass Recalcitrance: Deconstruction the Plant Cell Wall for Bioenergy. United Kingdom, UK: Blackwell Publishing Ltd, 1-6.

Himmel, M. E., 2009. Detecting cellulase penetration into corn stover cell walls by immuno-electron microscopy. *Biotechnol. Bioeng.,* 103, 480–489.

Hodge, D., Karim, M., Schell, D., Macmillan, J., 2008. Soluble and insoluble solids contributions to high-solids enzymatic hydrolysis of lignocellulose. *Bioresource Technology*, 99, 8940-8948.

Ishizawa, C. I., Davis, M. F., Schell, D. F., Johnson, D. K., 2007. Porosity and its effect on the digestibility of dilute sulfuric acid pretreated corn stover. *J. Agric. F. Chem.,* 55, 2575-2581.

Kelly, S., Rials, T., Snell, R., Groom, L., Sluiter, A., 2004. Use of near infrared spectroscopy to measure the chemical and mechanical properties of solid wood. *Wood Sci. Technol.*, 38, 257-276.

Kennedy, J. F., Rivera, Z. S., Lioyd, L. L., Warner, F. P., Silva, M. P. C., 1995. Determination of the molecular weight distribution hydroxyethylcellulose by gel permeation chromatography. *Carbohydrate polymers,* 26, 31-34.

Kirby, A. R., Gunning, A. P., Waldron, K. W., Morris, V. J., Ng, A., 1996. Visualization of plant cell walls by atomic force microscopy. *Biophys. J.,* 70, 1138–1143.

Klemm, D., Philipp, B., Heinze, T., Heinze, U., Wagenknecht, W., 1998. Comprehensive Cellulose Chemistry. Wiley-VCH Verlag GmbH: Berlin, Germany.

Klyosov, A. A., Mitevich, D. V., Sinitsyn, A. P., 1986. Role of the activity and adsorption of cellulose in the efficiency of the enzymatic hydrolysis of amorphous and crystalline cellulose. *Biochemistry,* 25, 540-542.

Kongruang, S., Han, M. J., Breton, C. I. G., Penner, M. H., 2004. Quantitative analysis of cellulose. *Appl. Biochem. Biotechnol.*, 213, 113-116.

Lakshmanan P, Geijskes R. J., Aitken K. S., Grof C. L. P., Bonnett G. D., Smith G. R. 2005. Sugarcane biotechnology: The challenges and opportunities. *In Vitro Cell. Dev. Biol. Plant*, 41, 345–363.

Laureano-Perez, L., Teymouri, F., Alizadeh, H., Dale, B. E., 2005. Understanding factors that limit enzymatic hydrolysis of biomass: characterization of pretreated corn stover. *Appl. Biochem. Biotechnol.,* 121-124, 1081-1099.

Liu, L., Ye, X. P., Womac, A. R., Sokhansanj, S., 2010. Variability of biomass chemical composition and rapid analysis using FT-NIR techniques. *Carbohydr. Polym.*, 81, 820-829.

Luo, X., Zhu, J. Y., 2011. Effects of drying-induced fiber hornification on enzymatic saccharification of lignocelluloses. *Enz. Microb. Technol.*, 48, 92-99.

Ma, J. F., Yang, G. H., Mao, J. Z., Xu, F., 2011. Characterization of anatomy, ultrastructure and lignin microdistribution in Forsythia suspensa. *Ind. Crop Prod.,* 33, 358–363.

McCann, M. C., Wells, B., Roberts, K., 1990. Direct visualization of cross-links in the primary plant cell wall. *J. Cell Sci.,* 96, 323–334

Masarin, F., Gurpilhares, D. B., Baffa, D. C. F., Barbosa, M. H. P., Carvalho, W., Ferraz, A., Milagres, A. M. F., 2011. Chemical composition and enzymatic digestibility of sugarcane clones selected for varied lignin content. *Biotechnology for Biofuels*, 4, 55.

Mazumder, B. B., Ohtani, Y., Cheng, Z., Sameshima, K., 2000. Combination treatment of kenaf bast fiber for high viscosity pulp. *Journal of Wood Science,* 46, 364-370.

Meder, R., Gallagher, S., Mackie, K. L., Bohler, H., Meglen, R. R., 1999. Rapide determination of the chemical composition and density of *Pinus radiate* by PLS modeling of transmission and difuse reflectance FTIR spectra. *Holzforschung,* 53, 261-266.

Moore, P. H., 1987. Anatomy and morphology. In: Heinz, D. J. (Ed.), Sugarcane Improvement Through Breeding. Elsevier, Amsterdam, pp. 273–311.

Morris, V. J., Gunning, A. P., Kirby, A. R., Round, A., Waldron, K., Ng, A., 1997. Atomic force microscopy of plant cell walls, plant cell wall polysaccharides and gels. *Int. J. Biol. Macromol.*, 21, 61–66.

Mosier, N., Wyman, C., Dale, B., Elander, R., Lee, Y. Y., Holtzapple, M., Ladisch, M., 2005. Features of promising technologies for pretreatment of lignocellulosic biomass. *Bioresour. Technol.*, 96, 673-686.

Nahzad, M. M., Ramos, L. P., Paszner, L., Saddler, J. N., 1995. Structural constraints affecting the initial enzymatic hydrolysis of recycled paper. *Enzyme Microb. Technol.*, 17, 68-74.

Nidetzky, N., Steiner, W., Hayn, M., Claeyssens, M. 1994. Cellulose hydrolysis by the cellulases from Trichoderma reesei: a new model for synergistic action. *Biochem. J.*, 298, 705–710.

Pala, H., Mota, M., Gama, F. M., 2007. Enzymatic depolymerisation of cellulose. *Carbohyd. Polym.*, 68, 101-108.

Pang, S., Rudin, A., 1992. Use of continuous viscometer and light scattering detector in characterization of polyolefin's: comparisons of data for individual and combined detectors. *Journal of Applied Polymer Science*, 46, 763-773.

Park. S., Venditti, R., Jameel, H., Pawlak, J. 2006. Changes in pore size distribution during the drying of cellulose fibers as measured by differential scanning calorimetry. *Carbohydr. Polym.*, 66, 97-103.

Perez, S., Mazeau, K. 2004. Conformation, structure and morphologies of cellulose. In: Polysaccharides: Structural Diversity and Functional Versatility, Dimitriu (ed), Macel Dekker, pp. 41-68.

Puri, V. P. 1984. Effect of crystallinity and degree of polymerization of cellulose on enzymatic saccharification. *Biotechnol. Bioeng.*, 26, 1219-1222.

Ramos, L. P., Breuil, C., Saddler, J. N. 1993. The use of enzyme recycling and the influence of sugar accumulation on the cellulose hydrolysis by *Trichoderma cellulases*. *Enzyme Microb. Technol.*, 15, 19-25.

Rezende, C. A., Lima, M. A., Maziero, P., Azevedo, E. R., Garcia, W., Polikarpov, I., 2001. Chemical and morphological characterization of sugarcane bagasse submitted to a delignification process for enhanced enzymatic digestibility. *Biotechnology for Biofuels*, 2011, 4:54.

Rigby, S. P., Fletcher, R. S., Riley, S. N. 2002. Determination of the multiscale percolation properties of porous media using mercury porosimetry. *Ind. Eng. Chem. Res.*, 41, 1205-1226.

Saddler, J. N., Brownell, H. H., Clermont, L. P., Levitin, N. 1982. Enzymatic hydrolysis of cellulose and various pretreated wood fractions. *Biotechnol. Bioeng.,* 24, 1389-1402.

Saha, B. C. 2003. Hemicellulose bioconversion. *Journal of Industrial Microbiology and Biotechnology,* 30,279-291.

Sant'Anna, C., Cost, L. T., Abud, Y., Biancatto, L., Miguens, F. C., De Souza, W. 2013. Sugarcane cell wall structure and lignin distribution investigated by confocal and electron microscopy. *Microsc. Res. Tech.,* 76, 829–834.

Segal, L., Creely, J. J., Martin, A. E. J, Conrad, C. M. 1962. An empirical method for estimating the degree of crystallinity of native cellulose using the X-ray diffractometer. *Tex. Res. J.,* 29, 786-794.

Sinitsyn, A. P., Gusakov, A. V., Vlasenko, E. Y. 1991. Effect of structural and physico-chemical features of cellulosic substrates on the efficiency of enzymatic hydrolysis. *Appl. Biochem. Biotechnol.*, 30, 43-59.

Springer, R., Raether, F., Caps, R., Manara, J. 2000. In situ measurement of light scattering in porous ceramics during sintering. *High Temperatures-High Pressures,* 32, 385-390.

Stone, J. E., Scallan, A. M. 1968. A structural model for the cell wall of water swollen wood pulp fibers based on their accessibility to macromolecules. *Cellulose Chemistry and Technology*, 2, 343-358.

Sun, Y., Cheng, J., 2002. Hydrolysis of lignocellulosic materials for ethanol production: A review. *Bioresource Technol.*, 83, 1-11.

Sun., Sun, X. F., Sun, R. C., Fowler, P., Baird, M. S., 2003. Inhomogeneities in the chemical structure of sugarcane bagasse lignin. *J. Agric. Food Chem.,* 51, 6719-6725.

Thompson, D. N., Chen, H. C., 1992. Comparison of pretreatment methods on the basis of available surface area. *Bioresour. Technol.*, 39, 155-163.

Ward, K. J. R., 1950. Crystallinity of Cellulose and Its Significance for the Fiber Properties. *Tex. Res. J.*, 20, 363-372.

Weimer, P. J., Hackney. J. M., French, A. D. 1995. Effects of chemical treatments and heating on the crystallinity of celluloses and their implications for evaluating the effect of crystallinity on cellulose biodegradation. *Biotechnol. Bioeng.,* 48, 169-178.

Ximenes, E., Kim, Y., Mosier, N., Dien, B., Ladisch, M. 2010. Inhibition of cellulases by phenols. *Enzyme Microb. Technol.*, 46, 170-176.

Xu, F., Zhong, X. C., Sun, R. C., Lu, Q., 2006. Anatomy, ultrastructure and lignin distribution in cell wall of Caragana Korshinskii. *Ind. Crop. Prod.,* 24,186–193.

Yang, B., Wyman, C. E. 2008. Pre-treatment: the key to unlocking low-cost cellulosic ethanol. *Biofuels Bioprod. Bioref.*, 2:26–40.

Yarbrough, J. M., Himmel, M. E., Ding, S. Y., 2011. Plant cell wall characterization using scanning probe microscopy techniques. *Bioenerg. Res.*, 102, 11097–11104.

Yoshida, M., Liu, Y., Uchida, S., Kawarada, K., Ukagami, Y., Ichinose, H., Kaneko, S., Fukuda, K., 2008. Effects of cellulose crystallinity, hemicellulose, and lignin on the enzymatic hydrolysis of *Miscanthus sinensis* to monosaccharides. *Biosci. Biotechnol. Biochem.*, 72, 805-810.

Yu, X., Minor, J. L., Atalla, R. H., 1995. Mechanism of action of Simons' stain. *Tappi J.,* 78, 175-180.

Yuri, A., Costa, L. T., De Souza, W., Sant'Anna, C. 2013. Revealing the Microfibrillar Arrangement of the Cell Wall Surface and the Macromolecular Effects of Thermochemical Pretreatment in Sugarcane by Atomic Force Microscopy. *Ind. Crops and Prod.,* 51,62–69.

Zhang, Y. H. P., Lynd, L. R., 2004. Toward an aggregated understanding of enzymatic hydrolysis of cellulose: Non-complexed cellulase systems. *Biotechnol. Bioeng*., 88, 797-824.

Zhao, X., Wang, L., Liu, D.-H., 2008. Peracetic acid pretreatment of sugarcane bagasse for enzymatic hydrolysis: a continued work. *J. Chem. Technol. Biotechnol.*, 83, 950-956.

In: Sugarcane
Editor: Eleanore Webb

ISBN: 978-1-61122-356-9

Chapter 2

SUGARCANE CROP MANAGEMENT IN BRAZIL: IMPACT ON SOIL ORGANIC CARBON DYNAMICS

***A. M. Silva-Olaya*[1], *L. A. Frazão*[2] *and F. F. C. Mello*[3]**
[1]Universidade de São Paulo - ESALQ-USP, Piracicaba, SP. Brazil
[2]Universidade Federal de Minas Gerais – ICA-UFMG, Montes Claros, MG, Brazil
[3]Instituto CNA, Módulo K, Brasília, DF, Brazil

ABSTRACT

Sugarcane (*Saccharum sp.*) is a C4 grass cropped in more than 70 countries. As a result of its photosynthetic cycle, this plant is highly efficient in turning solar radiation into biomass. In Brazil, sugarcane is cropped in about 10 Mha to obtain mainly sugar and ethanol, the latter of which is considered by international agencies an "advanced biofuel". However, as a monoculture with a crop cycle of six years, sugarcane is grown with agricultural practices that can potentially affect SOM dynamics and consequently interfere with the carbon balance of sugarcane ethanol. In this chapter we reviewed available data on management practices in sugarcane production, focusing on soil organic carbon (SOC) dynamics and greenhouse gas (GHG) emission impacts. Over the past 20 years there were significant improvements introduced in sugarcane agro-systems. One of most important improvements was the modification from the pre-harvest "burning management" to "green harvest" with maintenance of the dry leaves and tops in the field. This practice has potential benefits to the agricultural system and to SOM dynamics. Recent studies have verified a soil C accumulation rate of 1.5 Mg ha^{-1} $year^{-1}$ and a potential reduction in N fertilization by 36% - 40% within 30 and 45 years after implementing a green harvest system. In this way, from a GHG perspective, the "green harvest" of sugarcane could reduce GHG emissions from N fertilizers whose emission factor has been estimated at 1.11% in plant cane and 0.76% in ratoon cane cultivated in Brazil. Tillage operation can also affect key biogeochemical processes associated with soil C and N cycling. Despite the fact that soil tillage is an agricultural activity only performed every 5 to 6 years in sugarcane, these operations increase the mineralization of soil organic carbon (SOC) and the emission of CO_2. Conventional tillage implemented during the reformation of the field can cause losses equivalent to 80% of the C potentially accumulated in the soil during one year of "green harvesting". Meanwhile, reduced and minimum tillage practices have smaller effects on CO_2 emissions, accounting for losses of 12% and 2% of that rate of accumulation, respectively. In order to meet the increasing demand for ethanol, Brazil should increase the planting area of sugarcane in upcoming years. Thus it is crucial to implement sustainable management practices in this agro-system that supports carbon accumulation, improves soil quality and minimizes GHG emissions from soils, thereby reducing the carbon footprint of the ethanol and increasing the environmental benefits from fossil fuels replacement with sugarcane ethanol.

1. INTRODUCTION

Sugarcane (*Sacharum officinarum L*) is a crop originally from New Guinea that grows in tropical and subtropical regions of the world on both sides of the equator, between approximately 35° N and 35° S (Gomes & Lima, 1964).

With a C4 photosynthetic cycle highly efficient in converting solar radiation into biomass, sugarcane is cultivated in more than 70 countries occupying an area of 26 Mha with a total production of 1,832 million tons yearly (FAO, 2014).

In Brazil sugarcane has been cropped since the colonial period, beginning in 1515 when the plants were brought from Madeira Island (Cheavegatti-gianotto et al., 2011). After a long development process, Brazil has become the world's largest sugarcane producer; approximately 10 Mha are currently cultivated, accounting for 39% of the world production (FAO, 2014).

With a variable cycle, which depends of local climatic conditions, variety and cultural practices, sugarcane is replanted in Brazil every five to six years when yields decline and reach an economically unfeasible level (Matsuoka; Garcia & Calheiros, 1999). Generally, the first harvest occurs 12 or 18 months after planting. The following ratoon cane harvests are made once in a year until field renewal (Macedo; Seabra & Silva, 2008).

Due to its vigorous grown, photosynthetic efficiency and a production system that often includes the use of crop residues to generate power or processing mills, sugarcane is the most attractive feedstock for bioethanol production (Galdos et al., 2010). In Brazil production increased from about 70 million tons processed in 1974/1975 to 561 million tons in 2011/2012. This increment was a result from land expansion as well as from considerable improvements in the sugarcane production chain, including the development of plant varieties.

The production and use of sugarcane ethanol in Brazil began in 1975 with the launch of the Alcohol Program (Pro-Álcool), which resulted in the development of ethanol-fueled vehicles and, more recently, in flex-fuel vehicles. Currently sugarcane is the main source of renewable energy in Brazil, promoting the offset of about ~40% of gasoline needs (BEB, 2013) and contributing for the reduction of greenhouse gases (GHG) emissions.

Rapid growth of the sugarcane and ethanol industry has been accompanied by an increase in the area occupied by this crop. In the last five years almost three million hectares of sugarcane were added to the production system (CONAB, 2008, 2013) reaching a total planting area that represents 14% of total area harvested in the country (IBGE, 2014).

Two main production regions are recognized in Brazil: Northeast and South-Central. The latter region was responsible for the production of 550 million tons in 2010, which represented 90% of the national production of sugarcane that year. The South-Central region is also where land use change is more intensified. According to the most recent land use data for Brazil more than 90% of sugarcane expansion has been concentrated in the South-Central region with the replacement of pastures (71%) and annual cropland areas (13%). The conversion of natural vegetation into sugarcane has occurred in the past, but represents less than 1% of the expansion in this region (CONAB, 2010, 2013).

While Brazil continues to have opportunities to improve sugarcane production with additional land availability, ideal climate conditions, abundant water resources and solar radiation, sugarcane agricultural practices and land use change for sugarcane production can affect the GHG balance of this agricultural system, impacting the C intensity of the resulting ethanol.

The rapid expansion of ethanol production from sugarcane in Brazil has raised several questions regarding the sustainability of this biofuel. Positive impacts include the low production price and the reduction of GHG emission-mainly CO_2- by offsetting the use of fossil energy. However, burning of native vegetation, decomposition and oxidation of the soil organic matter (SOM) caused by land use change can result in GHG emissions to the atmosphere (Cerri et al., 2007; Fearnside et al., 2009) leading a decrease in soil C stocks (Lal and Kimble, 1997; Six et al., 2002) and consequently affecting the overall sustainability of the ethanol.

In contrast, the adoption of certain management practices can result in carbon sequestration, improving the C intensity of sugarcane, with less GHG emission per unit of energy produced. For example, changes in the harvest system from pre-harvest burning to green harvesting have shown to increase soil C stocks (Canellas et al., 2003; Galdos et al., 2009), making this a potential management option for mitigating GHG emissions from sugarcane production systems. In this context, the present chapter will briefly address the main impacts of sugarcane management practices on soil organic C dynamics.

2. Sugarcane Management Practices and Soil Organic C Dynamics

2.1. Sugarcane Harvest

In the traditional sugarcane production system two types of harvest are performed: i) sugarcane involving pre-harvest burning and subsequently manual cutting and ii) green harvesting without burning where trash is retained as an undisturbed layer on the soil surface.

Until the 1940s the majority of sugarcane sites had been harvested manually without burning, however due to increased labor costs burning trash (senescent leaves) before cutting was adopted (Resende et al., 2006). This practice, performed to facilitate manual harvest and transport operations, results in 100% loss of the straw (dead leaves) and 50% loss of the dry matter of tips (green leaves and the apical gem of the stalk) (Basanta et al., 2003), releasing GHG emissions to the atmosphere.

Burning sugarcane residues also releases other gases or even GHG precursors, including carbon monoxide (CO), methane (CH_4), non-methane volatile organic compounds (NMVOC) and nitrogen (N_2O, NO_x) (Levine, 2000). Figueiredo & Scala Júnior (2011) estimated that 941 kg CO_2eq ha^{-1} yr^{-1}- ~30% of total GHG emission in sugarcane production- are emitted to the atmosphere with the pre-harvest burning practice.

Crutzen & Andreae (1990) have linked biomass burning in the tropics to regional production of O_3 and photochemical smog, increased acid deposition, and a potential loss of fixed nitrogen (pyro-denitrification). Additionally, Pereira Netto, Cunha, & Krauss (2004) detected high concentrations of polycyclic aromatic hydrocarbons in soils located near to sugarcane burning areas, which represent a risk for human health since these compounds are often carcinogenic.

In order to improve environmental aspects of sugarcane production in Brazil the manual harvest of sugarcane has been gradually replaced by mechanical harvesting, also called green harvesting. Legal restrictions regarding sugarcane pre-harvest burning were imposed in São Paulo state, responsible for more than 50% of production in Brazil. State Law No. 11.241/2002 prescribed cessation of sugarcane burning by 2021 in mechanized areas (with slope $< 12\%$) and by 2030 for all areas cultivated with sugarcane. Later the Green Ethanol Protocol, an initiative from the sugarcane sector and the state government, anticipated the cessation of burning in mechanized areas by 2014 and on 2017 for all land cultivated with sugarcane. Currently more than 70% of the harvested area in the South-Central region and 60% of harvested areas across the country involve mechanical harvest (CONAB, 2013), where the leaves, tips and variable quantities of stalk pieces are retained in the field forming a thick mass of mulch varying from 10 to 30 Mg ha^{-1} (Trivelin, Rodrigues, & Victoria, 1996)

In contrast to the burning system, green harvest could contribute to soil conservation by influencing yields, weed control, fertilizer management, soil erosion, soil water infiltration rates and soil organic matter (SOM) dynamics, among other factors. Resende, et al. (2006) found that sugarcane yield was 25% higher in a system where trash was conserved compared with areas involving pre-harvest burning. This gain was attributed to extra nutrients recycled in slow-release form in the trash, and/or to changes in the soil physical condition created by the trash layer.

Retention of residues on the soil surface has resulted in increased SOM content, which influences the nutrient cycling and physical characteristics of the soil. Several studies have indicated modifications in soil density, structure, porosity, water flux among other physical characteristics resulting from the adoption of a green harvesting system in sugarcane. Ceddia et al. (1999) registered increases in the stability of microaggregates in sandy soil after five years of implementing mechanical harvest in Espírito Santo state in Brazil. Similar results were reported by Souza et al. (2005; 2008), Luca et al. (2008) and Machado et al. (2010) in experiments performed in the state of São Paulo. The decomposition of residues by microorganisms results in the formation of several composts, which contribute to the cementation and stabilization of the aggregates. However, the traffic of heavy machines during harvest operations in green harvesting systems can increase the soil bulk density and affect the porosity of the soil, with higher percent of micro-pores and lower percent of macro-pores than in burning systems (Souza, et al., 2005, 2008; Luca, et al., 2008).

Soil fertility is also influenced by the change in the harvest system. Slight acidification and decreases of exchangeable Ca, Mg, total P, Sum of Bases (SB) and Cation Exchange Capacity (CEC) levels in burned systems when compared to green harvest have been reported by different studies performed in Brazil (Pinheiro et al., 2010; Rachid et al., 2012).

Nevertheless, the major impact of ceasing pre-harvest burning is an increase of SOM input, which improves soil C stocks and soil C sequestration. A positive correlation between the maintenance of sugarcane trash and the increase of SOM content has been observed in several studies.

The rate of C sequestration varies with the climate, soil texture, soil management and time since adoption of a green harvesting system. Souza, et al. (2005) found an annual C stock change in the top soil layer (0-20 cm) of 1.45 Mg ha^{-1} $year^{-1}$ after conversion from burned to green harvest in a sandy soil located in São Paulo state, a value similar to that found by Luca et al. (2008) and higher than reported by Resende et al. (2006) in a long term experiment (0.15 Mg ha^{-1} $year^{-1}$) in a sandy soil at Pernambuco state representing a different climate and soil condition from the state of São Paulo. Studies performed in clay soils have also shown results highlighting the benefits of green harvesting system on SOC accumulation. Feller (2001) reported that an average of 0.32 Mg ha^{-1} $year^{-1}$ was accumulated in 12 years in the first 20 cm of depth of an Oxisol by omitting burning. Recently, higher values of annual C accumulation rates were reported, varying from 0.65 Mg ha^{-1} in 0-10 cm depth to 1.2 and 2.1 Mg ha^{-1} in 0-20 cm depth (Galdos et al. 2009; Luca et al. 2008; Razafimbelo et al. 2006).

Cerri, et al. (2011) used a dataset obtained from a literature review of soils cultivated with sugarcane in Brazil to estimate a mean rate of C sequestration equivalent to 1.5 Mg ha^{-1} $year^{-1}$ for the first 30 cm of soil depth. When considering the soil texture the authors indicated means at 2.04 Mg C ha^{-1} $year^{-1}$ for clayey soils and 0.73 Mg C ha^{-1} $year^{-1}$ for sandy soils. These values indicate higher potential of C sequestration than observed in soybean/maize cropland in Brazil, with a mean C accumulation rate for sandy and clayey soils estimated at 0.41 ± 0.06 Mg ha^{-1} $year^{-1}$ (La Scala Júnior, De Figueiredo, & Panosso, 2012).

The C sequestration potential from sugarcane fields is equivalent to 5.5 Mg CO_2 ha^{-1} $year^{-1}$, which corresponds to 5.14% of the mean annual CO_2 absorption of this crop (Ronquim, 2007). Regarding the balance of GHG emissions from sugarcane production in Brazil, the potential of soil C accumulation due to the adoption of green harvesting system can be sufficient to compensate all the emissions derived from the use of machinery and synthetic fertilizers- estimated at 2.8 Mg CO_{2eq} ha^{-1} $year^{-1}$ (Figueiredo & Scala Júnior, 2011)- and still contribute to an annual mitigation of 2.7 Mg CO_2 $ha^{-1}year^{-1}$.

2.2. Sugarcane N Fertilization

Proper nutrition is essential for crop growth and production. The efficient application of fertilizers is an important strategy to achieve profitable yields. In sugarcane systems planted on soils with regular fertility, yield decreases are estimated at 30% when fertilizers are not applied (Nunes Júnior, 1999).

Nitrogen (N) is an essential element in the crop formation period, which occurs normally from 3 to 5 months after planting (Orlando Filho, 1983). Compared with cereal crops, N in sugarcane has a different productivity response. Although both require N for vegetative phases, cereals transport N to a grain 'sink' as protein that is accumulated with maturation, while in sugarcane the maturation is delayed and the sugar content is reduced if an excess of N continues to accumulate in the crop until harvest (Thorburn et al., 2005).

A review performed by Azeredo et al. (1986) involving 135 experiments reported that only 19% of the studies responded to N fertilization. It is known that the extraction of N from the soil is around 100 to 130 Kg ha^{-1} and N deficiency causes a reduction in the synthesis of chlorophyll and amino acids as well as a decrease in the energy available for carbohydrate and carbon skeleton production, with a direct effect on crop growth and yield (Malavolta et al., 1997).

The efficiency of N fertilization in sugarcane is a widely questioned issue. Fertilization effects on productivity are variable and highlight differential behaviors between cultivars, soil type and historical land management (Azeredo et al., 1986; Franco et al., 2010; Korndörfer et al., 2002; Paes et al., 1997; Trivelin et al., 2002). The lack of fertilization response has been mainly reported in experiments assessing the productivity of cane plant. For regrowth (ratoon cane) most experiments show a response with fertilization. This tendency has been related to several factors, a standout among which is Biological atmospheric Nitrogen Fixation (BNF). Biological atmospheric Nitrogen Fixation is performed by diazotroph organisms, capable of forming associations by colonizing roots and internal plant tissue. A BNF contribution was first demonstrated by Ruschel, Henis, & Salati (1975) and later confirmed by the work of Lima, Boddey, & Döbereiner (1987) and Urquiaga, Cruz, & Boddey (1992). According to Carnaúba (1990) the presence of fixing microorganisms in the sugarcane rhizosphere is an indication that fixation must play an important role in N self-sufficiency. Urquiaga et al., (2012) assessing BNF contributions to nine commercial varieties under field conditions, found annual inputs of BNF ranging from 40 to 64 kg ha^{-1} of N. Resende et al. (2006) reported a positive N balance in an experiment that evaluated the effect of pre-harvest burning, N fertilizer and vinasse additions on the yield, N balance and soil fertility for the "CB 45-3" variety. The authors suggested an annual input of BNF of up to 58 kg ha^{-1} of N in the plots where no N fertilizer was added.

Biological atmospheric Nitrogen Fixation in sugarcane crops has been also related with low rates of N fertilization application in Brazil. Compared with other countries, sugarcane crops in Brazil use a low level of N fertilizers, ranging from 30 to 60 Kg ha^{-1} of N in plant cane and from 60 to 120 Kg ha^{-1} of N in ratoon cane. Higher amounts of N fertilizer are used in Australia, where the recommended N fertilizer applications for sugarcane plant and ratoon are generally from 120 to 200 Kg ha^{-1} $year^{-1}$ (Calcino, 1995).

The application of lower rates of N fertilization in Brazil can result in fewer GHG emissions, improving the carbon intensity of Brazilian sugarcane ethanol, with less GHG emission per liter produced. The manufacturing of N fertilizer often represents large emission of CO_2 due to the intensive energy requirement in the Haber-Bosch process; emissions of 0.86 kg CO_2-C per kg N was reported (Powlson, Riche, & Shield, 2005). Additionally, the enrichment of soil N by fertilizer application can promote N_2O emissions to the atmosphere, a powerful GHG with a Global Warming Potential (GWP) 298 times greater compared with CO_2 (Forster et al., 2007). The rate and type of fertilizer used affects N_2O emissions from the soil. The flux of this gas to the atmosphere depends on N_2O production and consumption by denitrification and nitrification processes in the soil, varying according to characteristics such as the concentration of inorganic N, soil temperature, soil water content, pH and land management among others indicated in Table 1.

Table 1. Key parameters influencing N_2O emissions from agricultural soils

Parameter	Effect on N_2O emissions
Soil aeration	• Intermediate aeration: high N_2O production • Low aeration: high denitrification rate, but mainly N2 production
Soil water content	• Increase soil water content: Increase N_2O emissions • Very wet conditions: Decrease N_2O emissions • Changing conditions (dry/wet): Higher N_2O production
Nitrogen availability	• Increase NO_3^-/NH_4^+ concentrations: Increase N_2O emissions
Soil Texture	• From sandy to clayey: Increase N_2O emissions
Tillage practice	• Ploughing: Lower N_2O emissions • No/low-tillage: Higher N_2O emissions
Soil pH	• Where denitrification is the main source of N_2O emission: increase pH values and decrease N_2O emissions • Where nitrification is the main source of N_2O emission: increase pH values and increase N_2O emissions
Organic material	• Increase OC content: Increase N_2O emission
Crops and vegetation	• Plants, specially their residues and remaining roots after harvest increase N_2O emission
Temperature	• Temperature increase: Increase N_2O emissions

Adapted from Brentrup et al., (2000).

Due to the complexity of N dynamics and multiple factors affecting N_2O emissions, the Intergovernmental Panel on Climate Change (IPCC) has proposed an emission factor in order to facilitate N_2O estimation in Life Cycle

Assessment (LCA) studies of different crops. This emission factor considers N_2O emissions to be equivalent to 1% of the amount of N applied as fertilizer (IPCC, 2007).

Despite studies performed in different crops showing that this IPCC Emission Factor (EF) can under and/or overestimate the emissions of N_2O, many GHG inventories performed in sugarcane production systems have used this factor due to the lack of field data. Only a few studies are available with direct N_2O emissions in sugarcane fields, and most of them were completed in Australia.

Lisboa et al. (2011) published the first results of N_2O from soils in typical sugarcane areas of São Paulo state in Brazil. The estimates were performed using the eddy covariance technique over 5 months and encompassed operations such as the harvest of a ratoon crop and the subsequent renovation of the field using tillage and organic and mineral fertilization. According to the authors 2.1 kg N_2O-N ha^{-1} were emitted to the atmosphere during the evaluation time. They also calculated an N_2O EF for N fertilizer application in sugarcane of 3.87%, but that estimate was performed based on studies in Australia and Hawaii. Due to the lack of measurements in control areas (e.g. natural vegetation and areas cropped with sugarcane without the application of N fertilizers), the evaluation with eddy-covariance technics are as yet inconclusive for the evaluation of N_2O emissions from sugarcane sites in Brazil.

More recently, Signor, Cerri & Conant (2013) indicated that N_2O emissions from N fertilizers in sugarcane are dependent on the N rate, with different responses according to the N source. The proportion of N emitted as N_2O is higher when ammonium nitrate is used, versus when urea is used. The emissions due to ammonium nitrate increase as the N dose applied to the soil increases. In contrast urea emissions reach a maximum point that occurs at an approximate dose of 114 kg N ha^{-1}. The emission factors estimated by the authors were different between the sites studied, varying from 0.80 – 12.94% and 2.84 – 6.67% due to ammonium nitrate and urea application in São Paulo state and from 1.22 – 1.53% and 0.31 – 1.10% in Goiás state due to ammonium nitrate and urea, respectively.

Carmo et al. (2013) reported EFs for ammonium nitrate and urea for sites in the state of São Paulo closer to the values found by Signor Cerri, & Conant (2013) in Goiás, which were not substantially different from the 1% emission factor proposed by the IPCC. The amount of fertilizer N emitted to the atmosphere as N_2O in plant cane (ammonium nitrate) and ratoon cane (urea) were 1.11 % and 0.76 % respectively.

Direct and indirect emissions of N_2O from urea fertilization were also quantified by Silva Paredes et al. (2014) in a study performed in the state of Rio de Janeiro under greenhouse and field conditions. In that experiment only 0.2% of the N added as urea (under field conditions) was emitted as N_2O. Under greenhouse conditions the total N_2O-N emitted during the monitoring period was 0.8% of the N applied as urea. The proportion of added N volatilized as NH_3 was 2.5% in field conditions and 31% in greenhouse conditions. Both the Carmo et al. (2013) and Silva Paredes et al. (2014) studies additionally quantified the emissions resulting from the addition of organic amendments to the soil, a common management practice used in sugarcane production systems in Brazil. Carmo et al. (2013) found that the use of organic fertilizer combined with mineral fertilizer resulted in higher emissions, with EF varying from 1.56% in plant cane to 1.8% in ratoon cane. A higher EF was reported by Silva Paredes et al., (2014) where vinasse was applied, with losses of 2.5% from the N added. According to the authors it seems that the form and timing of application of vinasse could influence on the emissions; therefore more studies are required in order to develop better management practices to minimize the GHG emissions due to the application of vinasse.

2.3. Soil Tillage

Tillage includes all operations of seedbed preparation that optimize soil and environmental conditions for seed germination, seedling establishment and crop growth (Rattan Lal, 1995). In sugarcane this practice can deeply influence the yield between ratoons if the operations are not completed with proper technology for each soil type.

Sugarcane production involves highly mechanized systems, with around 30 operations on the same area along the production cycle. These operations can potentially affect soil properties such as bulk density, pore size distribution and soil structure (Braunack & McGarry, 2006; Silva et al., 2009). The renovation of the field for new planting is performed after a typical frequency of four harvests (Macedo, Seabra, & Silva, 2008); therefore, appropriate tillage practices could prevent degradation of soil properties and allow maintenance of crop yield as well as agroecosystem stability.

The cultivation of the soil by tillage also affects SOM dynamics. The physical disturbance caused by soil tillage increases the mineralization of soil organic carbon (SOC) and CO_2 emissions (Reicosky et al., 1999) by breaking down macro aggregates and exposing carbon protected in their interiors to microbial processes (Cambardella & Elliott, 1992). Jacinthe & Lal (2005) have shown that protected C accounts for about 0.5% of the total organic carbon in the surface layer 0–5 cm of soils in no-till systems. The amount of C lost in the form of CO_2 due to soil-tillage practices is strongly correlated with the intensity of the disruption and the volume of soil disturbed by the implements used (Dao, 1998; La Scala, Bolonhezi, & Pereira, 2006; Rastogi, Singh, & Pathak, 2002; D. Reicosky & Archer, 2007). Thus, depending on the soil type and management system, soils under sugarcane may be important sources or sinks of atmospheric carbon (Bernoux et al., 2005; R Lal, Fausey, & Eckert, 1995).

Traditionally, conventional soil tillage is adopted during sugarcane field renovation, which consists of the mechanical removal of ratoon followed by the operations of subsoiling and harrowing. However, in recent years conventional tillage systems have been replaced by reduced tillage in some regions in Brazil. In reduced tillage, the destruction of the previous ratoon is performed either through the application of herbicide or mechanically, and the soil is lightly tilled in the planting row. Some of the potential advantages of reduced tillage include the reduction of soil erosion rates- since the fields are protected by the old crop stubble- as well as the reduction of manpower inputs, machinery and fuel requirements. In terms of biomass productivity, reduced tillage does not affect the yield when compared with conventional tillage (Carvalho et al., 2011; Tavares, Lima, & Zonta, 2010).

Tillage systems that cause less perturbation to the soil have been highlighted as potential sources of GHG mitigation within the agricultural sector. In sugarcane crops the adoption of reduced tillage systems could make substantial contribution to GHG emission reductions from the agrosystem. Short-term CO_2 losses from mineral soils resulting from deep tillage with a moldboard plow can be substantial when compared to losses from no-till or minimally tilled (shallow tillage) soils (Reicosky & Lindstrom, 1995; Reicosky & Lindstrom, 1993). La Scala, et al. (2006) found in an experiment performed in the state of São Paulo that conventional tillage (CT) practices in sugarcane increased CO_2 emissions 160% when compared with no-till treatments. Losses of 2.3 Mg ha^{-1} of C-CO_2 were caused by CT tillage in a period of 4 weeks. Minimum tillage (MT) involving chisel plowing reduced losses, quantified at 1.0 Mg ha^{-1} of C-CO_2 in the same study area. Considering the amount of C in the litter on the soil surface before the tillage procedure, the authors estimated that in 1 month 30% of litter C was transferred back to the atmosphere from soil respiration after conventional tillage.

Similar results were reported by Moitinho et al. (2013) in a clay oxisol located in Mato Grosso state. In their experiment, CO_2 emissions were quantified from CT on areas without residues on the soil surface for a period of 15 days. In this time period the mean CO_2 flux values were 49% higher in CT when compared with the No-till treatment. However cumulative emissions were lower than the value reported by La Scala et al. (2006), with losses of 33 Kg ha^{-1} of C-CO_2 due to CT practices. As the litter on the soil surface was removed before tillage, C losses were mostly caused by the exposure of additional labile C to microbes due to aggregate breakdown and changes in *k* factor induced by tillage. Rochette & Angers (1999) also pointed out a phenomenon caused by tillage described as "degassing", which involves the physical expulsion of CO_2 from the soil at the time of cultivation.

When residues are incorporated into the soil by tillage tools, the interaction between the mixing of organic residues and greater soil aeration may affect oxidation, thereby increasing the CO_2 flow. Teixeira et al. (2011) compared the effect of rotatory tillage in oxisol under sugarcane crop with and without residues, verifying that the incorporation of litter in the soil increased the emissions of CO_2 140%. Carbon losses were higher that the values reported by Moitinho et al. (2013) but lower than reported by La Scala et al. (2006). The cumulated CO_2 flow in no-till treatments over 2 weeks of measurements indicated C-CO_2 losses due to soil respiration as 373.2 kg ha^{-1}; meanwhile the losses under tillage treatments with and without residues were 681.1 and 500.1 Kg ha^{-1} C-CO_2 respectively. These authors also verified decreases in the mass of soil aggregates with diameters smaller than 4 mm and increases in the distribution of aggregates smaller than 4 mm, especially from 2 to 0.125 mm, after tillage operations.

The impact of tillage on SOC dynamics in sugarcane crops is variable. Silva-Olaya et al. (2013) quantified emissions of 954.79 kg ha^{-1} of C-CO_2 due to CT in an area using green harvesting. The C losses caused by reduced tillage (RT) and MT were smaller than CT, with values of 141 and 15.2 kg ha^{-1} of C-CO_2 respectively.

Differences among the results reported in the literature may be related to differences in soil type, soil texture, climatic conditions of the region during the period studied and the implements used for tillage practices. Several studies have found that soil moisture is a controlling factor in CO_2 emissions and that the sensitivity of CO_2 flux to soil moisture is greater under the conventional tillage compared to no-till and minimum tillage treatments (Jabro et al., 2008; La Scala, et al., 2006; Ussiri & Lal, 2009). Temporarily, the emissions were affected by precipitation events in all of the studies cited above. Emission increases were observed in the days with precipitations events. Meanwhile, the temperature seems to not be a control factor in tropical regions since it remains near to optimal conditions for microbial activity (La Scala et al. 2006; Panosso, Marques, Pereira, & La Scala, 2009).

While tillage is an agricultural practice only performed every 5 to 6 years in sugarcane production systems, it is still an important source of CO_2 to the atmosphere. The C losses are significant and comparable to the value, estimated by Galdos, Cerri & Cerri (2009), of potential annual C accumulation resulting from changing the harvesting system from burning plant residues to green cane harvesting adoption in Brazil.

According to Silva-Olaya et al. (2013) soil tillage for sugarcane reformation under conventional practices (CT) could generate a loss in a period of 44 days equivalent to 80% of the C that could potentially be accumulate in this soil layer during one year with mechanical harvesting. The impact of RT and MT is smaller than CT, with reported losses, respectively, at 12% and 2% from the accumulated C in one year. Data from La Scala et al. (2006) suggest that these losses are still higher than the annual C sequestration rates for sugarcane cultivation.

In order to guarantee the environmental sustainability of ethanol produced from sugarcane, C losses caused by tillage should be the minimized. Currently, 71.6% of the sugarcane production area in South-Central Brazil, responsible for 90 % of the country's production, involves green harvesting, and around 15 -20% of the plantations are renewed annually. If the consumption of diesel oil required by the machine operation for the renovation of the field is considered, the adoption of conservationist management practices could result in significant changes in GHG emissions. According to Oliveira Bordonal, Figueiredo & La Scala (2012) the adoption of CT and MT results in the consumption of 166.72 and 107.41 L ha^{-1} of diesel oil, respectively. When considering all agricultural operations- including soil tillage, planting, ratoon maintenance, harvest and transport to the mill- the mean annual rate of consumption of diesel oil is 189.03 L ha^{-1} under CT and 177.36 L ha^{-1} under MT management. In terms of CO_2 emissions, the use of diesel oil in sugarcane with green harvesting involving CT practices result in 750.2 Kg CO_{2eq} ha^{-1} yr^{-1}, meanwhile the adoption of MT results in 703.9 Kg CO_{2eq} ha^{-1} yr^{-1}.

The selection of sustainable management practices that support soil C accumulation, improve soil quality and minimize CO_2 emissions from soils could reduce the carbon footprint of ethanol, improving the environmental benefits of fossil fuels offset by sugarcane ethanol.

2.4. Sugarcane Expansion and Land Use Change

As land use and land use change (LULUC) has become one of the most important factors affecting the sustainability of sugarcane ethanol (Fargione et al., 2008; Gibbs et al., 2008; Lange, 2011; Lapola et al., 2010; Melillo et al., 2009), new research indicates that a sustainable path for sugarcane expansion is underway in Brazil.

Fargione et al. (2008) described a concept whereby land clearing for biofuel production would lead to a "biofuel carbon debt". To avoid such debt, biofuels would have to present a positive GHG offset when substituted for fossil fuels large enough to compensate emissions due to LULUC. In this study, Fargione et al. (2008) estimated a payback time of 17 years for sugarcane replacing wooded cerrado in Brazil (for ethanol) to up to 423 years for palm production in peatlands in Malaysia (for biodiesel).

This idea gained greater importance when indirect LUC (iLUC) was pointed by Melillo et al. (2008) and later by Lapola et al. (2010) as a major source of GHGs during biofuel production. The iLUC concept suggested that GHG emissions from bioenergy production went beyond land use change "in site". Rather, added to the carbon debt proposed by Fargione et al. (2008), the cultivation of crops for bioenergy production should include the GHG emissions resulting from the expansion of other land use activities replaced by sugarcane or any other biofuel crop.

As an example, Lapola et al. (2008) indicated that the substitution of natural systems to rangeland was necessary to offset the conversion from rangelands to sugarcane. The result of such activity would increase the payback time for sugarcane ethanol from 1 to 38 years if iLUC emissions occurred from the exchange of wooded cerrado or tropical forest, respectively.

However, the inclusion of GHG emissions resulting from iLUC is not a consensus among scientists. There are other major issues that drive land use change, such as commodity prices. Lower productivity of a specific commodity (e.g. maize) anywhere in the globe could increase LUC elsewhere to improve the production of the commodity. In such case, iLUC would be a global problem and not only for the place or country where that change in land use occurred.

In the case of Brazilian sugarcane, for example, more than 95% of expansion from 2000 to 2009 occurred over cultivated pasture (~70%), grain crops (~25%) and citrus (~1 %) (CONAB, 2009; Adami et al., 2012). The conversion of natural vegetation into sugarcane has occurred in the past, but represents less than 1% of the expansion in South-Central Brazil (Adami et al.,

2012), indicating a different pattern from those evaluated by Fargione et al. (2008) or Lapola et al. (2010).

In a recent study, Mello et al. (2014) completed an extensive evaluation of the soil C changes due to sugarcane expansion in South-Central Brazil. Their evaluation included major land use transitions into sugarcane across 5 states, totaling 135 study sites, and clarified soil C impacts with sugarcane expansion.

Their study showed that LUC from cultivated pastures into sugarcane decreased soil C stocks through the years, but indicated that only a short time span is necessary to recover CO_2 emissions due to LUC.

The payback time for soil C stocks losses from cerrado conversion to sugarcane were estimated at 8 years, versus 2-3 years for cultivated pastures. Adding C losses attributed to biomass removal, the payback time would increase to 17 years for cerrado and to 5-6 years for pastures conversion, respectively. On the other hand the authors indicated that the substitution of annual cropland with sugarcane could increase soil C stocks through the years, offsetting from 36 to 79 Mg of CO_2 after a 20 year time span, in contrast with substituting cerrado or cultivated pastures with sugarcane fields.

Therefore, the substitution of annual cropland with sugarcane rather than replacing cultivated pastures would mitigate GHG emissions. The authors indicated that around 3 Mha were converted into sugarcane in South-Central Brazil from 2000 to 2010 (73.04% from pastures, 25.08% from annual cropland and 0.52% from cerrado). Their results estimate net ecosystem emissions of0.7 to 1.0 Mg of CO^2 ha^{-1} $year^{-1}$, which in their words "reduces but does not negate the biofuel offset of 9.8 Mg of CO_2 ha^{-1} $year^{-1}$ (Fargione et al., 2008)". Finnaly, Goldemberg et al. (2014) proposed that ~7 Mha of sugarcane will be required by 2020 to meet increasing demand for ethanol in Brazil. The authors indicated that there are available, currently, ~200 Mha of pastures that could be converted into sugarcane. Therefore sugarcane expansion could occur without further conversion from natural ecosystems, due to specific policies that will allow the recovery of up to 15 Mha of pasture areas. These areas will be enough to promote the expansion of food and biofuel crops, while following a sustainable path towards the consolidation of a low C emission economy.

FINAL CONSIDERATIONS

Sugarcane is an important component of the Brazilian economy, supporting about 1.5% of the Gross Domestic Product (GDP) with the provision of sugar and ethanol. Currently, more than 50% of sugarcane crushed in Brazil is being distilled into ethanol (CONAB, 2013), one of the most sustainable biofuels currently produced at commercial scales. This fuel is experiencing a rapid increase in demand, leading to an expansion of sugarcane production in Brazil that reached 10 Mha during the last harvest season (CONAB, 2013).

The selection of adequate sugarcane production management practices plays an important role in the GHG balance and subsequently the sustainability of sugarcane ethanol. Among mitigation processes, the soil is an important natural reservoir of carbon (C). Cultivation of the soil with tillage increases the mineralization of SOC and the emission of CO_2 (Reicosky 1999). Thus, depending on the soil type and management system, sugarcane crop could be a source or sink of atmospheric C.

Conservation management practices in this agro-system can contribute to restoration of C lost through land cultivation. The conversion from manual harvesting involving pre-harvest burning to mechanical or green harvesting can input high quantities of dry matter on the soil surface, providing benefits to the system such as C accumulation and erosion control. The potential of soil C sequestration can be sufficient to compensate all the emissions derived from the use of machinery and synthetic fertilizers, and still contribute to the annual mitigation of 2.7 Mg CO_2 $ha^{-1}year^{-1}$

The selection of sustainable management practices that allow the increase of C accumulation, improve soil quality and minimize CO_2 emissions from soils in the sugarcane agro-system can help to reduce the C intensity of Brazilian sugarcane ethanol.

REFERENCES

Adami, M., Rudorff, B. F. T., Freitas, R. M., Aguiar, D. A., Sugawara, L. M., & Mello, M. P. (2012). Remote sensing time series to evaluate direct land use change of recent expanded sugarcane crop in Brazil. *Sustainability, 4*(4), 574-585.

Azeredo, D. d., Bolsanello, J., Weber, H., & Vieira, J. (1986). Nitrogênio em cana-planta, doses e fracionamento (Nitrogen in plant cane, rates and fractionation). *STAB, 4*(5), 26-32.

Basanta, M. V., Dourado-Neto, D., Reichardt, K., Bacchi, O., Oliveira, J., Trivelin, P., et al. (2003). Management effects on nitrogen recovery in a sugarcane crop grown in Brazil. *Geoderma, 116*(1), 235-248.

Bernoux, M., Cerri, C., Volkoff, B., Maria da Conceição, S., Feller, C., Cerri, C., et al. (2005). Gases do Efeito Estufa e Estoque de Carbono nos Solos: Inventário do Brasil (Greenhouse Gases and Soil Carbon Stocks: Brazilian Inventory). *Cadernos de Ciência e tecnologia (Books of science and technology), Brasília, 22*(1), 235-246.

BEB - Brazilian Energy Balance year 2012 (2013). Empresa de pesquisa energetica (Energy research company) , Rio de Janeiro, 288 p.

Brasil. Ministério de Agricultura, Pecuária e Abastecimento (Ministry of Agriculture, Livestock, and Supply) (2007).) Balanço nacional de cana-de-açúcar e agroenergia. (National balance of sugarcane and agroenergy) *MAPA, Secretaria de Produção e Agroenergia (Secretary of production and agroenergy)*, 139 p.

Braunack, M., & McGarry, D. (2006). Traffic control and tillage strategies for harvesting and planting of sugarcane (< i> Saccharum officinarum</i>) in Australia. *Soil and Tillage Research, 89*(1), 86-102.

Brentrup, F., Küsters, J., Lammel, J., & Kuhlmann, H. (2000). Methods to estimate on-field nitrogen emissions from crop production as an input to LCA studies in the agricultural sector. *The International Journal of Life Cycle Assessment, 5*(6), 349-357.

Calcino, D. (1995). *Australian sugarcane nutrition manual*: Bureau of Sugar Experiment Stations.

Cambardella, C. A., & Elliott, E. T. (1992). ParticulateE Soil Organic-Matter Changes Across A Grassland Cultivation Sequence. *Soil Science Society of America Journal, 56*(3), 777-783.

Canellas, L., Velloso, A., Marciano, C., Ramalho, J., Rumjanek, V., Rezende, C., et al. (2003). Propriedades químicas de um Cambissolo cultivado com cana-de-açúcar, com preservação do palhiço e adição de vinhaça por longo tempo (Chemical properties of a Inceptisol cultivated with sugarcane involving trash conservation and vinasse aplication for long term). *Revista Brasileira de Ciência do Solo (Brazilian journal of soil science), 27*(5), 935-944.

Carmo, J. B. d., Filoso, S., Zotelli, L. C., Neto, S., Eraclito, R., Pitombo, L. M., et al. (2013). Infield greenhouse gas emissions from sugarcane soils in Brazil: effects from synthetic and organic fertilizer application and crop trash accumulation. *GCB Bioenergy, 5*(3), 267-280.

Carnaúba, B. (1990). O nitrogênio en cana-de-açúcar (The nitrogen in sugarcane). *STAB-Açúcar, Álcool e Subprodutos (STAB-Sugar, alcohol and by-products), Piracicaba, 8*(3/4), 24-41.

Carvalho, L. A. d., Silva Junior, C. A. d., Nunes, W. A. G. d. A., Meurer, I., & Souza Júnior, W. S. d. (2011). Produtividade e viabilidade econômica da cana-de-açúcar em diferentes sistemas de preparo do solo no Centro-Oeste do Brasil (Productivity and economic feasibility of sugar cane in different systems of tillage in west-central Brazil). *Revista de Ciências Agrárias (Journal of Agricultural Sciences), 34*(1), 199-211.

Ceddia, M. B., Anjos, L. d., Lima, E., Ravelli Neto, A., & Silva, L. d. (1999). Sistemas de colheita da cana-de-açúcar e alterações nas propriedades físicas de um solo Podzólico Amarelo no Estado do Espírito Santo (Harvest systems in sugarcane and its effects on physical properties of an ultisol at Espirito Santo State). *Pesquisa Agropecuária Brasileira (Brazilian Journal of Agricultural Research), 34*(8), 1467-1473.

Cerri, C., Galdos, M., Maia, S., Bernoux, M., Feigl, B., Powlson, D., et al. (2011). Effect of sugarcane harvesting systems on soil carbon stocks in Brazil: an examination of existing data. *European journal of soil science, 62*(1), 23-28.

Cerri, C.E.P.; Easter, M.; Paustian, K.; Killian, K.; Coleman, K.; Bernoux, M.; Falloon, P.; Powlson, D.S.; Batjes, N.H.; Milne, E.; Cerri, C.C. (2007). Predicted soil organic carbon stocks and changes in the Brazilian Amazon between 2000 and 2030. *Agriculture Ecosystems & Environment*, v. 122, n. 1, p. 58-72.

Conab. (2010). Companhia Nacional De Abastecimento (National Supply Company). Perfil do setor de açúcar e álcool no Brasil (Sugar and ethanol profile in Brazil. Edição para a safra 2008-2009 (Edition for the harvest season 2008-2009). 74 p.

Conab. (2013). Companhia Nacional De Abastecimento (National Supply Company). Perfil do setor de açúcar e álcool no Brasil: safra 2011/2012 (Sugar and ethanol profile in Brazil: harvest season 2011/2012), *v. 5*, 74 p.

Crutzen, P. J., & Andreae, M. O. (1990). Biomass burning in the tropics: Impact on atmospheric chemistry and biogeochemical cycles. *Science, 250*(4988), 1669-1678.

Cheavegatti-Gianotto, A., de Abreu, H. M. C., Arruda, P., Bespalhok Filho, J. C., Burnquist, W. L., Creste, S., et al. (2011). Sugarcane (Saccharum X officinarum): A Reference Study for the Regulation of Genetically Modified Cultivars in Brazil. *Tropical plant biology*, 1-28.

Dao, T. H. (1998). Tillage and crop residue effects on carbon dioxide evolution and carbon storage in a Paleustoll. *Soil Science Society of America Journal, 62*(1), 250-256.

Figueiredo, E., & Scala Júnior, N. l. (2011). Greenhouse gas balance due to the conversion of sugarcane areas from burned to green harvest in Brazil. *Agriculture, Ecosystems & Environment, 141*(1/2), 77-85.

Fargione, J., Hill, J., Tilman, D., Polasky, S., & Hawthorne, P. (2008). Land clearing and the biofuel carbon debt. *Science, 319*(5867), 1235-1238.

Fearnside, P.M.; Righi, C.A.; Graça, P.M.L.D.A.; Keizer, E.W.; Cerri, C.C.; Nogueira, E.M.; Barbosa, R.I. (2009). Biomass and greenhouse-gas emissions from land-use change in Brazil's Amazonian "arc of deforestation": the states of Mato Grosso and Rondônia. *Forest Ecology and Management,* 258 (9), 1968-1978.

Feller, C. (2001). Efeitos da colheita sem queima da cana-de-açúcar sobre a dinâmica do carbono e propriedades do solo (Effects of green harvesting in sugarcane on soil carbon dynamics and soil properties). *Final Report Fapesp Contract, 98*, 12648-12643.

Forster, P., Ramaswamy, V., Artaxo, P., Berntsen, T., Betts, R., Fahey, D. W., et al. (2007). Changes in atmospheric constituents and in radiative forcing. *Climate change, 20*(7).

Franco, H. C. J., Trivelin, P. C. O., Carlos Eduardo, F., Vitti, A. C., & Otto, R. (2010). Stalk yield and technological attributes of planted cane as related to nitrogen fertilization. *Scientia Agricola, 67*(5), 579-590.

Galdos, M., Cerri, C., & Cerri, C. (2009). Soil carbon stocks under burned and unburned sugarcane in Brazil. *Geoderma, 153*(3), 347-352.

Galdos, M., Cerri, C.C., Bernoux, M., Cerri, C.E.P. (2010) Ethanol production from sugarcane and soil quality. In Lal, R., & Stewart, B. A. *Soil quality and biofuel production*: CRC Press.

Gibbs, H. K., Johnston, M., Foley, J. A., Holloway, T., Monfreda, C., Ramankutty, N., et al. (2008). Carbon payback times for crop-based biofuel expansion in the tropics: the effects of changing yield and technology. *Environmental Research Letters, 3*(3), 034001.

Goldemberg, J., Mello, F. F., Cerri, C. E., Davies, C. A., & Cerri, C. C. (2014). Meeting the global demand for biofuels in 2021 through sustainable land use change policy. *Energy Policy, 69*, 14-18.

Ibge (2014). Instituto Brasileiro De Geografia E Estatistica (Brazilian Institute of Geography and Statistics). Levantamento Sistemático da produção agrícola (Systematic reporting of agricultural production). Available in: <http://www.ibge.gov.br/home/estatistica/indicadores/agropecuaria/lspa/lspa_201404_4.shtm>. Accessed in: May 17 2014.

IPCC (Intergovernmental Panel on Climate Change) 2007 Climate Change: Synthesis Report. Contribution of Working Groups I, II and III to the

Fourth Assessment Report of the Intergovernmental Panel on Climate Change Core Writing Team, R K Pachauri and A Reisinger (Geneva: IPCC).

Jabro, J. D., Sainju, U., Stevens, W. B., & Evans, R. G. (2008). Carbon dioxide flux as affected by tillage and irrigation in soil converted from perennial forages to annual crops. *Journal of Environmental Management, 88*(4), 1478-1484.

Jacinthe, P.-A., & Lal, R. (2005). Labile carbon and methane uptake as affected by tillage intensity in a Mollisol. *Soil and Tillage Research, 80*(1), 35-45.

Korndörfer, G., Colombo, C., Chimello, M., & Leone, P. (2002). *Desempenho de variedades de cana-de-açúcar cultivadas com e sem nitrogênio (Performance of sugarcane varieties cultivated with and without nitrogen).* CONGRESSO NACIONAL DA SOCIEDADE DOS TÉCNICOS AÇÚCAREIROS E ALCOOLEIROS DO BRASIL. (Paper presented at the National Congress of Brazilian Association of Sugar and Ethanol Technicians).

La Scala Júnior, N., De Figueiredo, E., & Panosso, A. (2012). A review on soil carbon accumulation due to the management change of major Brazilian agricultural activities. *Brazilian Journal of Biology, 72*(3), 775-785.

La Scala, N., Bolonhezi, D., & Pereira, G. T. (2006). Short-term soil CO2 emission after conventional and reduced tillage of a no-till sugar cane area in southern Brazil. *Soil & Tillage Research, 91*(1-2), 244-248.

Lal, R.; Kimble, J.M. (1997). Conservation tillage for carbon sequestration. *Nutrient Cycling in Agroecosystems,* 49(1/3) 243-253.

Lal, R. (1995). *Tillage systems in the tropics: Management options and sustainability implications*: Food & Agriculture Org.

Lal, R., Fausey, N., & Eckert, D. (1995). Land use and soil management effects on emissions of radiatively active gases from two soils in Ohio. *Soil Management and Greenhouse Effect. CRC Press, Boca Raton, FL, 41*, 59.

Lange, M. (2011). The GHG balance of biofuels taking into account land use change. *Energy Policy, 39*(5), 2373-2385.

Lapola, D. M., Schaldach, R., Alcamo, J., Bondeau, A., Koch, J., Koelking, C., et al. (2010). Indirect land-use changes can overcome carbon savings from biofuels in Brazil. *Proceedings of the national Academy of Sciences, 107*(8), 3388-3393.

Levine, J. S. (2000). Global biomass burning: a case study of the gaseous and particulate emissions released to the atmosphere during the 1997 fires in

Kalimantan and Sumatra, Indonesia *Biomass Burning and its Inter-relationships with the Climate System* (pp. 15-31): Springer.

Lima, E., Boddey, R. M., & Döbereiner, J. (1987). Quantification of biological nitrogen fixation associated with sugar cane using a< sup> 15</sup> N aided nitrogen balance. *Soil Biology and Biochemistry, 19*(2), 165-170.

Lima, M., Ligo, M., Cabral, O., Boeira, R., Pessoa, M., & Neves, M. (1999). (Greenhouse gases emission from trash burning in Brazil) Emissão de gases de efeito estufa provenientes da queima de resíduos agrícolas no Brasil. *Jaguariúna: Embrapa Meio Ambiente.*

Lisboa, C. C., Butterbach-Bahl, K., Mauder, M., & Kiese, R. (2011). Bioethanol production from sugarcane and emissions of greenhouse gases–known and unknowns. *GCB Bioenergy, 3*(4), 277-292.

Luca, E. d., Feller, C., Cerri, C. C., Barthès, B., Chaplot, V., Campos, D. C., et al. (2008). Avaliação de atributos físicos e estoques de carbono e nitrogênio em solos com queima e sem queima de canavial (Evaluation of physical properties and soil carbon and nitrogen stocks as affected by burning or green trash management of sugarcane) . *Revista Brasileira de Ciência do Solo (Brazilian Journal of Soil Science), 32*(02), 789-800.

Macedo, I. C., Seabra, J. E., & Silva, J. E. (2008). Green house gases emissions in the production and use of ethanol from sugarcane in Brazil: the 2005/2006 averages and a prediction for 2020. *Biomass and bioenergy, 32*(7), 582-595.

Machado, W., Telles, T. S., Tavares Filho, J., Guimarães, M. d. F., Alves, G. B., & Borges, J. L. B. (2010). Physical properties of a rhodic haplustox under two sugarcane harvesting systems. *Revista Brasileira de Ciência do Solo, 34*(6), 1803-1809.

Malavolta, E., Vitti, G., & Oliveira, S. Avaliação do estado nutricional das plantas: princípios e aplicações. Piracicaba: Potafos, 1997. 201p. *Artigo Científico.*

Matsuoka, S., Garcia, A., & Calheiros, G. (1999). Hibridação em cana-de-açúcar (Artificail hybridization in sugarcane). *Hibridação Artificial de Plantas. Viçosa: Editora UFV*, 221-254.

Moitinho, M. R., Padovan, M. P., Panosso, A. R., & La Scala Jr, N. (2013). Efeito do Preparo do solo e resíduo da colheita de cana-de-açúcar sobre a emissão de CO_2 (Effects of soil tillage and sugarcane trash on CO2 emissions) *Revista Brasileira de Ciência do Solo (Brazilian Journal of Soil Science), 37*(6), 1720-1728.

Nunes Júnior, D. (1999). A redução da adubação ea produtividade (Fertilization rates reduction and productivity). *STAB, Piracicaba, 17*(3), 16.

Melillo, J. M., Reilly, J. M., Kicklighter, D. W., Gurgel, A. C., Cronin, T. W., Paltsev, S., et al. (2009). Indirect emissions from biofuels: how important? *Science, 326*(5958), 1397-1399.

Mello, F. F., Cerri, C. E., Davies, C. A., Holbrook, N. M., Paustian, K., Maia, S. M., et al. (2014). Payback time for soil carbon and sugar-cane ethanol. *Nature Climate Change*.

Oliveira Bordonal, R., Figueiredo, E. B., & La Scala, N. (2012). Greenhouse gas balance due to the conversion of sugarcane areas from burned to green harvest, considering other conservationist management practices. *GCB Bioenergy, 4*(6), 846-858.

Oliveira, M. E. D., Vaughan, B. E., & Rykiel, E. J. J. (2005). Ethanol as fuel: energy, carbon dioxide balances, and ecological footprint. Bioscience 55:593–602.

Oliveira, S. d. (1997). Avaliação do estado nutricional das plantas: princípios e aplicações (Evaluation of the nutritional state of plants: principles and applications). *Piracicaba: Potafós*.

Orlando Filho, J. (1983). Nutricao e adubacao da cana-de-acucar no Brasil [Saccharum officinarum, macroelemento, microelemento] (Nutrition and fertilization of sugarcane in Brazil [Saccharum officinarum, macroelemento, microelemento]).

Paes, J., Marciano, N., Brito, C. d., Cardoso, A., Martinez, H., & Mendes, A. (1997). Estudo de espaçamentos e doses de nitrogênio na produção e em algumas características biométricas de três variedades de cana-de-açúcar (Study of row space and nitrogen rates in the production and in some biometric characteristics of three varieties of sucarcane) . *STAB, 15*, 18-20.

Panosso, A. R., Marques, J., Pereira, G. T., & La Scala, N. (2009). Spatial and temporal variability of soil CO2 emission in a sugarcane area under green and slash-and-burn managements. *Soil & Tillage Research, 105*(2), 275-282.

Pereira Netto, A., Cunha, I., & Krauss, T. (2004). Persistence of polycyclic aromatic hydrocarbons in the soil of a burned area for agricultural purposes in Brazil. *Bulletin of environmental contamination and toxicology, 73*(6), 1072-1077.

Pinheiro, É. F. M., Lima, E., Ceddia, M. B., Urquiaga, S., Alves, B. J., & Boddey, R. M. (2010). Impact of pre-harvest burning versus trash conservation on soil carbon and nitrogen stocks on a sugarcane plantation in the Brazilian Atlantic forest region. *Plant and soil, 333*(1-2), 71-80.

Powlson, D. S., Riche, A., & Shield, I. (2005). Biofuels and other approaches for decreasing fossil fuel emissions from agriculture. *Annals of Applied Biology, 146*(2), 193-201.

Rachid, C. T., Piccolo, M. C., Leite, D. C. A., Balieiro, F. C., Coutinho, H. L. C., van Elsas, J. D., et al. (2012). Physical-chemical and microbiological changes in Cerrado Soil under differing sugarcane harvest management systems. *BMC microbiology, 12*(1), 170.

Rastogi, M., Singh, S., & Pathak, H. (2002). Emission of carbon dioxide from soil. *Current Science, 82*(5), 510-517.

Razafimbelo, T., Barthès, B., Larré-Larrouy, M.-C., Luca, E. F. D., Laurent, J.-Y., Cerri, C. C., et al. (2006). Effect of sugarcane residue management (mulching versus burning) on organic matter in a clayey Oxisol from southern Brazil. *Agriculture, ecosystems & environment, 115*(1), 285-289.

Reicosky, D., & Archer, D. (2007). Moldboard plow tillage depth and short-term carbon dioxide release. *Soil and Tillage Research, 94*(1), 109-121.

Reicosky, D., & Lindstrom, M. (1995). Impact of fall tillage on short-term carbon dioxide flux. *Soils and Global Change. Lal, R., J. Kimble, E. Levine, and BA Stewart (eds.). Lewis Publishers, Chelsea, Michigan*, 177-187.

Reicosky, D. C., & Lindstrom, M. J. (1993). Fall Tillage Method – Effect on Short-Term Carbon-Dioxide Flux from Soil. *Agronomy Journal, 85*(6), 1237-1243.

Reicosky, D. C., Reeves, D. W., Prior, S. A., Runion, G. B., Rogers, H. H., & Raper, R. L. (1999). Effects of residue management and controlled traffic on carbon dioxide and water loss. *Soil & Tillage Research, 52*(3-4), 153-165.

Rochette, P., & Angers, D. A. (1999). Soil surface carbon dioxide fluxes induced by spring, summer, and fall moldboard plowing in a sandy loam. *Soil Science Society of America Journal, 63*(3), 621-628.

Ronquim, C. (2007). Dinâmica espaço temporal do carbono aprisionado na fitomassa dos agroecossistemas do Nordeste do Estado de São Paulo . (Space-time dynamics of carbon phytomass in the agroecosystmes in Northeast of Sao Paulo state. *Documentos*.

Ruschel, A. P., Henis, Y., & Salati, E. (1975). Nitrogen-15 tracing of N-fixation with soil-grown sugarcane seedlings. *Soil Biology and Biochemistry, 7*(2), 181-182.

Signor, D., Cerri, C., & Conant, R. (2013). N2O emissions due to nitrogen fertilizer applications in two regions of sugarcane cultivation in Brazil. *Environmental Research Letters, 8*(1), 015013.

Silva, R. d., Lanças, K., Miranda, E., Silva, F., & Baio, F. (2009). Estimation and evaluation of dynamic properties as indicators of changes on soil structure in sugarcane fields of Sao Paulo State–Brazil. *Soil and Tillage Research, 103*(2), 265-270.

Silva-Olaya, A. M., Cerri, C. E. P., La Scala Jr, N., Dias, C. T. S., & Cerri, C. C. (2013). Carbon dioxide emissions under different soil tillage systems in mechanically harvested sugarcane. *Environmental Research Letters, 8*(1), 015014.

Six, J.; Feller, C.; Denef, K.; Ogle, S.M.; Sa, J.C.D.; Albrecht, A. (2002). Soil organic matter, biota and aggregation in temperate and tropical soils: effects of no-tillage. *Agronomie,* 22 (7/8), 755-775.

Resende, A. S., Xavier, R. P., de Oliveira, O. C., Urquiaga, S., Alves, B. J., & Boddey, R. M. (2006). Long-term effects of pre-harvest burning and nitrogen and vinasse applications on yield of sugar cane and soil carbon and nitrogen stocks on a plantation in Pernambuco, NE Brazil. *Plant and Soil, 281*(1-2), 339-351.

Silva Paredes, D., Lessa, A. C. d. R., de Sant'Anna, S. A., Boddey, R. M., Urquiaga, S., & Alves, B. J. (2014). Nitrous oxide emission and ammonia volatilization induced by vinasse and N fertilizer application in a sugarcane crop at Rio de Janeiro, Brazil. *Nutrient Cycling in Agroecosystems, 98*(1), 41-55.

Souza, Z. M., Beutler, A. N., Prado, R. M., & Bento, M. J. C. (2008). Efeito de sistemas de colheita de cana-de-açúcar nos atributos físicos de um Latossolo Vermelho (Effects of sugarcane harvest on physical properties of an Oxisol). *Científica, 34*(1), 31-38.

Souza, Z. M., de Mello Prado, R., Paixão, A. C. S., & Cesarin, L. G. (2005). Sistemas de colheita e manejo da palhada de cana-de-açúcar (Harvest system and residue management of sugarcane). *Pesq. agropec. bras., Brasília, 40*(3), 271-278.

Tavares, O. C. H., Lima, E., & Zonta, E. (2010). Sugarcane growth and productivity under different tillage and crop systems. *Acta Scientiarum. Agronomy, 32*(1), 61-68.

Teixeira, L. G., Fukuda, A., Panosso, A. R., Lopes, A., & Scala Jr, N. L. (2011). Soil CO2 emission as related to incorporation of sugarcane crop residues and aggregate breaking after rotary tiller. *Engenharia Agrícola, 31*(6), 1075-1084.

Trivelin, P., Vitti, A., Oliviera, M. d., Gava, G., & Sarriés, G. (2002). Utilização de nitrogênio e produtividade da cana-de-açúcar (cana-planta) em solo arenoso com incorporação de resíduos da cultura (Nitrogen utilization and sugarcane (plant cane) yield on a sandy soil with incorporated crop Residues). *Revista Brasileira de Ciência do Solo (Brazilian Journal of Soil Science), 26*(3), 637-646.

Trivelin, P. C. O., Rodrigues, J. C. S., & Victoria, R. L. (1996). Utilização por soqueira de cana-de-açúcar de início de safra do nitrogênio da aquamônia-15N e uréia-15N aplicado ao solo em complemento à vinhaça (Utilization by early harvest sugarcane ratoon of the nitrogen from ^{15}N-aqua ammonia and ^{15}N-urea applied to the soil as vinasse N complement). *Pesquisa Agropecuária Brasileira, 31*(2), 89-99.

Urquiaga, S., Cruz, K. H., & Boddey, R. M. (1992). Contribution of nitrogen fixation to sugar cane: nitrogen-15 and nitrogen-balance estimates. *Soil Science Society of America Journal, 56*(1), 105-114.

Urquiaga, S., Xavier, R. P., de Morais, R. F., Batista, R. B., Schultz, N., Leite, J. M., et al. (2012). Evidence from field nitrogen balance and 15N natural abundance data for the contribution of biological N2 fixation to Brazilian sugarcane varieties. *Plant and soil, 356*(1-2), 5-21.

Ussiri, D. A. N., & Lal, R. (2009). Long-term tillage effects on soil carbon storage and carbon dioxide emissions in continuous corn cropping system from an alfisol in Ohio. *Soil & Tillage Research, 104*(1), 39-47.

Valdes, C. (2011). Brazil's ethanol industry: looking forward. *ERS Report BIO-02, USDA*, June.

In: Sugarcane
Editor: Eleanore Webb
ISBN: 978-1-61122-356-9

Chapter 3

ECONOMIC IMPACT ASSESSMENT OF SILTING-UP AND EROSION PROCESSES: HOW SPATIAL DYNAMIC MODELS COUPLED WITH ENVIRONMENTAL VALUATION MODELS CAN CONTRIBUTE TO SUSTAINABLE PRACTICES IN SUGARCANE FARMING

***Rodrigo de Campos Macedo*[1], *Cláudia Maria de Almeida*[2], *João Roberto dos Santos*[2], *Bernardo Friedrich Theodor Rudorff*[2], *Britaldo Alves Soares Filho*[3], *Herman Rodrigues*[3] *and Wilson Cabral de Sousa Jr.*[4]**

[1]Brazilian Institute for Geography and Statistics (IBGE)
[2]National Institute for Space Research (INPE)
[3]Federal University of Minas Gerais (UFMG)
[4]Technological Institute of Aeronautics (ITA)

Abstract

This chapter approaches the economic valuation of environmental impacts related to soil erosion and silting-up of water streams, designed to allow the transfer of recovery costs to a policy of payment for ecosystem services. The aim of this study is to evaluate the contribution of silting-up mitigation to funding the environmental recovery of riparian areas found in sugarcane farms. The city selected for study is Arealva, located in the Central-West region of São Paulo State, Southeast of Brazil. Spatial dynamic models were conceived to simulate past land cover/land use changes (2005-2010) and future landscape scenarios (2010-2020) in the study area. The main observed changes that took place from 2005 to 2010 were: sugarcane expansion (6,012.71ha (49.68%)), mostly extending over grazing lands, and deforestation (3,107.16ha (22.33%)), predominantly converted into pastures. Three sets of scenarios were defined: i) stationary scenarios, in which the transition rates observed in former years were held constant (business as usual); ii) non-stationary scenarios with a partial recovery of environmentally protected areas along riversides (70% by 2015), and; iii) non-stationary scenarios with a full recovery of environmentally protected areas along riversides (100% by 2015). The regarded impacts are dependent on the estimated amount of lost soil, assessed by means of the Universal Soil Loss Equation (USLE). We also estimated the sediment accumulation rate in order to calculate siltation. The envisaged scenarios for environmental recovery can reduce environmental impacts up to 16% (US$41,479.29 to US$56,789.40) yearly. The riparian and alike vulnerable areas (prone to erosion and silting-up) can be recovered through a financing mechanism, relying either on water use charging or even on a taxation strategy implicitly considering the payment for ecosystem services. The silting mitigation would approximately contribute with US$13.83 to US$18.94 $ha^{-1}.year^{-1}$. In this way, sugarcane farms would have a financial incentive to restore and maintain the environmentally protected areas within their domain, reducing the environmental impacts related to silting-up processes.

Keywords: Silting-up, economic valuation, environmental impact, payment for ecosystem services

INTRODUCTION

The assessment of the landscape changes is key to efficiency in land management, playing an important role in the decision making related to land use and conservation of the environmental and natural resources.

Land-use and land-cover changes (LUCC) are directly translated into changes in marginal revenues, in the ecosystem services and in the environmental impacts generation. In order to assess the cause-and-effect relationship of such changes, it is necessary to consider environmental and socioeconomic aspects. However, it is difficult to compare disparate variables, involving unknown effects, distinct time scales, etc.

A suitable strategy for working with this complexity is converting the environmental impacts, the ecosystem services and the profitability into a common basis (monetary values, for example). Thus, it is possible to evaluate if an expansion of a kind of land use compensates for the reduction of a given land cover, i.e., if the marginal revenue increase offsets the environmental impacts associated to the new land use and the ecosystem services reduction associated to the altered land cover.

The environmental impacts and ecosystem services are converted into monetary values through several valuation methods. Among them, there are methods which are responsible for acting in the production function, i.e., they are based on changes in productivity or in production costs. These are simple methods, with high reduction of real phenomena. However, they are appropriate to modeling purposes.

The monetary values for the environmental impacts may be linked to a particular type of land use or land cover. This connection is conducted through a dose-response function. The dose is the magnitude of the LUCC, and the response is the consequence in monetary terms. By connecting them based on a spatial dynamic modeling, it is possible to integrate economic, social and environmental aspects. It is possible, then, to perform an environmental planning that expresses future consequences, enabling the comparison of several alternatives related to conservation, modification or conversion of a certain land use/land cover class, facilitating decision making.

The possibility of performing a spatial assessment of the main environmental impacts, comparing alternatives, emulating useful scenarios for environmental planning and monitoring and, especially, connecting data from different formats, is the motivation of this chapter.

The stated problem is related to the environmental impacts valuation associated to erosion and silting-up, linked to the replacement of pastures and native vegetation by sugarcane. The aim of this study is to evaluate the contribution of the silting-up mitigation in funding the riparian areas restoration.

THEORETICAL FRAMEWORK

Environmental Impacts

According to Brasil (1986), environmental impact is any change in the physical, chemical or biological properties of the environment, caused by any form of matter or energy, resulting from human activity which directly or indirectly affects:

- health, safety and welfare of the population;
- social and economic activities;
- biota;
- aesthetic and sanitary conditions of the environment;
- quality of environmental resources.

Studies related to environmental impacts are needed when we consider the assessment of some actions and projects consequences, in order to predict or minimize the quality or quantity loss of a specific environmental aspect (Donaire, 1995).

Impact assessment is seen as an environmental policy tool, formed by a set of procedures capable of ensuring that a systematic examination of its aspects and effects – considering alternatives – is adopted. The Environmental Impact Assessment (EIA) is one of the main legal instruments for conducting an evaluating impacts caused by anthropogenic activities (Gilbert, 1996; Romeiro, 2004).

Environmental Impact Statement/Report of Environmental Impact brings different ways of performing an assessment of environmental impacts, such as the approach PSR (Pressure, State, Response); SWOT analysis (Strength, Weakness, Opportunity, Threats); FMEA (Failure Mode and Effect Analysis); TCO analysis (Total Cost of Ownership); cost-benefit analysis; checklist; mathematical models; matrices or interaction diagrams (Leopold, Singer, etc.); matrices of weights; optimization; projections; and scenario planning, etc. (Carvalho, 2002; Romeiro, 2004).

Many techniques depend on weightings, making the assessment dependent on the reliability of the weights involved. As the environment involves physical (biotic and abiotic) and social (socioeconomic and cultural) factors, the impact assessment will be more assertive when involving as many experts as possible (Mirra, 1998; Silva, 2010).

Erosive Processes

The erosion processes are basically classified in natural and anthropic. The first one is related to the natural deterioration, which is responsible for sculpting the geomorphologic aspects of the landscape. The soil cover makes this removal very slow and it is offset by ongoing processes of soil formation. Under natural conditions, the deterioration cycle is usually balanced by the renewal (Bertoni; Lombardi Neto, 1990).

The anthropic erosion processes are fastest than the natural erosion ones. This accelerated erosion process could be technically defined as the removal of soil particles from the higher parts, by the action of rainwater or wind, and the transport and deposition of these particles into the lower parts of the relief, or into the bottom of lakes, rivers and oceans. Its most common variants are: water erosion and wind erosion (IPT; DAEE, 1997).

Water erosion is, in Brazil, more important than erosion caused by winds. It is composed of two stages: breakdown and transport. The breakdown is caused by the impact of raindrops, as well as by the water which flows across the surface. The raindrops touch the surface with a speed calculated in about 5 to 15 miles an hour, while the flood water speed is usually not more than 1 mile/h. The impact of the raindrops in a soil lacking in vegetation generates the particles disruption, the first step to erosion. When the soil surface is properly protected, the cover absorbs most of the kinetic energy of the raindrops (Lombardi Neto; Drugowich, 1995).

A large amount of soil can be removed, since its particles are disaggregated and suspended in the runoff water. The way the particle is transported depends on its size. Clay and silt are most easily carried by water due to the small size of their particles (Lombardi Neto; Drugowich, 1995).

There are, basically, three types of water erosion: the gradual removal of a thin surface layer of uniform thickness, covering practically all the relief, known as laminar erosion; the erosion in narrow bands along the largest slopes of land is called erosion in furrows; and the displacement of soil mass, forming large cavities or soil landslides is the process known as erosion in gullies (Carey; Silburn, 2006).

Considering these three types, laminar erosion is the most important. Soil losses related to this type of erosion often outweigh the other two forms, those arising from ridges and gullies. All immediately affect the production capacity of the land in a property, while the slow nature of the degradation process causes many problems which are not noticed by farmers before reaching large proportions in order to be corrected (Pierzynski et al., 2000).

The increased pace of erosion has produced noticeable conditions, such as presence of gullies, stunted roots exposed, fallen roadblocks, deep paths in pastures, water reservoir siltation, floods in arable fields, muddy waters in rivers and streams. Dragging of soils and fertilizers to rivers and lakes changes the aquatic micro-fauna and, consequently, the overall fauna, with serious losses.

The greatest or the lowest susceptibility of land to erosion by water depends on a number of factors. Four of these factors are considered the main ones: the climate in the region; the soil type; the terrain slope; and the soil management. The most important climate factors related to erosion are the distribution and the amount and intensity of rainfall. Regarding soils, susceptibility to erosion depends primarily on its physical characteristics, especially its texture, its permeability and its deepness. Flood speed, marked by greater or lesser soil particles dragging, depends on the slopes of the terrain. Finally, the management, or how the land is being used, determines the soil mobility (Sparovek et al., 2007).

Covering the ground with a dense layer of vegetation or debris from previous crops, it is possible to notice that the direct impact of the raindrops is absorbed and there is greater water infiltration. The presence of vegetation minimizes flood. Moreover, the roots are wrapped, holding the ground. The disaggregation and transport of particles vary according to the system of cultivation. Some crops become more susceptible to soil erosion than others. In general, soils with annual crops are more exposed than those grown with perennial or semi-perennial plants (Drugowich et al., 2010).

The way how crops are planted have a great influence. In any culture there are some precautions which must be observed to protect the soil, such as planting level, terracing and no-tillage. These practices could be divided into edaphic, vegetative and mechanical (Almeida et al., 2000):

- *edaphic*: usability adjustment, burning control or elimination, fertilization, crop rotation;

- *vegetative*: zoning, forestry, alley cropping with interception of water runoff through vegetation, planting grass on the slopes of roads, windbreakers, adequate control of weeding, mulching, tillage;
- *mechanical*: soil preparation and contour farming; subsoiling; terraces of the ridge type; terracing, rational arrangement of carriers, and structures for infiltration and deviation of water from the roads, structures to control gullies, retention basins.

In the State of São Paulo (Brazil), there are about 6,700 erosive focuses, and most of them are medium and large (IPT; DAEE, 1997). In rural areas, it is estimated that about 80% of cultivated land is suffering erosion beyond the limits of natural soil recovery. According to Scanavaca Jr. (2011), the State loses 200 million tons of soil per year, with about 50 million being disposed in rivers and streams. This situation generates less fertile soil and silting increasing, as well as it reduces the farm value and increases water treatment costs. The balance between consumption and production is unfavorable, reaching 10kg of soil to 1kg of food. Besides the loss of the resource itself, most of the carbon is fixed in the soil (Bustamante; Oliveira, 2008).

Deposition is the amount of accumulated sediment in a delimited period of time which did not exceed the limit of a given area in question. In order to occur such deposition, there must be transmission or downward movement of water and solids in suspension by superficial flow in the areas between the furrows (Ritter; Shirmohammadi, 2001). This deposition is segmented, with part of the sediment being carried by waterways, part deposited near the source of sediment, and also redistributed in a large extension of the floodplain downstream of the basin or in water reservoirs (Bertolini et al., 1993).

Part of the sediment originated in erosive events is carried downslope and can be placed in the slope itself, while another part can reach waterways (Douglas, 1990). The loose sediment which does not reach watercourses is placed in depressions or in concavities, under vegetation or in other places where the surface flow loses its ability to transport (Rhoton et al., 1982). The sedimentation occurs after and/or during rainfall events, when many soil particles are detached and transported downslope, being retained by plants, depressions or any other obstacle located downstream (Bryan, 2000).

The erosion promotes the generation of sediments which may start processes of turbidity and/or sedimentation of water bodies (Andrade, 2009). Machado et al. (2003) relate erosion and siltation to land use and land cover changes.

Ecosystem Services

Besides the minimization of environmental impacts, there is the possibility of recovering the remaining native vegetation through environmental adaptation. Thus, there could be an increasing of their ecosystem services, those able to support and meet the conditions of human life, such as air purification, protection of soil and natural pest control (De Groot, 1992). Also:

"They are those services that nature provides to absorb, filter and promote water quality; to recycle nutrients and provide soil structure; to maintain climate stability, minimizing disasters such as floods, droughts and storms; to ensure and increase agricultural and industrial production, providing the needed biodiversity and genetic diversity for crop improvement or for drugs, cosmetics or new materials, supplementing processes that human technology does not dominate or replace such as pollination, photosynthesis and waste decomposition" (John, 2008, p 459.).

The characterization of ecosystem services is derived from studies of environmental valuation and inclusion of environmental factors in business negotiations and international agreements. At the beginning, services were considered environmental costs and were associated to evaluations of impacts. This negative characterization related to cost became a positive concept of services and, generally, it was not adequately paid. Costanza et al. (1997) show that, considering the services provided by all existing biomes, the estimated average annual value of these services is US$ 33 trillion, almost the double of the entire world economy GDP.

The Millennium Ecosystem Assessment (MEA, 2003; 2005) suggests the existence of a large number of ecosystem functions and their associated goods and services, gathering ecological functions into four main categories:

- regulatory functions: related to a natural capacity of ecosystems and ecological processes in controling the maintenance of biotic processes through biogeochemical cycles of benefit to living beings, such as clean air, water balance, soil conservation, pollination, sanitary and epidemiological control;
- habitat/support functions: natural ecosystems provide habitat for animal and plant species breeding processes, contributing to biodiversity conservation in situ and genetic diversity. It is the maintenance of ecological and biological processes (nutrient cycling, soil formation, primary production, etc.);

- supply functions: corresponding to the processes of photosynthesis and autotrophic processes that convert carbon dioxide, water and nutrients in carbohydrate structures, which are used for generating higher biomass (provision of raw materials, such as water, food, fiber, genetic resources, biochemical, forestry and fisheries, etc.);
- information functions: resulting from the moments in which the natural ecosystem contributes to the maintenance of human health by providing active ingredients for the pharmaceutical industry, or when promoting functions of reflection, spiritual enrichment and recreation/tourism.

According to Burstein et al. (2002), there is a basic typology of ecosystem services that differentiate into:

- carbon sequestration, which includes the conservation of existing stocks, as well as the increase of fixed carbon in products from forests and other areas where these stocks exist and where they are increased;
- water services and performance monitoring of watersheds, incorporating services such as water supplying and groundwater aquifers recharging, life extension and hydraulic infrastructure, prevention and mitigation of disasters caused by meteorological phenomena of excess or lack of precipitation;
- conservation of biological diversity, including conservation of niches and reduction of habitat fragmentation in the regional landscape, through the creation of ecological corridors;
- scenic beauty, considered as an enhancement factor of natural properties and as a component of recreation services provision.

The ecosystem services performed by riparian forests are associated with the quantity (permanent) and the quality (purity) of water, such as protecting the soil from raindrop impact, reducing the erosive susceptibility, infiltration, vertical intercept, reducing the risk of flooding, landslides, i.e., all the variables which could affect the hydrological cycle (Tonhasca JR., 2004).

The integrated management should consider the risks of reducing the supply of ecosystem services resulting from changes in land use and land cover, which are related to conversions of ecosystems. The remnants of native vegetation shelter ecosystem functions which originate ecosystem services such as springs and watercourses protection; soil cover; nutrient cycling; retention of soil on steep slopes; food, fiber and energy provision; maintenance of genetic resources for the development of industrial, pharmacological and agricultural products; provision of wood and minerals; climate stabilization; pests and diseases control; air and water purification; regulating the water flow and quality; controlling sedimentation; maintenance of soil fertility and nutrient cycling; decomposition of organic waste; aesthetic and cultural benefits and opportunities for leisure (Vilar, 2009). On the other hand, if the ecosystem is inadequate as a degraded area, with restrict environmental attributes and low resilience, the service scenario performance is inversely shown, and, thus, the ability to generate environmental services is subject to the ecosystem integrity and conservation status (Daily, 1997).

Ecosystem Services Payments

There are many examples of mechanisms for capturing the values related to services provided by nature - environmental taxes, green protocols, green taxes, fines, etc. The valuation and policies for ecosystem services payments are strategies for matching economic growth and natural benefits maintenance, adopted by the most relevant environmental agendas (Fearnside, 2004).

Lee and Mahanty (2009) point out that payment for ecosystem service is a political attitude. By adopting the principle of "protector-receiver", the objective is to provide financial incentives for contributing to the maintenance or for increasing the supply of ecosystem services. This policy recognizes the role of protector-receiver and provides the adjustment of the conventional production model to a more sustainable system which ensures both environmental improvements and income generation. This is not compensation, nor the interpretation of environmental conservation as onus (ISA, 2008).

Study Area

The study area is Arealva, a city located in the state of Sao Paulo, Southeast region of Brasil. Its limitrophe planimetric coordinates are: 22º01'44,40" S, 48°54'39,60" W and its average altitude is 445m. Figure 1 shows the location of the municipality in relation to Brazil and the State of São Paulo, including the biomes, rivers and roads. Arealva has 505km² and 7,842 inhabitants (Macedo et al., 2013).

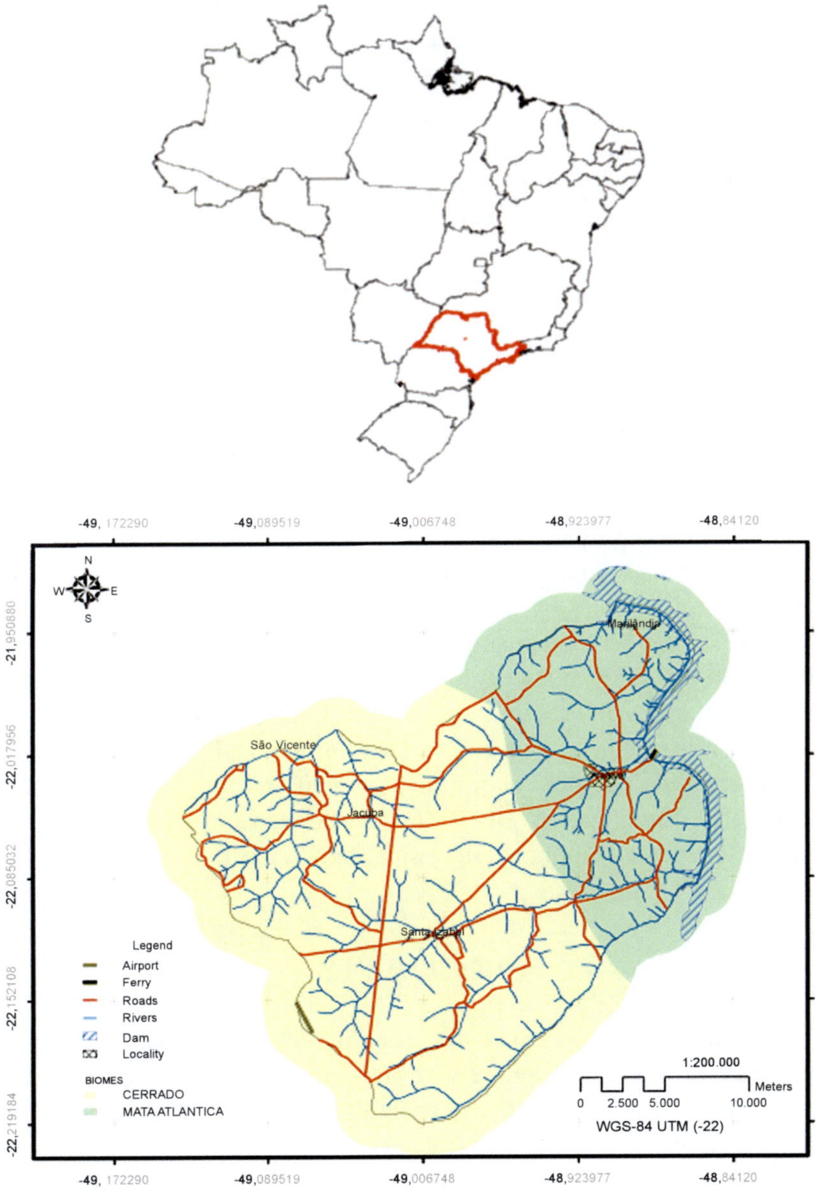

Source: Macedo et al., 2013.

Figure 1. Study area. The city is in a contact zone of two biomes: Cerrado (orange) and Atlantic Forest (green). It is possible to see the municipality (Arealva) and the regions of Marilândia, Jacuba, Santa Izabel and São Vicente, as well as the roads (in red) and the rivers (in blue), including part of the Ibitinga dam.

The climate is considered high-altitude tropical (Aw), sub humid (C), mesothermal (B'), marked by a dry season during the year. Summer's main features are high humidity and high temperature (~24.3° C). Winter is cold and dry (~18.9° C), not exceeding nine rainy days throughout the season. This climate allows the development of all tropical climate cultures which develop their vegetative growth between September and February. The registered pluviometric average between 2000 and 2010 was 1,500 mm a year (Arealva, 2010).

Regarding the hydrography, Tiete is the main river, to which most of the streams and rivers flow. In general, water used for human consumption and livestock watering comes from common and semi-artesian wells. The irrigation is provided by surface water catchment, mainly from Tiete River (São Paulo, 1983; 2008; Arealva, 2010).

The city is located on the east plateau of Sao Paulo. The relief is slightly hilly, with a predominant declivity of 3-8%, enabling mechanized and semi-mechanized agricultural practices. Its eastern side is flatter, while its western side is marked by the presence of hills with a higher dissection rate (Arealva, 2010).

Considering the geological and pedological aspects, Arealva is inserted in the Bauru Group (Vale do Rio do Peixe formation), a region formed by sedimentary rocks, mainly sandstones. The mineral produced different soil types, such as distrophic red oxysoils, red-yellow oxisoil, red-yellow argisoil and abrupt argisoil. The presence of shallow soil and the slightly hilly relief determine the classification of 75% of the territory as high erosion susceptibility, especially in the western side (São Paulo, 2000).

PROCEDURES

Table 1 summarizes the data used, including their type, base year, purpose and reference to the generation of land use and land cover maps (2005 and 2010), as well as to the simulations. Table 2 shows the applications used and their main features.

All procedures related to the field survey, to the preparation and validation of land use/land cover maps (2005 and 2010), to the detection of changes which occurred (between 2005 and 2010), to the calibration, to the parameterization and validation of the LUCC model, to the simulation of past scenarios (2006 to 2010) and the generation of future scenarios (till 2020) are described in Macedo et al. (2013), whose methodological procedure is illustrated in Figure 2.

Table 1. Used data, format, reference date and purpose

Data	Format	Base Year	Purpose	Source
Land use and land cover maps	Raster	2005 and 2010	Obtain land use/cover classes.	Macedo et al., 2013
Watersheds	Polygon	2010	Calculating the rate of sediment delivery.	Ana, 2010
Slope – degrees	Raster	2000	Calculating the Topographic Factor.	Valeriano, 2008; Valeriano; Rossetti, 2011
Elevation (Topodata)	Raster	2000	Calculating the Catchment Area.	Valeriano, 2008; Valeriano; Rossetti, 2011
Pluviosity	Table	1990-2010	Calculating the Erosivity.	Ciiagro, 2013; DAEE, 2013
Pedological map	Polygon	1982	Calculating the Erodibility.	Almeida et al., 1982
Silting-up events	Table with information based on water body/stream	2005-2010	Water quality monitoring.	CETESB, 2012; SABESP, 2012
Values related to dredging and alike silting-up mitigation actions	Table	2004-2013	Value related to the cost of silting-up mitigation.	Nóbrega, 2004; Bigaran; Tizato, 2009; Moreira, 2011

Table 2. Applications used

Software	Purpose
ARCGIS v. 10	Variables standardization, tables, datum and projection; vector editing; rasterization; raster editing; accumulation area assessment.
Erosividade Brasil	Calculating the erosivity.
Erodibilidade Brasil	Calculating the erodibility.
IDRISI v. 14 (Kilimanjaro)	Calculating the soil loss for comparative purposes.
USLE-2D	Calculating the topographic factor, for comparative purposes.

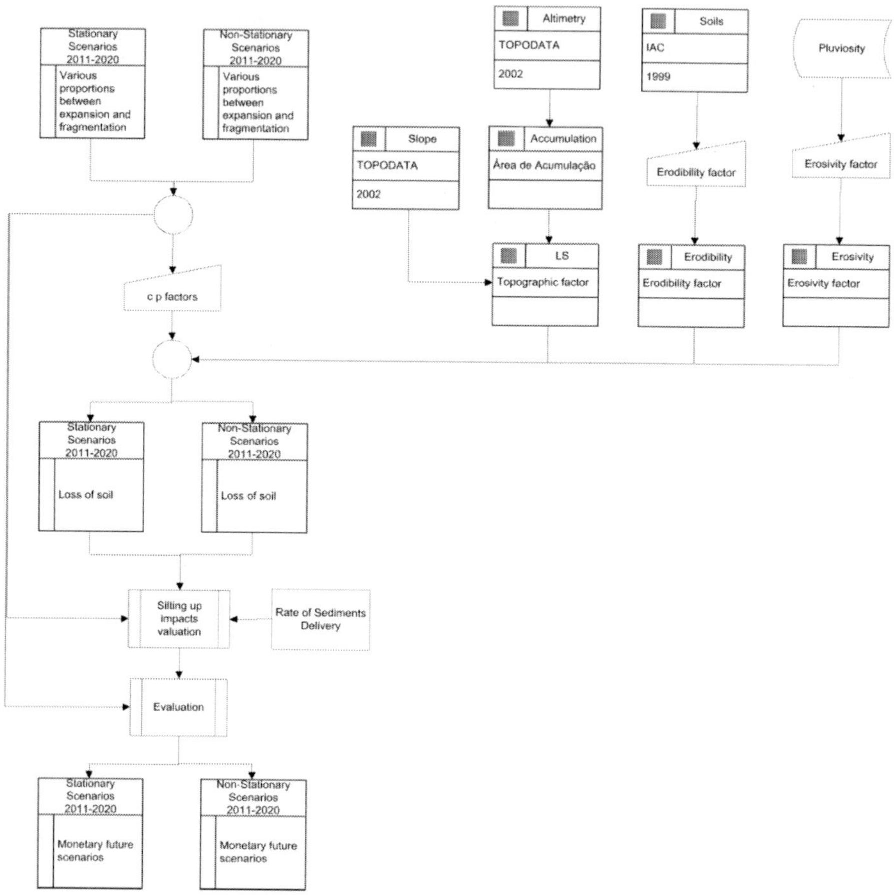

Figure 2. Methodological procedure for generating future land cover/land use scenarios and its valuation of environmental impacts produced by the silting-up process.

The scenarios were proposed considering the revegetation of riparian areas as a strategy for minimizing environmental impacts and for recovering environmental services which have been already committed. According to São Paulo (2011), the riparian area of Arealva is about 3,000 inhabitants. It is estimated that around 35% of this area is irregularly occupied. Moreover, there are no remaining areas under recovery.

Two sets of scenarios were defined:

- Stationary scenarios, which maintained the transition rates and reproduced the features detected (business as usual). The historical trend is the replacement of pasture by sugarcane;

- Non-stationary scenarios of readjustment of riparian areas. In this scenario, two rates of revegetation were simulated: i) 70% by 2015 and; ii) 100% by 2015.

Non-stationary scenarios of environmental readjustment simulated the increasing native vegetation and the reduction of the other classes in the riparian area. Outside them, the stationary transition rates were maintained, i.e., the detected changes (historical trend), including the reduction of native vegetation were kept for calculations.

The selected environmental impacts were related to erosion and silting-up. They are influenced not only by the type of land use and the management practices performed, but also due to the topography. Therefore, we estimated the potential soil loss based on the universal soil loss equation (USLE), considering the annual stationary and non-stationary scenarios (Equation 1).

$$A \square R * K * LS * CP \quad (1)$$

USLE

A = rate of erosion per unit area, in $t.ha^{-1}.year^{-1}$;

R = erosive power of the rain, in $MJ.mm.ha^{-1}.h^{-1}.year^{-1}$;

K = soil erodibility, in $t.h.MJ^{-1}\ mm^{-1}$;

LS = land slope and length, dimensionless;

CP = degree of soil cover (C) and conservation practices (P), dimensionless.

Considering the average rainfall in the last 20 years and the erosivity data presented in the Erosividade-Brazil database (Silva, 2004; Silva et al., 2006), we measured the erosivity (R) for the entire municipality, since there is no significant spatial variation in the annual rainfall height in the study area (Cataneo et al., 1992).

According to each type of soil, we adopted the erodibility factor suitable for the observed features (Bertoni; Lombardi Neto, 1990; Marques, 1996). Table 3 shows the respective values.

The equation used in the Erodibilidade-Brasil application (Silva; Alvares, 2005) is cited in Mitchell and Bubenzer (1980), converted into the international system, according to Foster et al. (1981).

Table 3. Erodibility factor by soil class

Class	Erodibility (K)
Dystrophic red latosoils + dystrophic red-yellow oxisoils, A moderate, medium texture, flat and slightly-wavy relief.	0.042
Eutrophic red-yellow argisoils + dystrophic and eutrophic red argisoils, both sandy / medium texture, mild slightly relief.	0.02
Dystrophic red latosoils + dystrophic red-yellow and red argisoils, A moderate, medium texture, flat and slightly-wavy relief.	0.0162
Eutroferric and dystroferric red latosoils, A moderate, clayey, flat and slightly-wavy relief.	0.013

Source: Lombardi-Neto and Bertoni (1975); Bertoni and Lombardi-Neto (1990); Marques (1996).

Table 4. Land use factors (C) and the conservation practices (P) land use/land cover class

Class	C	P
Urban	0	0
Water body	0	0
Sugarcane	0.18	0.5
Other cultures	0.25	0.7
Grassland	0.3	0.5
Silviculture	0.1	0.2
Riparian area (non-vegetated)	0.26	0.06
Native vegetation	0.05	0.2
Riparian area (vegetated)	0. 012	0.1

Source: Bertoni and Lombardi-Neto (1990); Lepsch et al. (1991).

The topographic factor (LS) was calculated based on Equation 2 (Azim-Zade, 2010), using the digital elevation model resampled to 20m of spatial resolution (Cowen, 1993; Desmet; Govers, 1996; Salgado, 2011, Salgado et al., 2011;. Salgado et al., 2012).

$$LS = (CA * \frac{PS}{22.13})^{0.6} * (\frac{sin(d*0.01745)}{0.09})^{1.3} \qquad (2)$$

LS = Topographic factor;
CA = Catchment area;
PS = Pixel Size;
d = Slope.

We calculated the land use factors (C) and conservation practices factors (P) (Bertoni and Lombardi-Neto, 1990; Lepsch et al., 1991), as shown in the Table 4.

The desilting costs are proportional to the amount of suspended matter, which is related to the erosion processes. In Arealva, both the company which is responsible for the local hydroelectric power plant and the company which holds the concession of the Alcohol Waterway carry the costs of dredging, estimated at around US\$9.00.t^{-1} (Nóbrega, 2004; Bigaran; Tizato, 2009; Moreira, 2011). The amount of sediments depends on several factors, including the watershed area and the length of rivers (Sousa Jr., 2011). It is necessary to estimate the rate of sediment delivery (RS), described in Equation 3 (Roehl, 1962).

$$RS = log2.88753 - 0.83291.colog\frac{R}{L} \tag{3}$$

RS = rate of sediment delivery, dimensionless;
R = range between the highest and the lowest elevation in the basin, in meters;
L = length of the main stream of water from the basin, in meters.

We considered the watersheds for estimating the RS. Based on it, we calculated the environmental impact related to siltation (Equation 4) (SOUSA Jr., 2011).

$$ISILT = A * RS * 0.5 * PDRED \tag{4}$$

ISILT = Impact related to siltation, in US\$.ha^{-1}.$year^{-1}$;
A = Soil loss, in t.ha^{-1}.$year^{-1}$;
RS = Rate of sediment delivery;
PDRED = Average price of dredging and desilting, in US\$.t^{-1}.

The coupling of the environmental impacts valuation in LUCC models allows the design of several scenarios which integrate not only plausible future landscape patterns, but mainly the derived consequences in terms of environmental impacts.

Assessment of the Environmental Impacts Resulting from Erosion and Silting-Up for the Envisaged Land Use/Land Cover Scenarios

The environmental impacts derived from erosion and silting-up, caused by the replacement of grassland and native vegetation by sugarcane culture, are dependent on the estimated amount of lost soil. This estimate can be seen in Figure 3.

There is a clear reduction of soil loss in the scenarios of environmental readjustment. In the stationary scenario, there is a reduction in soil loss due to the replacement of grasslands by sugarcane, since the sugarcane fields own a c factor accounting on average for half of the c factor belonging to the other land use/cover classes. Scenarios with environmental readjustment presented lower values than stationary scenarios, ranging from 1 to 16%. Even with the recovery of riparian areas, there will be an erosive potential, since it is also conditioned by erosivity, erodibility and topographic factors, which are the most relevant ones in calculating soil loss (Valeriano, 2003; Chaves, 2010; Salgado, 2011).

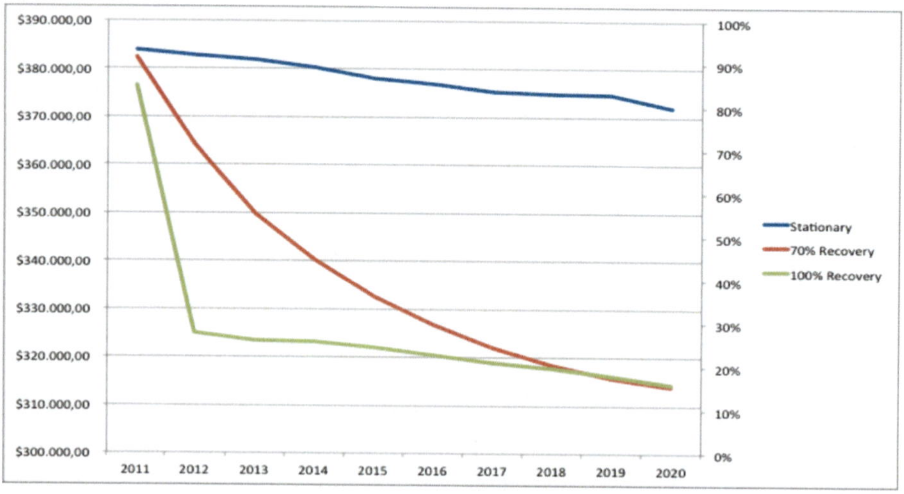

Figure 3. Dynamics of the soil loss estimate related to the considered scenarios, from 2011 to 2020 ($year^{-1}$).

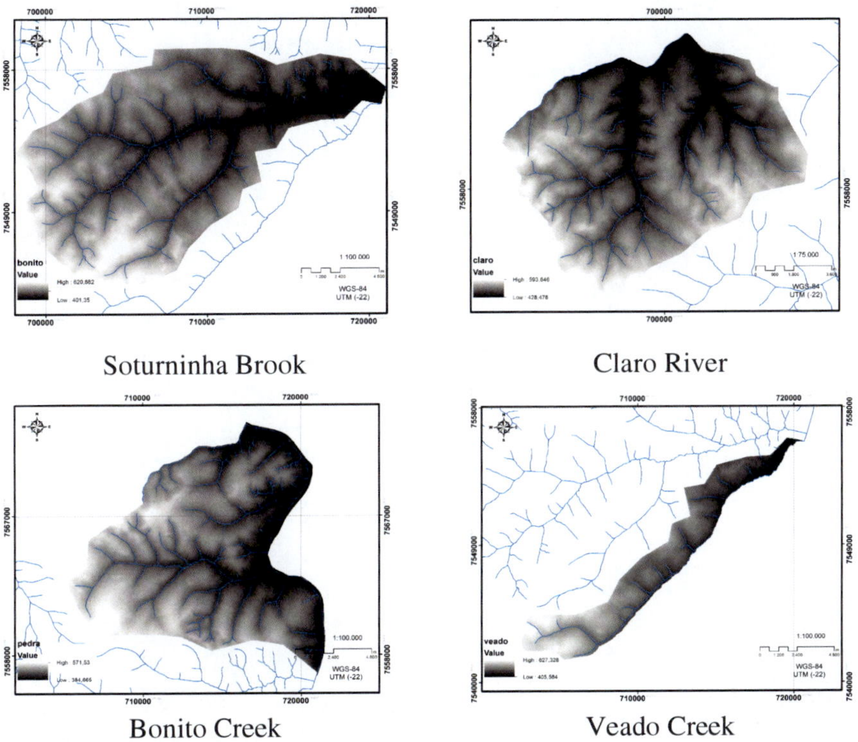

Source: Adapted from ANA (2010); Valeriano (2008).

Figure 4. Watersheds considered for the calculation of the rate of sediment (RS) delivery in Arealva.

Table 5. Calculation of the sediment delivery rate (RS), altimetric range, name and length of the main watercourse and the watersheds area of Arealva

Watershed	Area (ha)	Main waterstream	Length (m)	Elevation range (m)	RS
Soturninha Brook (1)	15,771.09	Soturninha Brook	14,521.70	186.86	0.272248022
Claro River	10,866.10	Claro River	13,121.62	165.17	0.266164605
Bonito Creek	19,241.34	Bonito Creek	26,333.29	219.53	0.144882799
Veado Creek	4,652.99	Veado Creek	24,316.37	221.74	0.172569649

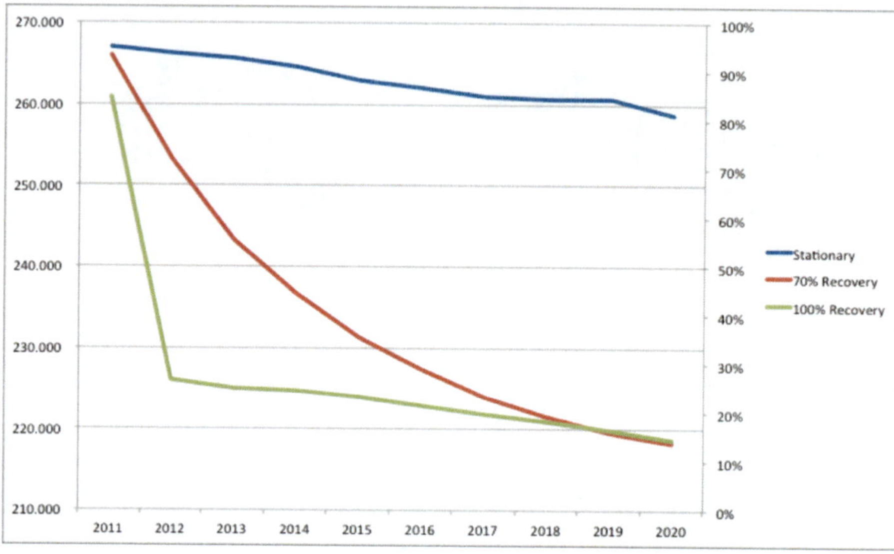

Figure 5. Dynamics of the environmental impact estimate related to the dredging and desilting of the watersheds in Arealva for the considered LUCC scenarios, from 2011 to 2020 (US$.year^{-1}).

From time to time (nearly in a one-year span), investments associated with dredging for reducing sediments are made in the Ibitinga reservoir, so as to enable the continuity in electrical energy generation and also to minimize the risk of low flow and draughts in the Tietê River, which could jeopardize the operationality of the Alcohol Waterway. These investments can be transferred to the final consumer, but they are presently regarded as a control measure against a situation of widespread environmental impact, although such impact is in fact largely originated upstream and unfairly beared by the agricultural sector.

In the particular case of dredging, the estimated cost depends on the rate of sediment delivery. This rate was calculated by assessing the proportial area of all watersheds partially or entirely contained within the municipal boundaries of Arealva, as it can be seen in Figure 4 and in Table 5.

Considering that 50% of the sediments will have to be effectively dredged, it can be observed in Figure 5 that the dynamics of the economic impacts related to dredging expenses are directly proportional to soil loss.

In the analyzed period, the environmental readjustment scenarios can reduce the impacts related to dredging and desilting in 16%, being US$414,793.00 (70% recovery) and US$567,895.00 (full recovery). If we consider the total riparian area (about 3,000 ha), the mitigation of environmental impacts could sum up approximately US$13.83 $ha^{-1}.year^{-1}$ or US$18.94 $ha^{-1}.year^{-1}$. These values are lower than the reported payment of main environmental services, but these financial resources could certainly be allocated for the recovery of riparian areas.

A reduction in profitability was noticed, despite there was a reduction in the selected environmental impacts, since the environmental readjustment includes the replacement of certain types of land use that are not in compliance with the current legislation (Brasil, 2012). However, if we consider that all sectors must be in accordance with the laws and that many ecosystem services and environmental impacts were not considered, reducing profitability should not be taken as a hindrance to the recovery of riparian areas. Moreover, some agricultural activities such as dairy and cattle production are underpaid and cause severe and large extent environmental impacts. This requires investments in infrastructure, such as drilling wells, for example.

Therefore, the environmental impacts reduction coupled with an increased productivity compensates for any decrease in a productive area caused by environmental readjustment initiatives.

CONCLUSION

The assessment of environmental impacts related to agricultural activities and ecosystem services in wild environments by LUCC monitoring is essential for an integrated and comprehensive analysis. The economic variables coupled with the spatial variables enable more representative models, allowing greater effectiveness in their applications.

The environmental readjustment in riparian areas is not only a legal issue. Ecosystem services are important for the sustainability of agricultural activities. Any reduction in terms of aggregate profitability should be offset by an increase in the ecosystem services provision and by a reduction in the risk of environmental impacts.

Environmental readjustments normally reflect in reduced profitability (sacrificed income). However, it brings a dual benefit: it increases the supply of ecosystem services and decreases the risk of negative environmental impacts. Both benefits are external to the market and difficult to be detected by the agents.

The environmental readjustment scenarios are able to reduce environmental impacts in up to 16% (US$41,479.29 - US$56,789.40) annually. This calculation is exclusively related to the selected study area. Such scenarios are also able to increase the wealth of the concerned municipality, depending on a two-fold strategy of: i) assigning the due importance to ecosystem services, including actions for rendering environmental resources readily available; and of ii) internalizing environmental impacts, including a fair reallocation of dredging costs upstream, where sediment delivery predominantly takes place.

The provision of an economic tool designed for the valuation of ecosystem services, coupled with the internalization of potential environmental impacts, is able to render the environmental readjustment feasible and competitive. The riparian and other fragile areas (prone to erosion and silting-up) could be recovered through a funding mechanism, which could be based on procedures of charging for water use and/or even on a strategy for ecosystem services payment. Such mechanism could also rely on a progressive discount on rural property taxes proportional to the percentage of recovered riparian area inside the property. The mitigation of silting-up would approximately account for savings of US\$13.83 - US\$18.94 $ha^{-1}.year^{-1}$.

Obviously, there are other impacts that could be mitigated, such as plagues, diseases, biodiversity loss, fire, etc., what would contribute to the reduction of the total value associated with them,. It is worth highlighting that this particular work concerns a reduced analysis as to its geographical, temporal and thematic dimensions. The span of time considered for this analysis ranges from 2005 to 2010 for the purpose of calibration, and from 2010 to 2020 for the sake of prediction. Furthermore, the considered environmental impacts do not regard further implications in the medium- and long-run and neither do they take into account a geographical extent beyond the Arealva municipality boundaries. In other words, we can state that the erosion and silting-up processes produced in Arealva will have implications outside this municipality, and this has not been taken into account in our analyses. Lastly, we must clarify that the land cover and land use classes adopted in this work comprise a wide range of environmental impacts and ecosystem services, which could not be dealt with in a thorough manner, since this would make this research unfeasible in view of constraints related to the lack of technical and financial resources, human capital, data availability, computational processing capacity and execution deadlines. Nevertheless, we strong recommend that local stakeholders promote further valuations of environmental impacts and ecosystems services as comprehensive as possible in the future, what would certainly result in more substantial savings derived from the mitigation of other environmental impacts found in the study area.

ACKNOWLEDGMENTS

The authors thank the Brazilian National Agency for the Upgrade of Graduate Personnel - CAPES, National Institute for Space Research - INPE / PG-SERE and the São Paulo State Foundation for Research Support - FAPESP, Grant n. 08/56252-0, for their valuable financial support. We also thank the Brazilian Institute for Geography and Statistics (IBGE), the Campinas Environmental Institute (IAC), the Vidágua Environmental Institute, the Sao Paulo Technological Research Institute (IPT), and the Agrosatélite enterprise for the provision of additional thematic data. The authors are as well particularly grateful to Dr. Joshua Farley (University of Vermont and Gund Institute).

REFERENCES

Almeida, C. L. F.; Oliveira, J. B.; Prado, H. Semi-detailed soil survey of the São Paulo: Jaú Grid (SF22 Z-B-II). 1:100.000. Campinas: Agronomic Institute, 1982 (*in portuguese*).

Almeida, R. G. Degradation, recovery and sustentability of cultivated pastures. Seminar presented to University of Viçosa, Viçosa/MG. 2000 (*in portuguese*).

ANA (Agência Nacional Das Águas, National Water Agency). Georeferenced data bases: microbasins. Ministry of Environment. Available in: <http://www.ana.gov.br/bibliotecavirtual> (*in portuguese*).

______. Program for improvement of quality and quantity of water in rural basins through the financial incentives for producers: The Water Producer Program. Available in: <http://www.ana.gov.br/produagua/> (*in portuguese*).

Andrade, J. M. F. Construction of an environmental sustainability index for agricultural industry of sugarcane. Dissertation in Agroenergy. Fundação Getúlio Vargas – FGV. São Paulo/SP, 2009. 260p. (*in portuguese*).

AREALVA. Municipal plan for rural sustainable development from 2010 to 2013. Municipality of Arealva. Rural Development Council. 2010 (*in portuguese*).

Azim-Zade, Z. GIS assessment of Pipeline corridor erosion along Baku-Tbilisi-Ceyhan route. Available in: <http://proceedings.esri.com/library/userconf/pug10/papers/gis_assessment_of_erosion_corridor.pdf>. 2010.

Bertolini, D.; Lombardi-Neto, F.; Lepsch, I.; Oliveira, J. B.; Drugowich, M. E.; Andrade, N. O.; Galeti, P. A.; Bellinazzi Jr., R.; Dechen, S. C. F. Handbook of management and conservation of soil and water: available technologies to control the soil runoff. Campinas/SP, CATI. 1993. 65p. (*in portuguese*).

Bertoni, J.; Lombardi Neto, F. Soil conservation. Ícone Ed., São Paulo/SP, 1990. 355p. (*in portuguese*).

Bigaran, J. T.; Tizato, L. H. G. Tietê-Paraná waterway. Piracicaba/SP. 2009. 19p. (*in portuguese*).

Brasil. Law n. 12.727, 2012. Regulates the protection of native vegetation. 2012 (*in portuguese*).

______. Resolution CONAMA n. 001, 1986 (*in portuguese*).

Bryan, R. B. Soil erodibility and process of water erosion on hillslope. *Geomorphology*, Amsterdam, v. 32, p. 385-415, 2000.

Burstein, J.; Mendoza, G. C.; Aguilar, J.; León, E. Report about the Environmental Services Payment in Mexico In: Rosa, H.; Kandel, S. FORD Foundation and PRISMA. 2002. Available in: <http://www.undp.org.cu/eventos/aprotegidas/Pago_Serv_Amb_Mexico_Fund_Ford.pdf> *(in spanish)*.

Bustamante, M. M. C.; Oliveira, E. L. Impact of agricultural, forestry and livestock activities on natural resources. In: II International Symposium on Savanna Agricultural; IX National Symposium of Cerrado, 2008. Proceedings... Brasília/DF, 2008, p.647-669. (*in portuguese*).

Carey, B.; Silburn, M. Erosion control in grazing lands. *Queensland, Natural Resource Sciences*, L91, 2006.

Carvalho, P. F. The environmental issue requiring a new world order. São Paulo, Ed. HUCITEC, p. 111-145. 2002 (*in portuguese*).

Cataneo, A.; Carvalho G. M.; Vasques Filho, B. S.; Lombardi Neto, F. Correlation between the average monthly rainfall erosivity index and the rain coefficient of the Mococa/SP. Científica, v. 19, n. 1, p. 1-7. 1992 (*in portuguese*).

CETESB (Companhia de Tecnologia de Saneamento Ambiental, Company of Environmental Sanitation Technology). Monitoring of water quality. Available in: <www.cetesb.sp.gov.br>. 2012 (*in portuguese*).

Chaves, H. M. L. Uncertainties in the erosion prediction with USLE: impacts and mitigation. *Brazilian Journal of Soil Science*, v. 34, p. 2021-2029, 2010 (*in portuguese*).

CIIAGRO (Centro integrado de informações agrometeorológicas, Center of agrometeorological information). Rainfall data. Available in: <http://www.ciiagro.sp.gov.br/ciiagroonline/> (*in portuguese*).

Costanza, R.; D´Arge, R.; Groot, R.; Farber, S.; Grasso, M.; Hannon, B.; Limburg, K.; Naeem, S.; O´Neill, R. V.; Paruelo, J.; Raskin, R. G.; Sutton, P.; Van Den Belt, M. The value of the world´s ecosystem services and natural capital. *Nature*, v. 387, p. 253-260, 1997.

Cowen, J. A proposed method for calculating the LS factor for use with the USLE in a grid-based Environment. In: Thirteenth Annual ESRI User Conference. Proceedings... San Francisco, p. 65-74, 1993.

DAEE (Departamento de Águas e Energia Elétrica do Estado de São Paulo, Department of Water and Energy, São Paulo State). Rainfall database. Available in: <http://www.sigrh.sp.gov.br/cgi-bin/bdhm.exe/plu?lig= podfp> (*in portuguese*).

Daily, G. C. (Ed.). Nature's services: Societal dependence on natural ecosystems. Island Press, Washington, DC, USA, 1997.

De Groot, R. S. Functions of nature. Evaluation of nature in environmental planning, management and decision making. 315p. 1992.

Desmet, P. J. J.; Govers, G. A GIS procedure for automatically calculating the USLE LS factor on topographically complex landscape units. *Journal of Soil and Water Conservation*, v. 51, p. 427-433, 1996.

Donaire, D. Environmental Management in the Company. São Paulo: Ed. Atlas. Cap. 3. 1995 (*in portuguese*).

Douglas, I. Sediment transfer and siltation. In: Turner II, B. L.; Clark, W. C.; Kates, R. W.; Richards, J. F.; Mathews, J. T.; Meyer, W. B. (Eds.). The Earth as transformed by human action. Cambridge: Cambridge University Press, 1990. p. 215-234.

Drugowich, M. I.; Savastano, S.; Savastano, S. A. A. L. Erosion in pastures in the state of São Paulo. 2010. Available in: <http://www.cati.sp.gov.br/new/acervo/recursos_naturais/ErosaoPastagensSP.pdf> (*in portuguese*).

Fearnside, P. M. Environmental services as a basis for sustainable use of tropical forests in the Brazilian Amazon. p. 9-23. In: IV Workshop Advances in Energy Studies: Ecology - Energy in Latin America. Proceedings... Unicamp, 2004 (*in portuguese*).

Foster, G. R.; McCool, D. K.; Renard, K. G.; Moldenhauer, W. C. Conversion of the Universal Soil Loss Equation to SI metric units. Journal of Soil and Water Conservation. Ankeney, Iowa, v. 36, n. 6. p. 355-359. 1981.

Gilbert, M. J. ISO 14001/BS 7750 - Environmental management system. IMAM, 1996 (*in portuguese*).

IPT (Instituto de Pesquisa e Tecnologia de São Paulo, Institute of Research and Technology of São Paulo); DAEE (Departamento de Água e Esgoto de São Paulo, Department of Water and Sanitation of São Paulo). Diagnosis of erosion in São Paulo State. 1997 (*in portuguese*).

ISA (Instituto Socioambiental, Socioenvironmental Institute). Environmental services: knowing, valuing and caring. Subsidies for the watersheds protection. São Paulo. 2008. 119p. (*in portuguese*).

John, L. Environmental services. In: ISA. Brazil-Environmental almanac. A new perspective to understand the situation of Brazil and our contribution to the planetary crisis, 2008. São Paulo. 553p. (*in portuguese*).

Lee, E.; Mahanty, S. Payments for environmental services and poverty reduction: Risks and opportunities. The Center for People and Forests. Issues Paper. Bangkok 10903, Thailand, 2009. 40p.

Lepsch, I. F.; Belinazzi Jr., R.; Bertolini, D.; Espindola, C. R. Handbook to survey of the physical environment and land classification. Ed. 4. Campinas: SBCC, 1991. 175p. (*in portuguese*).

Lombardi Neto, F.; Bertoni, J. Erodibility of soils from São Paulo. Technical Bulletin, n. 27. Campinas: Agronomic Institute, 1975. 12p. (*in portuguese*).

Lombardi Neto, F.; Drugowich, M. I. (coord.). Technical handbook on soil and water management. Campinas/SP. Department of Agriculture and Food Supply, 1995 (*in portuguese*).

Macedo, R. C.; Almeida, C. M.; Santos, J. R.; Rudorff, B. F. Spatial dynamic modeling of land use and land cover changes related to sugarcane expansion. *Bulletin of Geodetic Sciences*, v. 19, n. 2, p. 314-339. 2013 (*in portuguese*).

Machado, R. E.; Vettorazzi, C. A.; Xavier, A. C. Simulation of alternative land use scenarios in a watershed using GIS and modeling techniques. 2003. *Brazilian Journal of Soil Science*, v. 27, n. 4, p. 727-733. Viçosa/MG. 2003 (*in portuguese*).

Marques, J. J. G. S. Estimates and comparisons of the rainfall factors and soil erodibility in Brazil. Master's degree in Agronomy. Federal University of Lavras, Lavras/MG, 119p. 1996 (*in portuguese*).

MEA (Millennium Ecosystem Assessment). Ecosystem and human well-being: a framework for assessment. Washington, DC: Island Press, 2003.

MEA (Millennium Ecosystem Assessment). Ecosystem and human well-being: synthesis. Washington, DC: Island Press, 2005. Available in: <http://www.millenniumassessment.org/en/index.aspx>.

Mirra, A. L. V. Environmental impact: aspects of brazilian legislation. São Paulo: Ed. Oliveira Mendes, 1998 (*in portuguese*).

Mitchell, J. K.; Bubenzer, G. D. Soil loss estimation. In: Kirkby, M. J.; Morgan, R. P. C. *Soil erosion*. John Wiley & Sons Ltd., 1980.

Moreira, M. C. A. Evaluation of the influence of the Barra Bonita dam in the morphodynamics of the Tietê River and its impacts to navigation. PhD thesis in Civil Engineering. Faculty of Civil Engineering, Architecture and Urbanism. 2011 (*in portuguese*).

Nóbrega, C. A. W. National Agency for Waterway Transportation. Dredging. Foreign Trade Chamber. Rio de Janeiro, 2004 (*in portuguese*).

Pierzynski, G. M.; Sims, J. T.; Vance, G. F. Soils and environmental quality, 2. ed. CRC Press, Boca Raton, FL. 459p. 2000.

Rhoton, F. E.; Meyer, L. D.; Whisler, F. D. A laboratory method of predicting the size distribution of sediment eroded form surface soils. *Soil Science Society America Journal*, Madison, v. 46, p. 1259-1263, 1982.

Ritter, W. F.; Shirmohammadi, A. Agricultural drainage and water quality. p. 59-90. In: Ritter, W. F.; Shirmohammadi, A. (eds.) Agricultural Nonpoint Source Pollution: Watershed Management and Hydrology. Lewis Publishers, Boca Raton, FL. 2001.

Roehl, J. W. Sediment source areas, delivery ratios and influencing morphological factors. International Association of Hydrological Sciences, v. 59, p. 202-213, 1962.

Romeiro, A. R. (Org.). Evaluation and accounting of the environmental impacts. Campinas. Ed. Unicamp. Sao Paulo/SP, 2004 (*in portuguese*).

SABESP (Companhia de Saneamento Básico do Estado de São Paulo, Company of Sanitation of the São Paulo State). Water treatment. Available in: <http://site.sabesp.com.br/site/interna> (*in portuguese*).

Salgado, M. P. G. Evaluation of SRTM and GDEM data, applied to modeling the topographic factor of USLE. Master's degree in Remote Sensing. INPE, São José dos Campos/SP. 2011. 151p. (*in portuguese*).

______; Formaggio, A. R.; Rudorff, B. F. T. Refinement of the digital elevation model applied to erosion modeling of watersheds. Geografia, v. 36, p. 201-211. Rio Claro/SP. 2011.

______; Formaggio, A. R.; Rudorff, B. F. T. Validation of SRTM data applied to modeling the topographic factor of USLE. *Brazilian Journal of Cartography*, v. 64, n. 4, p. 429-442. 2012 (*in portuguese*).

São Paulo. Government. Municipal strategy of industrial and habitational development - Arealva/SP, 1983 (*in portuguese*).

______. IPT. Diagnosis of the current situation of water resources and establishing technical guidelines for the preparation of the Tietê-Jacaré Basin Plan - Final Report. São Paulo: CBH-TJ/FEHIDRO/IPT. 2000 (*in portuguese*).

______. Technical Report n. 402/08. Review of the Tietê-Jacaré Basin Plan (UGRHI 13), 2008 (*in portuguese*).

______; Fehidro; Vidagua. Consolidated regional atlas of the land use in riparian areas of the Tietê-Jacaré river watershed. Strategic plan for the evaluation of environmental status of riparian areas. Bauru/SP. 78p. 2011 (*in portuguese*).

Scanavaca Jr., L. Preserve for profit: environmentally and economically. Plasticulture magazine, Campinas/SP, p. 18 - 19, 2011 (*in portuguese*).

Silva, A. F.; Ferreira, A. C. S. A theoretical study on the accounting of environmental impacts in the biofuels industry. Journal of Accounting and Organizations – FEA-RP/USP, v. 4, n. 8, p. 139-159, 2010 (*in portuguese*).

Silva, A. M. Rainfall erosivity map for Brazil, Catena, v. 57, n. 3, p. 251-259. 2004.

______; Alvares, C. A. Erodibility of soils from São Paulo: information gathering and structuring of a database. *Geosciences*, v. 24, p. 33-42. 2005 (*in portuguese*).

______; Rodrigues, B. B.; Marques, M. A. Erosivity Brazil - software in preparation. In: XVI Brazilian Meeting on Management and Conservation of Soil and Water. Proceedings... Aracaju/SE. 2006 (*in portuguese*).

Sousa Jr., W. C. de. Ecosystem services payments: riparian forests, erosion, silting on and water quality. Project of Riparian Forest Recovery - Technical Products, n. 01, 2011. 44p. Available in: <http://www. ambiente.sp.gov.br/mataciliar> (*in portuguese*).

Sparovek, G.; Correchel, V.; Barretto, A. G. O. P. The risk of erosion in Brazilian cultivated pastures. *Scientia Agricola*, v. 64 n. 1, Piracicaba, 2007.

Tonhasca Jr., A. The ecological services of the atlantic forest. *Science Today Magazine*, v. 35, n. 205. p. 64- 65. 2004 (*in portuguese*).

Valeriano, M. M. Vertical curvature of strands in watersheds by digital elevation model analyzis. *Brazilian Journal of Agricultural and Environmental Engineering*, v. 7, n. 3, p. 539-546, Campina Grande/PB. 2003 (*in portuguese*).

______. Topodata: Geomorphometric data base in Brazil. Guide to using local geomorphometric data. São José dos Campos: INPE, 2008. 44p. On line.

Available in: <http://www.dpi.inpe.br/topodata/data/guia_enx.pdf>. 2011 (*in portuguese*).

______; Rosseti, D. F. Topodata: brazilian full coverage refinement of SRTM data. *Applied Geography*, v. 32, n. 2, p. 300–309, 2011.

Vilar, M. B. Economic valuation of environmental services on farms. Dissertation. Federal University of Viçosa. Viçosa/MG, 2009. 171 p. (*in portuguese*).

In: Sugarcane
Editor: Eleanore Webb
ISBN: 978-1-61122-356-9

Chapter 4

ENVIRONMENTAL IMPLICATIONS OF USING WASTE FROM SUGARCANE INDUSTRY IN AGRICULTURE

Rafael G. Botellho[1]*, Cintya A. Christofoletti[2], Jorge E. Correia[3] and Valdemar L. Tornisielo[1]

[1]Laboratório de Ecotoxicologia, Centro de Energia Nuclear na Agricultura, Universidade de São Paulo - CENA/USP. Piracicaba-SP, Brasil
[2]Centro Universitário Hermínio Ometto - FHO/UNIARARAS, Av. Dr. Maximiliano Baruto, Araras-SP, Brazil
[3]Universidade Estadual Paulista Júlio de Mesquita Filho – UNESP, Rio Claro-SP, Brazil

ABSTRACT

The ethanol industry is of great importance to the Brazilian economy since the sugarcane is one of the most important monocultures in the country. Although its activities are regulated by numerous rules in order to minimize the environmental impacts, the sector is worried about the amount of waste that results from the sugar and alcohol production process. Among the waste from this process, the bagasse, filter cake and vinasse can be highlighted. The environmental impacts of using these waste/byproducts in the agriculture are still not completely elucidated and

---

* Corresponding author: Rafael Grossi Botelho. Email: rbotelho@cena.usp.br.

have been emphasized, mainly due to the adverse effects on the aquatic and soil environments. Thus, the present chapter will gather the available data regarding the use of bagasse, filter cake and vinasse, highlighting their chemical characteristics, the effects on the soil and aquatic environments due to its use in agriculture.

Keywords: Aquatic environments, Bagasse, Filter cake, Vinasse

INTRODUCTION

General Aspects, Importance of Sugarcane in Brazil and Production Data

Sugarcane is a semi-perennial plant grown in tropical and semi-tropical regions that belongs to the Poaceae family and *Saccharium* genus. Although it is an alogamous plant that reproduces by scattering pollen, in commercial crops, it usually reproduces by vegetative propagation (Ming et al. 2006). The *Saccharium* genus includes two wild species *(S. spontaneum* and *S. robustum)* and four cultivated species *(S. officinarum, S. baberi, S. sinense* and *S. edule)* (Grivet, 2006; Aitken and McNeil, 2010). Sugarcane is originally from New Guinea (Miranda, 2008) and was introduced in Brazil in 1532 by Martin Afonso de Souza in the present city of São Vicente, São Paulo State, where the first sugar mill named São Jorge was built (Miranda, 2008).

Sugarcane is the main raw material for the production of ethanol and sugar. The sugarcane cultivation expanded rapidly due to favorable climate and soil, thus allowing Brazil to become the world leader in sugarcane production for the first time in 1980. Currently, Brazil still ranks first in world production followed by India, China and Thailand, according to the Food and Agriculture Organization (FAO, 2012). In addition, Brazil still ranks among the largest producers of sugar and ethanol, continuously expanding into foreign markets with the use of biofuel as an alternative energy source.

The sugarcane production in Brazil aims to meet the needs of foreign and domestic markets for sugar. In the domestic market, it also needs to meet the demand of ethanol production for fuel and, thus, sugarcane planted area coupled with good management practices and climate conditions are decisive for the pursuit of an economically competitive market. Data of the Sugarcane Industry Association (União da Indústria de Cana-de-Açucar, UNICA) show that in 1980 when Brazil became the largest sugarcane producer, the planted area was 2,768,514 acres; 6,983,814 acres less than the planted area in 2012 (9,752,328 hectares). Figure 1 shows the evolution of the area planted with sugarcane in Brazil from 1980 to 2012 according to (UNICA).

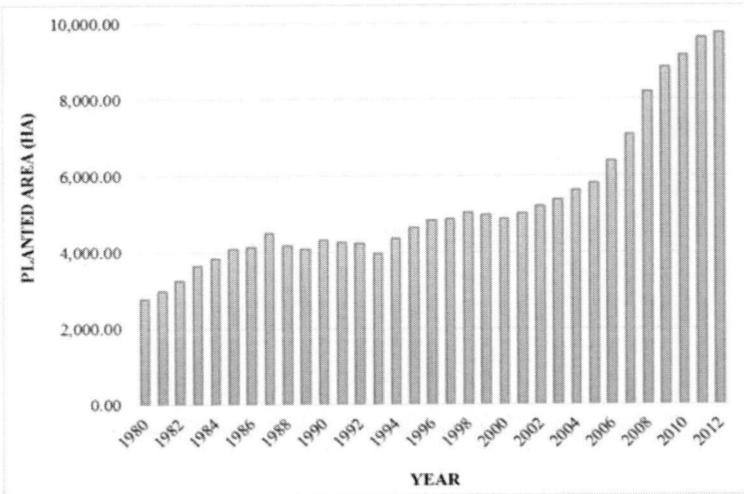

Figure 1. Area planted in hectares (ha) with sugarcane in Brazil from 1980 to 2012.

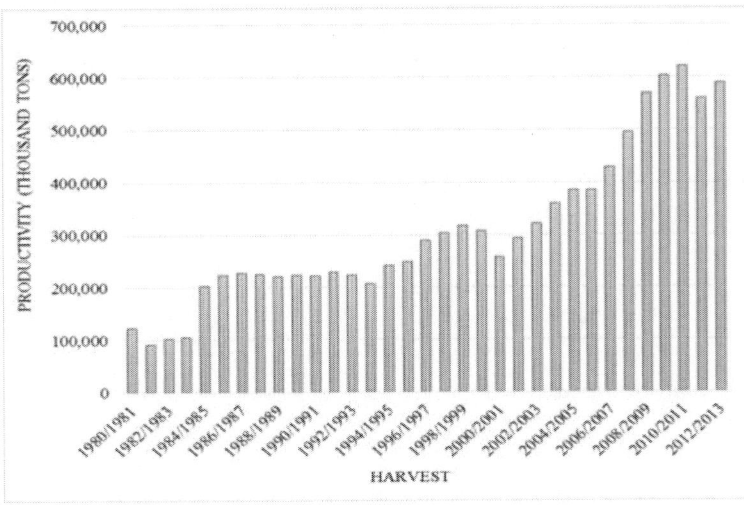

Figure 2. Production of sugarcane in Brazil between 1980 and 2013 according to (UNICA).

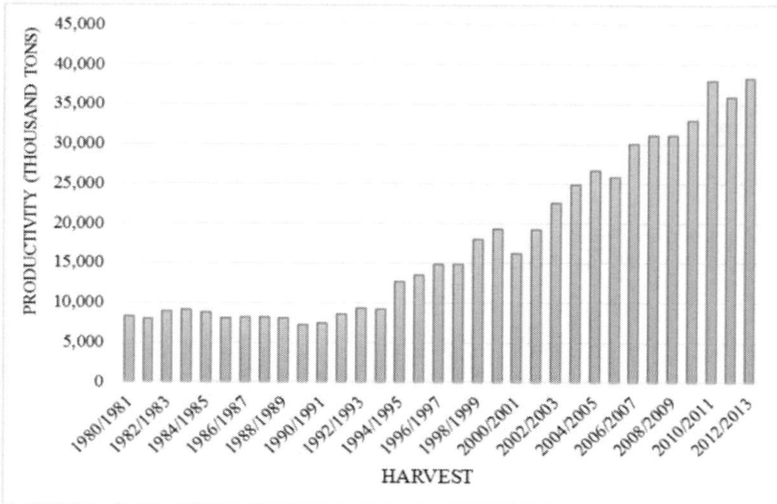

Figure 3. Sugar yield in Brazil between 1980 and 2013.

According to the National Supply Company (Companhia Nacional de Abastecimento, CONAB) the area planted with sugarcane in Brazil continues to expand, with an estimated increase of approximately 314 thousand hectares in the 2013/2014 season, a growth of 3.7% compared to the 2012/2013 season. Sao Paulo is the state with the largest planted area (51.31%), followed by Goiás (9.3%), Minas Gerais (8.0%), Mato Grosso do Sul (7.09 %), Paraná (7.04%), Alagoas (5.02%) and Pernambuco (3.25%), and less than 3.0% in the other states (CONAB, 2013).

Due to the increase in planted area over the years, the productivity of sugarcane (Figure 2) and, consequently, sugar (Figure 3) and alcohol increased.

According to CONAB estimates, productivity is expected to increase by 6.8% for the 2013/2014 harvest season from 69,407 kg/ha in 2012/2013 season to 74,100 kg/ha. Increasing of the planted area and investing in field maintenance are the main reasons for this increase. The main producer is the Center-South region with an estimated production of 594.1 million tons, corresponding to 11.5% increase compared to the previous harvest.

In the 2013/2014 harvest, the Brazilian sugar production is estimated at 40.97 million tons, an increase of 6.88% from the previous year (38.34 million tons) according to CONAB. This increase is directly related to increasing sugarcane productivity, primarily in the South Central Region (CONAB). The largest sugar producer regions in Brazil, in descending order, are the Southeast (70.03%), Midwest (10.21%), Northeast (9.87%), and South (8.83%).

Ethanol production is also expanding. In the 2012/2013 harvest, 23.64 billion liters of ethanol were produced, 3.53 billion liters or 14.94% less than what is planned for the 2013/2014 harvest, which is 27.17 billion liters. From this total, 12.02 billion liters is anhydrous alcohol, which is blended with gasoline, and the remainder is hydrated alcohol that is sold as fuel in gas stations throughout Brazil. The Southeast region stands out as the largest ethanol producer and it is responsible for 92.13% of the total ethanol produced in the country.

In Brazil, in 1975, in order to overcome the 1973 oil crisis and to meet the demand of domestic and foreign markets for fuel, the National Alcohol Program (PROÁLCOOL) was created and introduced as an alternative to oil-derived fuel. The great expansion seen in the sugarcane industry starting this same year has begun to play an important part in environmental pollution. This is the result of the little attention that was given to the utilization of byproducts of the industry since the introduction of the sugarcane culture in the country.

The processing of sugarcane into sugar and ethanol produces byproducts known as filter cake, vinasse and bagasse, which have been highlighted for their commercial importance (Cortez et al., 1992). However, inappropriate and indiscriminate disposal of such byproducts in the soil and/or water bodies has been discussed in the past decades to address the environmental problems associated with this practice (Christofoletti et al., 2013).

The Ethanol Industry: From Planting to the Byproducts

According to Alvarenga and Queiroz (2009), the sugarcane industry is very important from the viewpoint of economics and job creation; however, problems can result from the sugarcane processing, from planting to harvesting, if the practices associated with these activities are not conducted properly, respecting the environment and the environmental laws, as well. Among the main problems caused by the sugarcane agribusiness are: the loss of biodiversity, caused by deforestation and the tendency for sugarcane monoculture; contamination of soil and surface water through excessive use of fertilizers, such as liming and mineral agrochemical use; soil compaction due to heavy machinery transit during planting; cultural practices and harvesting; siltation of water bodies due to soil erosion in renewal areas; removal of soot and greenhouse gases from the sugarcane burning during the harvest period

(Alvarenga and Queiroz, 2009). For several years now, vinasse and filter cake have been used in the fertigation of the sugarcane plantations, applied through long transport channels and also directly in the soil.

Filter Cake: Composition, Use and Environmental Impact

The filter cake results from the clarification (purification) of sugarcane juice that resulted from the crushing process. The sugarcane juice is heated, and a suspension of calcium hydroxide is added to it to increase the pH and promote the flocculation of organic colloids, which are precipitated as calcium and phosphate salts. The precipitate is mixed with sugarcane bagasse and filtered in a vacuum rotary drum filter to extract the remainder juice. The filter cake is the product collected from the filter screens, after the filtration process.

Filter cake composition varies according to variety and maturity of the sugarcane, soil type and the process used to clarify the juice in the distilleries (Cortez et al., 1992). According to Gupta et al. (2011), the principal components of the filter cake are nitrogen (1.9%), phosphorus (1.8%), potassium (0.9%), calcium (4.3%), magnesium (0.7%), sulfur (3.2%), sodium (0.1%), manganese (0.034%), zinc (0.008%) and copper (0.053%).

The filter cake is rich in organic matter and nutrients, and is potentially a major source of carbon and nitrogen, especially for poor soils (Badole et al., 2001). It is, therefore, used as a natural fertilizer to replace the chemical (Nardin, 2007). In addition to its use as fertilizer, filter cake can be used to produce biogas by anaerobic fermentation. In India, where this practice is adopted, about 3.4×10^8 m^3 of gas are produced per year, serving as a probable source of energy. Other possibilities also include using it as a supplement in fish feeding in semi-intensive cultures and for citric acid production via fermentation of the solid residue by the fungus *Aspergillus niger* (Gupta et al., 2011; Shankaranand et al., 1993).

According to Yadav (1995), 4 million tons of filter cake are produced when 134 million tons of sugarcane are crushed.

The filter cake is applied in the off season in the sugarcane crop, particularly in the pre-planting, in the furrow between the rows of the entire planted area (Cortez et al., 1992). Although this residue is potentially a powerful fertilizer, some substances with significant toxic potential can be present in the filter cake due to its frequent use, especially in the very culture of sugarcane.

Effects on Soil Properties

It is extremely relevant to know how filter cake use affects soil properties due to the extensive use of this residue as fertilizer in many countries such as South Africa, Argentina, Australia, Brazil, India, Pakistan, Switzerland and Taiwan (Blackburn, 1984; Barry et al., 2001).

Both filter cake and bagasse are rich in organic matter and applying it *in natura* in the soil is limited to small amounts per area (Cortez et al., 1992). Therefore, according to Cortez et al. (1992), it is necessary to induce the degradation of this organic matter through composting in order to make the use of this residue technically and economically feasible.

When applied to the ground, the filter cake increases the levels of organic carbon, total nitrogen, phosphorus and potassium (Kaur et al., 2005). Barry et al. (2001) observed a large increase in the quantities of phosphorus above those required by the crop, resulting in increased acidic levels of phosphorus extractable from the soil (up to 1,260 mg/kg). The authors estimated that from 4 to 5 filter cake applications are necessary, for a soil that has never received this residue, to reach the phosphorus levels comparable to a soil with a history of long-term use.

On the other hand, the use of filter cake increases the pH levels of the soil, especially fine texture soils, and reduces aluminum saturation percentage. The increase in pH also reduces cadmium availability, although its concentration increases with filter cake application (Barry et al., 2001).

Toxicity

The addition of organic and inorganic compounds to improve soil fertility and plant growth is one of the most common practices in agriculture. The physicochemical properties of the filter cake, as well as its effects on vegetative growth of higher plants and consequently its use as natural fertilizer are well known. However, there are few studies in the literature on the quantity and the way it should be applied, stored, and how toxic this waste can be to both humans who handle it in the field and especially to the terrestrial fauna of the places it fertilizes.

Many fertilizers, whether organic or inorganic, affect not only the target plants but also the entire fauna and micro-fauna dynamics of the soil where the fertilizer is applied, interfering for example, with the development of nematodes beneficial to the soil (Muller and Gooch, 1982; Rodriguez-Kabana, 1986).

Rodrigues et al. (2005) assessed the effect of filter cake added to the soil on the migration and persistence of *Steinernema glaseri,* an entomopathogenic nematode, while using larvae of *Galleria mellonella* as host indicator for the activity of infective juveniles of the nematodes. The authors observed a significant decrease in the number of infected larvae and living adult nematodes in the larvae in sandy soils treated with filter cake. This result demonstrates the negative impact of this residue in the life cycle of this species of roundworm when applied in proportions greater than 25%, although it has not affected their persistence.

Bagasse: Composition, Use and Environmental Impact

Sugarcane bagasse is a fibrous lignocellulosic material that results from the crushing of the sugarcane. It is estimated that for every ton of crushed sugarcane approximately 250 kg of bagasse are generated (UDOP, 2012), making this the most abundant lignocellulosic byproduct in Brazil.

Sugar and alcohol industries lacked large storage area for bagasse when the first mills were built. Therefore, the bagasse was placed in outdoor areas and, quite often, near aquatic environments. As the sugarcane business expanded, the amount of bagasse produced also increased and, thus major industries had to devise better ways to store this byproduct and provide an environmentally friendly destination.

The sugarcane bagasse consists of two carbohydrates, cellulose and hemicellulose, which are embedded in a matrix called lignin (Rezende et al., 2011). Cellulose represents 40% of the bagasse, while hemicelluloses and lignin account for 35% and 15%, respectively (Teixeira et al., 2008). However, according to Rodrigues and Camargo (2008) this composition may vary according to soil type, sugarcane variety, harvesting techniques and material used.

Bagasse has been used to produce energy by burning it; paper; concrete for civil construction; second-generation ethanol, cosmetics and bioplastics; and, to feed animals, especially ruminants. However, undoubtedly, the most important use is to produce all the energy necessary to run both sugar and alcohol processes in the industry. The importance of being self-sufficient lies in the fact that the ethanol industry is independent of the energy generated by the hydroelectric plants, whose production oscillates according to environmental conditions.

Using bagasse to produce energy is very important not only for the sugar industry but also for the environment, since it is used to replace fossil fuels which are primarily responsible for the generation of greenhouse gas emissions, and, therefore, cause global warming. In this sense, the production of "biofuel" by sugarcane industries is carried out through a renewable energy or clean energy, provided that the gases emitted during the burning of bagasse are controlled.

Lora and Teixeira (2001) highlight the advantages of using bagasse; low cost, low emissions of carbon dioxide, nitrogen and sulfur oxides, and particulate matter compared to fossil fuels. The disadvantage, according to these authors, is that however small the emission of particulate matter and greenhouse gases, the industry has to invest in equipment such as scrubbers for removal, which generates a cost. Other authors emphasize some disadvantages of the use of bagasse for power generation. To Cortez et al. (1992), the problems regarding the use of bagasse for energy purposes include low energy density (214,800 kcal/m³), high humidity of bagasse *in natura*, difficulty to store, transportation costs and its decomposition over time.

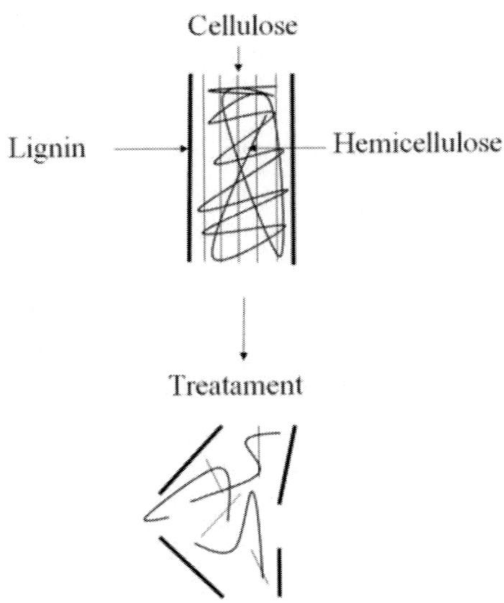

Figure 4. Schematics of treatment acting on the lignocellulosic material (adapted from Mosier et al. (2005).

Bagasse is also used as animal feeding, especially for ruminants, because it is produced in large quantities and at a time when there is a shortage of forage (Virmond, 2001). However, bagasse is low in protein and high in fiber and, therefore, to improve its nutritional value and digestibility, it is necessary to break the lignin to release the carbohydrates using biological, physical, chemical (Cardoso, 2006) and enzymatic treatments. Figure 4 shows the action mechanism of bagasse treatments. The biological treatment uses fungi and bacteria to produce the enzyme, lignin peroxidase, which is able to decompose the lignin present in the bagasse (Ogeda and Petri, 2010). On the other hand, according to Van Soest (1994), the physical and chemical treatments aim to break the bonds between lignin, cellulose and hemicellulose, in which the lignin acts as a substrate for its own degradation by microorganisms. The enzymatic treatment is based on the enzymatic hydrolysis by enzymes of the class of cellulases.

The physical treatment consists of using irradiation (Taherzadech and Karimi, 2008) or reducing the particle size through grinding and sieving (Ogeda and Petri, 2010). Furthermore, chemical treatments are based on the use of ammonia (Gollapalli et al. 2002), solvents (Matsumara et al. 2006) or diluted acids and bases (Betancur and Pereira, 2010).

The bagasse burning process generates ashes, approximately 24 kg for every ton of bagasse burnt. The ashes must be disposed of in an environmentally friendly way or used in the cement production process as described below (FIESP/CIESP, 2001). Ashes contain high silica levels (about 60% of the total weight), plus minor amounts of oxides of potassium, magnesium, phosphorus and calcium, which may vary according to the type of sugarcane cultivated, the fertilizers and herbicides used in the culture, as well as the climatic conditions (Cordeiro et al. 2009).

Currently, the ashes that result from burning the sugarcane bagasse seem to be specially used in the cement manufacture due to its high silica content, and according to Massazza (1988), ground silica possesses binding properties in the presence of water and calcium hydroxide. Therefore, its use for this purpose becomes economically and environmentally interesting, since this byproduct can be sold by sugar mills that do not have storage capacity, without causing any impact on human health (respiratory diseases) and the environment that could arise from incorrect storage.

It is also noteworthy that bagasse ashes can replace sand in the production processes of cement-derived materials. It is estimated that 100-120 tons of sand are removed annually from the rivers in Brazil while four million tons of ashes are produced from burning the bagasse (UNICA, 2011). Also according to UNICA (2011), for each 1 m^3 of concrete, ashes can replace up to 50% of sand.

Conventional plastics are made from petroleum-based raw material, i.e., a natural non-renewable resource. There is a great concern about the environment due to the time required for degradation of some components, since some may take hundreds of years to degrade. It is, therefore, necessary to replace the conventional plastic with a product made from renewable natural resources. Thus, bioplastics draw attention, since they are produced from renewable resources and, when discarded, are easily degraded.

According to Viveiros (2002), the manufacture of bioplastics has been researched worldwide, using several raw materials such as castor oil, sugarcane, beet, corn, soy protein and lactic acid. The most satisfactory results were obtained with castor oil and sugarcane by producing polyhydroxybutyrate and polyurethane, respectively, via microorganisms' action. Polyhydroxybutyrate (PHB) is a thermoplastic polymer belonging to the class of polyhydroxyalkanoates obtained from bacteria, which, when fed with sugar or other carbon source, store energy in the form of polyester (Rehm and Steinbüchel, 1998). This material after processing can be used in various applications such as packaging, containers and disposables (Gatenholm and Mathiasson, 1994).

Sugarcane bagasse can also be used for second generation ethanol production, which yields liquid or gaseous fuels derived from biomass, such as the biodiesel produced from organic oils, and bioethanol, from the fermentation of carbohydrates. There are two processes to produce bioethanol, bagasse treatment (as discussed earlier in this chapter) with the goal of breaking the long chains of cellulose and hemicellulose via enzymatic and/or chemical hydrolysis yielding smaller sugars followed by fermentation.

Ferreira-Leitão et al. (2010) emphasize that using agro-industrial residues to produce renewable fuels can prevent accumulation of waste, thus avoiding public health and environmental problems.

Vinasse: Characterization and Environmental Impact

Vinasse is an effluent from the distillation of ethanol. It is estimated that every liter of ethanol produced generates 8-18 liters of vinasse (Parnaudeau et al., 2008) whose composition varies according to the equipment and materials used in the distillation process (Kumar et al., 1998; Naik, Jagadeesh, Alagawadi, 2008). However, it is generally known that vinasse is composed of approximately 93% water and 7% minerals (Laime et al. 2007) with high concentrations of organic matter as organic acids and cations (potassium, calcium, and magnesium) and low concentrations of phosphorus and nitrogen (Laime et al., 2011). Vinasse has a dark color and low pH (3.5-5.0) (España-Gamboa et al., 2011); therefore, these features plus the amount of organic matter present, makes vinasse a hazardous byproduct that can contaminate aquatic environments. Table 1 shows the composition of vinasse as determined by Botelho et al. (2012) and Christofoletti et al. (2013).

Table 1. Physico-chemical parameters and metals present in sugarcane vinasse

Parameter s	Prices	Reference
pH	4.0	Botelho et al. (2012)
Total carbon	3.17 $g \cdot L^{-1}$	
Total organic matter	5.70 $g \cdot L^{-1}$	
Total nitrogen	0.39 $g \cdot L^{-1}$	
Total phosphorus (P_2O_5)	0.03 $g \cdot L^{-1}$	
Potassium (K_2O)	1.64 $g \cdot L^{-1}$	
Calcium	0.38 $g \cdot L^{-1}$	
Magnesium	0.16 $g \cdot L^{-1}$	
Total copper	1.00 ppm	
Total manganese	6.00 ppm	
Total zinc	1.00 ppm	
Total iron	35.00 ppm	
Sulfur	0.29 $g \cdot L^{-1}$	
Barium	0.41 $mg \cdot L^{-1}$	Christofoletti et al. (2013)
Copper	0.35 $mg \cdot L^{-1}$	
Electrical conductivity	13.530 µs/cm	
Chrome	0.04 $mg \cdot L^{-1}$	
Mercury	0.0019 $mg \cdot L^{-1}$	
Molybdenum	0.008 $mg \cdot L^{-1}$	
Nickel	0.03 $mg \cdot L^{-1}$	
Nitrate	1.30 $mg \cdot L^{-1}$	
Nitrite	0.008 $mg \cdot L^{-1}$	
Kjeldahl nitrogen	267 $mg \cdot L^{-1}$	
Total K	2,056 $mg \cdot L^{-1}$	
Zinc	1.66 $mg \cdot L^{-1}$	
Total sodium	50.2 $mg \cdot L^{-1}$	
Total sulfur	1,219 $mg \cdot L^{-1}$	
Total magnesium	237 $mg \cdot L^{-1}$	
pH	3.9	

As discussed earlier in this chapter, sugarcane production increased greatly since the 80s, and thus, it can be stated that the vinasse production accompanied this growth. Researchers have focused their efforts on trying to find a suitable use and treatment for this byproduct due to its composition and the large volume generated. The literature cites recycling in fermentation; fertigation; evaporative concentration; yeast and energy production; and, feedstock production for livestock and poultry (Robertiello, 1982). However, the use of vinasse in fertigation will be discussed in this chapter due to the sheer size of the area planted with sugarcane in Brazil (Figure 1). It is noteworthy that, in the past, due to little knowledge and lack of storage areas, vinasse was discharged in aquatic environments, causing serious pollution problems (Santos et al., 1981; Demattê et al., 2004).

Vinasse used in fertigation of the sugarcane planted area is a technology that uses rationally the natural resources, preventing vinasse from being released into rivers and allowing fertilization of arable areas (Christofoletti et al., 2013).

In Brazil, the first studies about applying vinasse in the soil date to the 50s, and 30 years later, in 1980, the use of vinasse as a fertilizer became quite widespread throughout the country.

The use of vinasse as fertilizer was beneficial for the sugar and alcohol industry, since purchase and application of synthetic chemicals decreased, reducing costs, as well as for the environment, which was no longer directly contaminated by this residue.

The use of vinasse to fertilize sugarcane plantations has become a widespread activity because the initial investment is low. The practice requires basically pipes, pumps, trucks for transportation and settling tanks/clarifiers. Furthermore, the application is quick and increases sugarcane productivity.

Vinasse is also produced in many other countries as a byproduct of alcohol produced from different raw materials such as beets, wine and fruits in Europe and corn in North America. The vinasse obtained from different raw materials also has different properties.

Vinasse used as fertilizer is beneficial to the sugar industry; however, it can contaminate aquatic environments if the practice is not conducted according to the Norm P4.231 of 2006 issued by the São Paulo State Environmental Company (Companhia Ambiental do Estado de São Paulo, Cetesb, 2006), which contemplates some parameters such as type of soil to be fertilized; distance from rivers, streams; soil field capacity (water retention) and percentage of salts in the soil (Laime et al., 2011). Moreover, the climatic conditions are factors that should be observed during this activity will be discussed later.

Impact on Soil

The use of vinasse as liquid fertilizer on the soil generates numerous benefits when it does not exceed the ion retention capacity of the soil, as reported by Ramalho and Sobrinho (2001) and Silver (2001). Therefore, the physical and chemical parameters of both the soil to be fertilized and the vinasse must be determined. If this residue is applied in concentrations above the soil retention/holding capacity, it may seriously impact the biota of this environment as well as aquatic ecosystems, since this residue can reach the water through either leaching or runoff.

It has been reported by some researchers that the addition of vinasse can change soil physico-chemical and biological characteristics. Canella et al. (2003) observed an increase in the concentration of organic matter in the soil after application of vinasse while Zolin et al. (2011) reported increased organic carbon and potassium. Santos et al. (2009) reported a change in the composition of fungi and bacteria in the soil after applying vinasse. In a study conducted in the 50s, Camargo and colleagues reported increased microorganisms with prevalence of the fungi *Neurospora* spp, *Aspergillus* spp, *Penicillium* spp, *Mucor* spp, and also the bacteria *Strepytomices* ssp. Importantly, the increase of microorganisms in the soil changes the biological and chemical processes of this environment, notably organic matter decomposition, nitrification (Resende et al., 2006), denitrification (Leal et al., 1983), N_2 fixation and pH increase (Silva et al., 2007; Oliveira et al., 2013).

Other beneficial impacts on the physical, chemical and biological properties of both soil and plant have also been reported (Camargo et al., 1983; Andreoli, 1986; Canellas et al., 2003; Zolin et al., 2011).

The deleterious effects of vinasse application in the soil are related to low and/or no seed germination as reported in several studies.

However, according to Pant and Adholeya (2007), this effect is related to the concentration applied to the soil. Still, depending on the concentration, vinasse can cause soil salinization, change soil quality due to nutrients imbalance, especially Mn (Agrawal and Pandey, 1994), reduce alkalinity and cause loss of crops (Kumar and Viswanathan, 1991), increase phytotoxicity and produce uncomfortable odor (Navarro et al., 2000; Santana and Machado, 2008).

Impact on Aquatic Environments

The impact of vinasse on aquatic ecosystems is known since the 70s, when it was disposed of directly in these environments. This residue has high contaminant potential that can greatly impact aquatic environments due to its physical and chemical characteristics, i.e., high organic matter content, acid pH and highly corrosive. It is known that once the organic matter, present in large quantities in an effluent or waste, is released into a receiving water body, it will be consumed (degraded) by the microorganisms. However, these microorganisms utilize molecular oxygen in the respiratory process to degrade the organic matter, thus resulting in an anoxic environment and, consequently, in the death of aquatic organisms.

Another important feature of vinasse, which contributes to deterioration of aquatic environment, is the pH. As reported in Table 1, the vinasse has an acidic pH and it is known that most aquatic organisms tolerate near neutral pH. The National Environmental Council (Conselho Nacional do Meio Ambiente, CONAMA) through Resolution 357 of 2005 requires that pH of Brazilian rivers should be between 6 and 9 to protect the aquatic biota (BRAZIL, 2005), and thus, if vinasse is released directly into the water, the pH might change and result in elimination of aquatic organisms.

Several studies have been conducted to evaluate the toxicity of vinasse to aquatic organisms. However, due to the sheer size of the sugarcane industry and what it represents to Brazil and the world, coupled with the increase of cultivated lands, sugarcane and vinasse production, further studies are still necessary. Kumar and Gopal (2001) reported a reduction of protein in the liver, muscle, brain and kidneys, as well as increased mucus production after exposing the fish *Channa punctatus* to various dilutions of vinasse.

Some studies were conducted aiming to evaluate the toxicity of vinasse before and after treatment. One of the environmental problems related to vinasse, besides those already mentioned, is its dark brown color. Wilkie et al. (2000) stated that colored effluents such as vinasse can inhibit the growth of aquatic plants since by inhibiting light penetration in their environment. Thus, Ferreira et al. (2011) treated vinasse with *Pleurotus sajor-caju* in order to lighten the color and conducted toxicity tests with aquatic organisms before and after treatment. The results showed that the lighter color reduced toxicity to the algae, *Pseudokirchneriella subcapitata;* the microcrustaceans, *Daphnia magna* and *Daphnia similis;* and the cnidarian, *Hydra attenuata.*

Another study was conducted by Botelho et al. (2012) to assess the toxicity of vinasse to the microcrustaceans, *Ceriodaphnia dubia* and *D.*

magna; and the fish *Danio rerio*, before and after adjusting the pH to 7.0. The authors concluded that after correcting the pH, vinasse was 4, 7 and 3 times less toxic to the organisms mentioned above, respectively.

Final Considerations

The sugar and alcohol industry is of great importance to the Brazilian economy as well as to other countries. However, most efforts were directed towards technological development of the industry, process improvements and energy production. Given the need to reconcile sustainability and production, mitigating the impacts caused to the environment by the production processes became necessary in order to secure a better market position. Thus, an effective and beneficial alternative to both sugar and alcohol sector and the environment was to utilise the byproducts in various activities as mentioned in this chapter.

Three main byproducts bagasse, filter cake and vinasse are generated during the processing of sugarcane to produce alcohol and sugar (Figure 5).

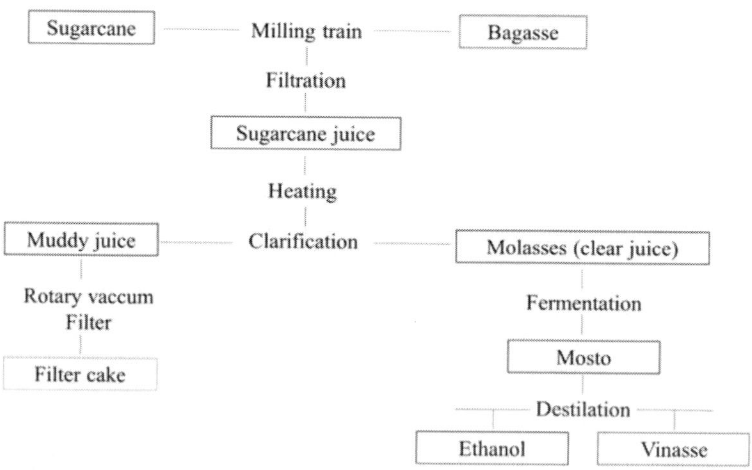

Figure 5. Byproducts generated during the sugarcane processing.

Unarguably, the correct use of vinasse and filter cake in fertigation of soil and other activities, and of bagasse for energy production or animal feed production, among others, tend to minimize the direct environmental impacts on soil-water-air systems. However, as presented, the physical and chemical composition of these byproducts can indirectly impact the environment, and, thus, further studies involving the assessment of toxicity to organisms should be conducted. Moreover, vinasse and filter cake application should take into consideration the climatic conditions, such as precipitation, while following the environmental requirements in order to maintain the balance between productivity and sustainability.

ACKNOWLEDGMENTS

The authors thank the Fundação de Amparo à Pesquisa do Estado de São Paulo for financial support (process 2012/50197-2 and 2013/08143-5). We are also thankful to our friend Nikolas Camilo Gomes, a student of Biological Sciences, Escola Superior de Agricultura Luiz de Queiroz/Universidade de São Paulo, for his friendship and companionship, who unfortunately left us this year.

REFERENCES

Agrawal, C.S.; Pandey, G.S. Soil pollution by spent wash discharge: depletion of manganese (II) and impairment of its oxidation. *J. Environ. Biol.,* 1994, 15, 49–53.

Aitken, K.S.; McNeil, M. Diversity Analysis. In Genetic, Genomic and Breeding of Sugarcane; Henry, R.J., Kole, C., Ed., New Hemisphere: *Science Publishers,* 2010, p.19-42.

Alvarenga, R.P.; Queiroz, T.R. Book of abstracts, 2nd International Workshop – Advances in Cleaner Production, São Paulo, SP, 1-9, 2009.

Andreoli, I. Effects of vinasse on some chemical and physical properties of a Dark-Red Latossol medium texture. Ph.D. Dissertation, Luiz de Queiroz *College of Agriculture*, SP, 1986.

Badole, S.B.; More, S.D.; Raut, M.M. Residual effect of different organic and inorganic nutrient sources on yield and quality of groundnut. Ann. Plant Physiol. 2001, 15, 54-57.

Barry, G.A.; Rayment, G.E.; Jeffrey, A.J.; Price, A.M. Changes in cane soil properties from applications of sugar mill by-products. In Proceedings of the 2001 Conference of the Australian Society of Sugar Cane

Technologists held at Mackay, Queensland, Australia., Hogarth, D.M., Ed., Cab direct: Brisbane, 2001, p. 185–191.

Bassanta, M.V.; Dourado-Neto, D.; Reichardt, K.; Bacchi, O.O.S.; Oliveira, J.C.M.; Trivelin, P.C.O.; Timm, L.C.; Tominaga, T.T.; Correchel, V. Cássaro, F.A.M.; Pires, L.F.; de Macedo, J.R. Management effects on nitrogen recovery in a sugarcane crop grown in Brazil. *Geoderma,* 2003, 116, 235-248.

Betancur, V.G.J.; Pereira, N.J. Sugarcane bagasse as feedstock for second generation ethanol production. Part I: diluted acid pretreatment optimization. *Eletron. J. Biotechn.,* 2010, 13, 1-9.

Blackburn, F. Sugarcane. Tropical Agriculture Series. Longman: London and New York, 1984.

Botelho, R.G.; Tornisielo, V.L.; Olinda, R.A.; Maranho, L.A.; Machado-Neto, L. Acute toxicity of sugarcane vinasse to aquatic organisms before and after pH adjustment. *Toxicol. Environ. Chem.,* 2012, 10, 2035-2045.

BRAZIL. Ministry of Environment. National Environmental Council (CONAMA). N.357 resolution of 17 March 2005. Provides the classification of water bodies and environmental guidelines for its framework and establishes the conditions and standards for effluent discharge, and other measures. http://www.mma.gov. br / port / conama / res / res05 / res35705.pdf (accessed April 09, 2014).

Camargo, O.A.; Valadares, J.M.A.S.; Geraldi, RN. Chemical and physical features of soil which received vinasse for long period. Campinas, SP, 1983.

Canellas, L.P.; Velloso, A.C.X.; Marciano, C.R.; Ramalho, J.F.G.P.; Rumjanek, V.M.; Rezende, C.E.; Santos, G.A. Chemical soil properties of an inceptisol under long-term sugarcane crops with vinasse application and without slash burning. *Rev. Bras. Ciênc. Solo,* 2003, 27, 935-944.

Cardoso, M. G. Brandy production quality. UFLA: Lavras, 2006.

Christofoletti, C.A.; Pedro-Escher, J.; Correia, J.E.; Marinho, J.F.U.; Fontanetti, C.S. Sugarcane vinasse: environmental implications of its use. *Waste Manage,* 2013, 33, 2752-2761.

National Supply Company. Monitoring of the Brazilian crop: sugarcane, according to a survey, August 2013. http: //www.conab. gov.br (accessed June 03 2014).

Company of Environmental Sanitation Technology. Vinasse - Criteria and procedures for application to agricultural soil. http://www.cetesb.sp. gov.br/Tecnologia/camaras/P4_231.pdf (accessed June 12 2014).

Cordeiro, G.C.; Toledo Filho, R.D.; Tavares, R.M.; Fairbairn, E.M.R. Ultrafine grinding of sugarcane bagasse ash for application as pozzolanic admisture in concrete. *Cement. Concrete. Res.,* 2009, 39, 110-115.

Cortez, L.; Magalhães, P.; Happi, J. Major byproducts from sugarcane agribusiness and its valuation. Brazilian Journal of Energy, 1992, 2, 111-146.

Demattê, J.A.M.; Gamaa, M.A.P.; Coopera, M.; Araújo, J.C.; Nanni, M.R.; Fiorioa, P.R. Effect of fermentation residue on the spectral reflectance properties of soils. *Geoderma,* 2004, 120, 187-200.

de Resende, A.S.; Xavier, R.P.; de Oliveira, O.C.; Urquiaga, S.; Alves, B.J.R.; Boddey, R.M. Long-term effects of pre-harvest burning and nitrogen and vinasse applications on yield of sugar cane and soil carbon and nitrogen stocks on plantation in Pernambuco, *N.E. Brazil. Plant Soil.,* 2006, 281, 339–351.

España-Gamboa, E.; Minagos-Cortes, J.; Barahona-Perez, L.; Dominguez-Maldonado, J.; Hernández-Zarate, G.; Alzate-Gaviria, L. Vinasse: characterization and treatments. *Waste Manag. Res.,* 2011, 29, 1235-1250.

Ferreira, L.F.R.; Aguiar, M.M.; Messias, T.G.; Pompeu, G.B.; Lopez, A.M.Q.; Silva, D.P.; Monteiro, R.T. Evaluation of sugar-cane vinasse treated with Pleurotussajor-caju utilizing aquatic organisms as toxicological indicators. *Ecotoxicol. Environ. Saf.,* 2011, 74, 132–7.

Ferreira-Leitao, V.; Gottschalk, L.M.F.; Ferrara, M.A.; Nepomuceno, A.L.; Molinari, H.B.C.; Bon, E.P.S. Biomass residues in Brazil: availability and potential uses. *Waste Biomass Valor*, 2010, 1, 65-76.

FIESP/CIESP. Expanding the supply of energy from biomass (sugarcane bagasse), FIESP/CIESP: São Paulo, 2001.

Food and Agriculture Organization. Top production – Sugar cane – 2012. http://faostat.fao.org/site/339/default.aspx (accessed Jun 9, 2014).

Gatenholm, P.; Mathiasson, A. Biodegradable natural composites. II. Synergistic effects of processing cellulose with PHB. *J. Appli. Polym. Sci.,* 1994, *51*, 1231-1237.

Gollapalli, L.E.; Dale, B.E.; Rivers, D.M. Predictive digestibility of ammonia fiber explosion (AFEX) treated rice straw. *Appl. Biochem. Biotechnol.,* 2002, 98-*100*, 23-35.

Grivet, L.; Glaszmann, J.C.; D´Hont, A. Molecular evidence for sugarcane evolution and domestication. In Darwin´s Harvest. New Approaches to the Origins, Evolution and Conservation of Crops; Motley, T., Zerega, N., Cross, H., Ed., Columbia University Press: New York, 2006, p.49-66.

Gupta, N.; Tripathi, S.; Balomajumder, C. Characterization of pressmud: A sugar industry waste. *Fuel,* 2011, 90, 389-394.

Kaur, K.; Kapoor, K.K.; Gupta, A.P. Impact of organic manures with and without mineral fertilizers on soil chemical and biological properties under tropical conditions. *J. Plant. Nutr. Soil Sci.,* 2005, 168, 117-122.

Kumar, S.; Viswanathan, L. Production of biomass, carbon dioxide, volatile acids, during distillery waste treatment by bacterial strains. *Enzyme Microb. Technol.,* 1991, 13, 179-187.

Kumar, V.; L. Wati, P.; Nigam, I.M.; Banat, B.S.; Yadav, D.; Singh, Marchant, R. Decolorization and biodegradation of anaerobically digested sugarcane molasses spent wash effluent from biomethanation plants by white-rot fungi. *Process Biochem.,* 1998, 33, 83–88.

Kumar, S.; Gopal, K. Impact of distillery effluent on physiological consequences in the freshwater teleost Channa punctatus. *Bull. Environ. Contam. Toxicol.,* 2001, 66, 617–22.

Laime, E.M.O.; Fernandes, P.D.; Oliveira, D.C.S.; Freire, EA. Technological possibilities for the disposal of vinasse: a review. R. Trop.: Ci. Agr. Biol. 2011, 5, 16-29.

Leal, J.R.; Amaral Sobrinho, N.M.B.; Velloso, A.C.X.; Rossielo, R.O.P. Potencial redox e pH: variação em um solo tratado com vinhaça. *Rev. Bras. Ciênc. Solo,* 1983, 7, 257–261.

Lora, E.E.S.; Teixeira, F.N. Energy and Environment. In Conservation of Energy: Energy Efficiency for Facilities and Equipment; Marques, M., Haddad, J., Martins, A.R.S., Ed.; Unifei: Itajubá, 2001; Vol.1; p.30-93.

Massazza, F. Em Lea's Chemistry of Cement and Concrete; 4th ed.; Wiley: New York, 1988.

Matsumura, Y.; Sasaki, M.; Okuda, K.; Takami, S.; Ohara, S.; Umetsu, M.; Adschiri, T. Supercritical water treatment of biomass for energy and material recovery. *Combust. Sci. Tech.,* 2006, 178, 509-536.

Melo, A.S.S.A.; Silva, MP Estimating the value of the positive externality of using vinasse on sugarcane production: a case study. http://www.ecoeco.org.br/conteudo/publicacoes/encontros/iv_en/mesa4/2.pdf (accessed Aug 20, 2011).

Ming, R.; Moore, P.H.; Wu, K.K.; D´Hont, A.; Glaszmann, J.C.; Tew, T.L.; Mirkov, T.E.; Da Silva, J.; Jifon, J.; Rai, M.; Schnell, R.J.; Brumbley, S.M.; Lakshmanan, P.; Comstock, J.C.; Paterson, A.H. Sugarcane improvement through breeding and biotechnology. Plant Breed Rev. 2006, *27,* 15-100.

Miranda, J. R. History of cane sugar. Komedi: Campinas, 2008.

Mosier, N.; Wyman, C.E.; Dale, B.E.; Elander, R.T.; Lee, Y.Y.; Holtzapple, M.; Ladisch, M. Features of promising technologies for pretreatment of lignocellulosic biomass. *Biores Technol.,* 2005, 96, 673-686.

Muller, R.; Gooch, P.S. Organic amendments in nematode control. An examination of the literature. *Nematotropica,* 1982, 12, 319-326.

Naik, N.M.; K.S.; Jagadeesh, and A.R. Alagawadi. Microbial decolorization of spent wash: A review. *Indian. J. Microbiol.,* 2008, 48, 41–48.

Nardin, RR Filter cake applied in ultisol and agronomic effects in two varieties of sugar-cane at two times of harvest. Ph.D. Dissertation, Agronomic Institute, Campinas, SP, 2007.

Navarro, A.R.; Sepúlveda, M.C.; Rubio, M.C. Bio-concentration of vinasse from the alcoholic fermentation of sugar cane molasses. *Waste Manage,* 2000, 20, 581-585.

Ogeda, T.L.; Petri, D.F.S. Enzymatic hydrolysis of biomass. Quím Nova. 2010, 33, 1549-1558.

Oliveira, B.G.; Carvalho, J.L.N.; Cerri, C.E.P.; Cerri, C.C.; Feigl, B.J. Soil greenhouse gas fluxes from vinasse application in Brazilian sugarcane areas. *Geoderma,* 2013, 200–201, 77–84.

Pant, D.; Adholeya, A. Enhanced production of ligninolytic enzymes and decolorization of molasses distillery wastewaster by fungi under state fermentation. *Biodegradation,* 2007, 18, 647–59.

Prata, F.; Lavorenti, A.; Regitano, J.B.; Tornisielo, V.L. Degradation and sorption of ametryne in two soils with vinasse application. *Pesq. Agropec. Bras.,* 2001, 36, 975–981.

Ramalho, J.F.G.P.; Sobrinho, N.M.B.A. Heavy metals accumulation due to the use of residues in soils cultivated with sugarcane. *Floresta Ambient.,* 2001, *8,* 120–129.

Rehm, B.H.A.; Steinbüchel, A. Book of abstract, Proceedings of the International Symposium on Natural Polymers and Composites, Biotechnological Production of Polyester: Novel Production Organisms and metabolic routes to novel polyester, Atibaia, SP, 1998.

Rezende, C.A.; Lima, M.A.; Maziero, P.; Azevedo, E.R.; Garcia, W.; Polikarpov, I. Chemical and morphological caracterization of sugarcane bagasse submitted to delignification process for enhanced enzymatic digestibility. *Biotechnol. Biofuels,* 2011, 4, 54.

Robertiello, A. Upgrading of agricultural and agroindustrial wastes: the treatment of distillery effluents (vinasses) in Italy. *Agric. Wastes.,* 1982, 4, 387–395.

Rodrigues, R.C.D.; Aguillera, M.M.; Gobbi, N.; Pizano, M.A. Influence of different doses of filter cake in the soil on the migration and persistence of *Steinernema glaseri* (Steiner, 1929) (Nematoda: Rhabditida). *Holos Environ.*, 2005, *5*, 52-64.

Rodrigues, C.P.; Camargo, J.A. Sugarcane bagasse for potential co-generation of electricity and cellulosic ethanol. São Joaquim da Barra: College Iara, Coimbra, 2008.

Rodriguez-Kabana, R. Organic and inorganic nitrogen amendments to soil as nematode suppressants. *J. Nematol.,* 1986, 18, 129-135.

Sangwan, P.; Kaushik, C.P.; Garg, V.K. Vermiconversion of industrial sludge for recycling the nutrients. Bioresour Technol. 2008, *99*, 8699- 8704.

Santana, V.S.; Machado, N.R.C.F. Photocatalytic degradation of the vinasse under solar radiation. *Catal. Today,* 2008, 133–135, 606–610.

Santos, G.A.; Rossiello, R.P.; Fernandes, M.S. Effects of stillage on soil ph, germination rate and potassium accumulation in corn. *Pesq. Agropec. Bras.,* 1981, 16, 489–493.

Santos, T.M.C.; Santos, M.A.L.; Santos, C.G.; Santos, V.R. Effect of fertirrigation with in nature vinasse on the microorganism of the soil. *Rev. Caatinga.*, 2009, 22, 155–160.

Shankaranand, V.S.; Lonsane, B.K.; Sugarcane – pressmud as a novel substrate for production of citric acid by solid-state fermentation. *World J. Microbiol. Biotechnol.,* 1993, 9, 377-380.

Silva, M.A.S.; Griebeler, N.P.; Borges, L.C.. Use of stillage and its impact on soil properties and groundwater. *Rev. Bras. Eng. Agríc. Ambient.,* 2007, 11, 108–114.

Taherzadech, M.J.; Karimi, K. Pretreatment of lignocellulosic wastes to improve ethanol and biogas production: a review. *Int. J. MolSci.,* 2008, 9, 1621-1651.

Teixeira, F.A.; Veloso, C.M.; Pires, A.J.; Silva, F.F.; Nascimento, P.V.N. Losses in ensiling of elephantgrass added with cocoa meal and sugarcane. *Arq. Bras. Med. Vet.,* 2008, 60, 227-233.

Union producers of bioenergy. http://www.udop.com.br (accessed May 10, 2014).

The Sugarcane Industry Union. Sugarcane bagasse can earn value by replacing sand in construction. http://www.unica.com.br/noticias/show.asp ?nwsCode=%7B4E794FD3-7EC2-403C-a66d-32,501,604,600 0% 7D (accessed September 3, 2012).

The Sugarcane Industry Union. http://www.unica.com.br (accessed Mar 18, 2014).

Van Soest, P.J. Nutritional ecology of the ruminant. Cornell University Press: Ithaca, 1994.

Virmond, M. Evaluation of sugarcane bagasse treated with different chemicals through kinetics ruminal games and degradability test. Ph.D. Disseration. University of São Paulo, Pirassununga, SP, 2001.

Viveiros, M. Bagasse turns biodegradable plastic. Folha de São Paulo, Folha Online [Online] November 3, 2002, p A15. Caderno Ambiente. http://www1.folha.uol.com.br/folha/ cotidiano/ult95u62132.shtml. (accessed Nov 20, 2009).

Zolin, C.A.; Paulino, J.; Bertonha, A.; De Freitas, P.S.L.; Folegatti, M.V. Exploratory study of the stillage use along the time. I. Characteristics of the soil. *Rev. bras. eng. agríc. ambient.*, 2011, 15, 22-28.

Waliszewski, K.N.; Romero, A.; Pardio, V.T. Use of cane condensed molasses solubles in feeding broilers. *Anim. Feed Sci. Technol.*, 1997, 67, 253-258.

Wilkie, A.C.; Riedesel, K.J.; Owens, J.M. Stillage treated characterization and anaerobic treatment of ethanol stillage from conventional and cellulosic feed stocks. *Biomass Bioenerg.*, 2000, 19, 63–102.

Yadav, R.L. Soil organic matter and NPK status as influenced by integrated use of green manure, crop residues, cane trash and urea in sugarcane-based crop sequences. *Bioresour. Technol.*, 1995, 54, 93–98.

In: Sugarcane
Editor: Eleanore Webb

ISBN: 978-1-61122-356-9

Chapter 5

Valorization of Sugarcane Bagasse Ash Waste to Produce Sustainable Clay-Based Ceramics: A Brief Review

J. N. F. Holanda
Northern Fluminense State University,
Laboratory of Advanced Materials/Group of Ceramic Materials,
Campos dos Goytacazes, RJ, Brazil

Abstract

The sugarcane industry generates huge amount of sugarcane bagasse ash waste worldwide. The management of this solid waste material has resulted in increased economic, social and environmental concerns in the world. Over the years, the sugarcane ashes have been mainly disposed as soil fertilizer. More recently, the recycling of such solid waste into clay-based ceramics appear to be a viable economic and environmental option. The prospective benefits of using sugarcane bagasse ash waste to produce clay-based ceramics include conservation of natural resources and use of costless raw materials. This chapter presents a brief review on the valorization and reuse of sugarcane bagasse ash from the sugarcane industry in the production of sustainable clay-based ceramics for civil construction.

Keywords: Sugarcane bagasse ash, waste, reuse, valorization, clay ceramics

1. Introduction

The agribusiness focused on the production of food and energy sources is currently of the highest importance for the economic and social development of nations. Despite this, the agribusiness activities also generate huge amounts of wastes which, if not reused, could cause damage to the environment and public health. Currently emphasis is given to sugarcane industry that is primarily based on the production of sugar and ethanol. In fact, the sugarcane industry generates huge amounts of wastes, including the sugarcane bagasse. In general, the sugarcane bagasse is used in the mills to produce electricity [1]. As a result, the sugarcane industry generate large amount of sugarcane bagasse ash (SCBA) worldwide [2], which is referred hereafter as SCBA waste. The management of this solid waste material has resulted in increased economic, social, and environmental concerns.

The SCBA waste is a non-biodegradable solid waste material composed mainly of silica (SiO_2) [2]. Currently, significant amount of this waste material has been used as soil fertilizer. However, this practical has been very questionable due to modify the physical and chemical properties of the soils, resulting in environmental impacts.

The field of traditional ceramic materials (bricks, ceramic blocks, roofing tiles, floor tiles, and others) is considered a viable technological approach for the valorization of solid wastes as a source of low-cost alternative raw materials [3-11]. The raw materials used in the clayey formulations are mainly common clays, kaolin, feldspars, talc, and quartz [12]. The formulations consist basically of mixtures of plastic and non-plastic components with wide variability of the chemical, mineralogical and physical compositions. This means that the clayey formulations have great potential to absorb significant amounts of solid wastes.

A large amount of information is available on the valorization of solid waste materials in the production of clay-based ceramics. However, the same cannot be said about the SCBA waste. This chapter presents a brief review on the valorization of SCBA waste from the sugarcane industry to produce sustainable clay-based ceramics for civil construction. Is worth mentioning that, the prospective benefits of using SCBA waste to produce clay-based ceramics include conservation of natural resources and use of costless raw materials.

2. Application of SCBA Waste in Clay-Based Ceramics

2.1. Characteristics of SCBA Waste

SCBA waste is produced as a combustion by-product from energy co-generation in the sugarcane mills. It presents black color due to carbon content, and is rich in mineral particles (quartz) and porous plates of sugarcane bagasse not burnt (Figure 1).

The chemical composition and loss on ignition of SCBA wastes originated from different places are presented in Table 1. As can be observed, the SCBA wastes exhibit broad variability in terms of chemical composition. The reasons for this are: i) the use of different soils in which sugarcane grow; ii) fertilization method; and iii) soil management. As can be seen in Table 1, the SCBA waste is predominantly composed of SiO_2 (38.31 – 72.74 wt. %), followed by Al_2O_3, Fe_2O_3, CaO, and K_2O (minor amounts of Ti, Mg, Mn, Na, S, and P oxides). The loss on ignition (LOI) of SCBA waste is relatively high, and is mainly due the organic matter and calcite. In fact, the SCBA wastes present high content of organic matter [2].

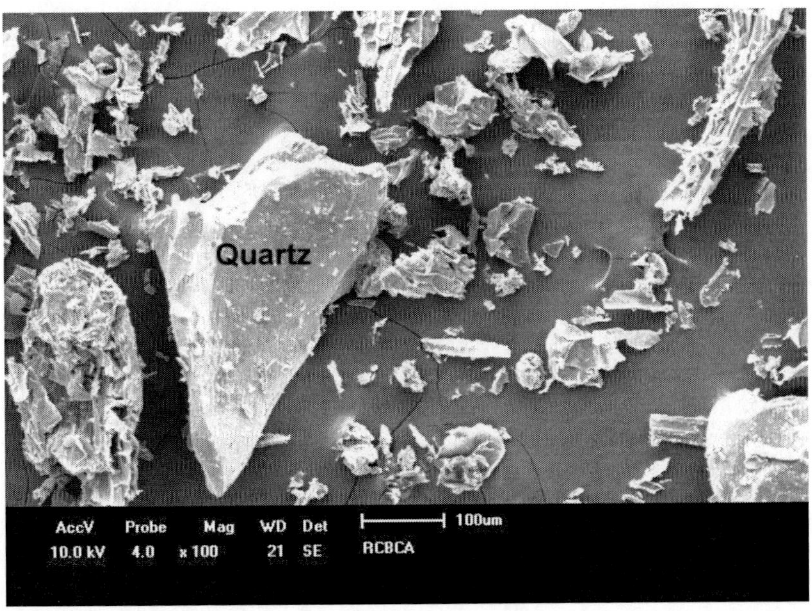

Figure 1. Morphology of the SCBA waste particles.

Table 1. Chemical composition and loss on ignition of several SCBA wastes (wt.%)

Oxides	Ref. [13]	Ref. [14]	Ref. [15]	Ref. [16]	Ref. [17]	Ref. [18]
SiO_2	61.59	72.74	59.87	60.97	38.31	51.66
Al_2O_3	5.92	5.26	20.69	0.09	2.84	9.92
Fe_2O_3	7.36	3.92	5.76	0.09	3.36	2.32
TiO_2	1.46	0.32	-	-	0.21	0.74
CaO	5.00	7.99	3.36	5.97	10.76	2.59
MgO	1.17	2.78	1.87	8.65	0.94	1.44
MnO	0.10	-	-	0.42	-	0.14
K_2O	6.22	3.47	1.37	9.02	1.77	2.10
Na_2O	-	0.84	1.11	0.70	-	1.23
P_2O_5	0.98	1.59	-	8.34	0.90	0.90
SO_3	0.42	-	1.06	-	0.45	-
LOI^+	9.78	-	0.63	5.70	40.21	24.15

+ Loss on ignition.

Table 2. Mineralogical phases identified in the SCBA wastes used to produce clay ceramics [13-15, 19-21]

SCBA waste	Crystalline phases
SCBA (Colombia)	Quartz and mullite
SCBA (Nigeria)	Quartz, cliftonite, moissanite, and titanium oxide
SCBA (northeastern Brazil)	Quartz, potassium silicate, calcium silicate, calcium oxide, and magnesium oxide
SCBA (southeastern Brazil)	quartz, potassium carbonate, Cristobalite, hematite, calcium phosphate, and mullite
SCBA (Thailand)	Quartz, calcite, and microcline
SCBA (India)	Quartz, cristobalite, and calcite

The mineralogical phases present in SCBA wastes originated from different places are presented in Table 2. In particular, the X-ray diffraction pattern of SCBA waste from southeastern Brazil is shown in Figure 2. The SCBA waste samples have different mineralogical compositions. However, for all SCBA waste samples the silica (SiO_2) is considered to be the majority mineral phase. This is in agreement with the chemical composition data (Table 1).

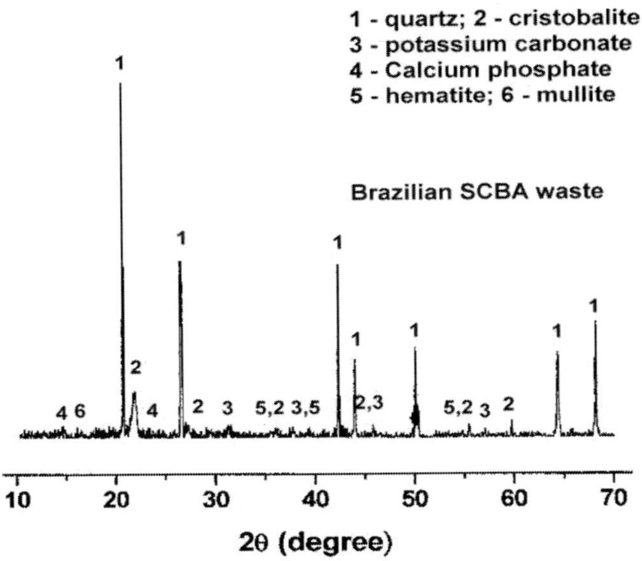

Figure 2. X-ray diffraction pattern form Brazilian SCBA waste.

The SCBA wastes (after combustion process) have a wide particle size range, being rich in particles with size above 63 μm. In fact, the SCBA wastes present high content of silica particles (Table 2). As a consequence, the SCBA wastes have a non-cohesive nature. Thus, the SCBA wastes could be used into plastic clayey formulations as a non-plastic component, which provides structural support that helps to retain shape during drying and firing processes.

2.2. Clay-based Ceramic Incorporated with SCBA Waste

Borlini (2006) investigated the influence of the incorporation of SCBA waste on the workability and technological properties of clay-based ceramics. The clay pieces containing up to 20 wt.% of SCBA waste with different particles sizes (< 44 μm, < 75 μm, < 149 μm, and < 840 μm) were prepared by pressing at 20 MPa and fired between 900 and 1200 °C. The experimental results showed that the SCBA waste is composed mainly of silica (SiO_2) and potassium oxide (K_2O). The incorporation of the SCBA waste tends to improve the workability of the clayey formulations by reducing its global plasticity. It was also established that the particle size of the SCBA waste did not appreciably affect the technological properties of the ceramic pieces, except the sample with fraction < 44 μm. Finally, it was established that the clay-based ceramics field is a suitable alternative for final disposal of SCBA waste with benefits for the ceramic processing.

Teixeira et al., (2008) reported on the reuse of SCBA waste as potential quartz replacement in clay ceramic. Clay ceramic pieces containing up to 10 wt.% of SCBA waste were prepared by pressing and fired between 800 and 1,200 ºC. The linear shrinkage and flexural strength were determined. It was found that the SCBA waste has a very high quartz concentration and low concentration of fluxing oxides (Fe_2O_3, CaO, MgO, and K_2O). The incorporation of the SCBA waste tends to improve the linear shrinkage of the fired pieces. However, the incorporation of SCBA waste had a detrimental effect on the flexural strength. They also concluded that the clayey formulation used can incorporate up to 10 wt.% of SCBA waste to produce clay bricks.

Paranhos (2010) reported on the use of SCBA waste to produce clay-based ceramic. Five clayey formulations containing up to 40 wt.% of SCBA waste and three formulations containing up to 30 wt.% of calcined SCBA waste were prepared. The SCBA wastes were used replacing feldspar. The clay ceramic pieces were prepared by pressing and fired between 1,100 and 1,250 ºC. The following technological properties were determined: linear shrinkage, water absorption, apparent porosity, apparent density, and flexural strength. Mineralogical analysis of the fired pieces was done via XRD. The results showed that it is possible to use up to 10 wt.% of SCBA waste as a replacement of feldspar in the manufacture of ceramic floor tiles.

Aigbodian et al., (2010) investigated the reuse of SCBA waste in the metallurgical and materials industry. It was found that the SCBA waste is rich in silica (SiO_2) particles. The experimental results also showed that the SCBA waste present chemical and mineralogical compositions suitable for the manufacture of ceramic products such as insulation, membrane filters, and structural ceramics.

Faria et al., (2012) reported on the reuse of SCBA waste in the manufacture of clay bricks, replacing natural clay material by up to 20 wt.%. Clay bricks were prepared by uniaxial pressing at 21 MPa and fired at 1,000 ºC. The technological properties (linear shrinkage, water absorption, apparent density, and mechanical strength) were determined. Microstructural analysis was done by scanning electron microscopy. It was found that the SCBA waste is rich in crystalline silica (SiO_2), which acts to reduce the plasticity of the clayey formulation. The results also showed that up to 10 wt. % of SCBA waste could be incorporated into clay brick formulation to result in good technological properties.

Alvaréz-Ramírez et al., (2012) reported on the use of SCBA waste and lime as chemical stabilizers (i.e., as replacement for cement) in compacted soil blocks. The soil blocks containing up to 10 wt.% of SCBA waste were prepared by pressing and cured in a curing room at 90 % relative humidity until the time of test at 7, 14, and 28 days of age. The experimental results showed that the incorporation of 10 wt.% of SCBA waste in combination with 10 wt.% of lime significantly improves the mechanical (flexural and compressive strengths) and durability properties of the compacted soil blocks. The X-ray diffraction analysis of the SCBA waste + lime mixture indicated the formation of chemical stable compounds such CSH and CAH.

Tonnayopas (2013) reported on the reuse of SCBA waste into a clay body. The clay bricks containing up to 50 wt.% of SCBA waste were uniaxially pressed and fired in air at 1,050 ºC. The following technological properties were determined: water absorption, apparent density, and compressive strength. The sintered microstructure was evaluated via scanning electron microscopy. The SCBA waste sample used is a non-plastic material rich in SiO_2 and CaO. In addition, the SCBA waste sample presented high value of loss on ignition (40 wt.%), which is credited to the calcite decomposition. The experimental results showed that the SCBA waste sample has high potential to be used as an alternative raw material to produce clay bricks. In addition, the clay-30wt.% SCBA waste composition fired at 1,050 ºC was optimal for meeting the brick quality of Thai specification.

3. Discussion

Figure 3 shows the processing flow diagram with the methodologies used to produce clay-based ceramics incorporated with SCBA wastes originated from different places. It can be seen that all research works followed the conceptual flow diagrams for the manufacture of different clay-based ceramics such as clay bricks, soil blocks, and ceramic tiles: i) clay bricks and ceramic tiles (raw materials, preparation of clayey body, shaping, drying, and firing); and ii) soil blocks (raw materials, preparation of clayey body, shaping, and curing). However, as shown in Tables 1 and 2, the researchers have used different SCBA waste samples to produce clay-based ceramics.

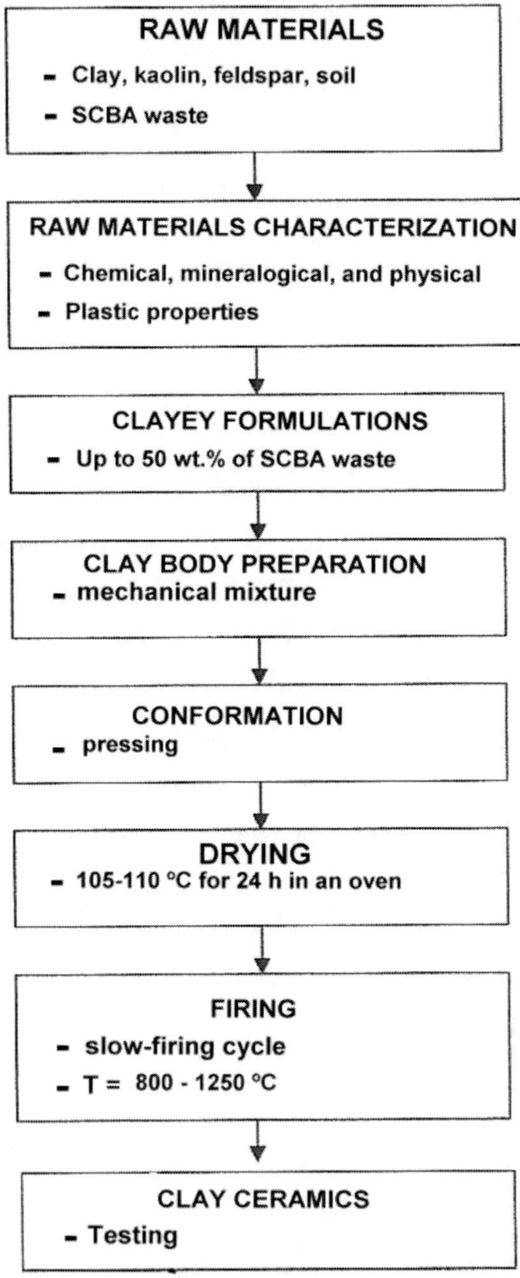

Figure 3. Methodology used for manufacturing SCBA waste bearing clay-based ceramics.

The SCBA waste samples used to produce clay-based ceramics exhibit wide chemical and mineralogical variability (Tables 1 and 2). The SCBA wastes contain, as a rule, an appreciable amount of silica (SiO_2) and moderate amounts of Al_2O_3, Fe_2O_3, CaO, and K_2O. These compounds also are present in the conventional raw materials used to manufacture of clay-based ceramics. Thus, a good chemical and mineralogical compatibility between the SCBA waste and clayey raw materials should be expected. This means that the SCBA wastes have a great potential to be used as low-cost alternative raw material to produce clay-based ceramics.

According to literature data (Table 3) different ceramic processing conditions have been used in the development of new clay-based ceramics containing SCBA waste. In general, the clayey formulations could tolerate the incorporation of SCBA waste in moderate amounts.

Table 3. Processing conditions used in the manufacture of clay-based ceramics

SCBA waste	SCBA waste amount (wt.%)	Forming method	Drying	Firing/curing process	Clay ceramic
Ref. [23]	Up to 10	Pressing at 18 MPa	-	Fired in an electrical kiln between 800 and 1,200 °C with heating rate of 10 °C/min	Clay bricks
Ref. [20]	Up to 40	Pressing at 45 MPa	Dried at 110 °C for 24h in an oven	Fired in an electrical kiln between 1,100 and 1,250 °C with heating rate of 5 °C/min	Ceramic tiles
Ref. [13]	Up to 20	Pressing at 21 MPa	Dried at 110 °C for 24h in an oven	Fired in an electrical kiln at 1,000 °C (24 h cold to cold)	Clay bricks
Ref. [17]	Up to 10	Pressing at 24 ton load	-	Curing room at 90 % relative humidity during 7, 14, and 28 days	Compacted soil blocks
Ref. [18]	Up to 50	Pressing at 100 MPa	Air-dried at room temperature for 24 h, and then oven dried at 105 °C for 8 h	Fired in an electrical kiln at 1,050 °C with heating rate of 2 °C/min until 500 °C, and then 5 °C/min until the maximum temperature	Clay bricks

Table 4. Technological properties of fired clay-based ceramics incorporated with different SCBA wastes

Properties	SCBA waste samples				
	Ref. [23]	Ref. [20]	Ref. [13]	Ref. [17]	Ref. [18]
Liner shrinkage, %	~ 0 – 3.5	3.05 – 10.01	2.0 – 2.8	-	0 – 4
Apparent density, g/cm^3	-	1.41 – 2.37	1.62 – 1.70	-	1.4 – 1.9
Water absorption, %	-	0.10 – 16.22	22.88 – 25.66	-	11 -18
Mechanical strength, MPa	~ 2.5 – 10*	9.89 – 32.85**	0.5 – 3.0+	1.40 – 1.45** 17.7 – 21.3*	23 – 43*
Weight loss, %	-	-	-	-	9 -13

* Compressive strength; ** flexural strength; +tensile strength.

Clay bricks have been produced with common clay being replaced with up to 50 wt.% of SCBA waste. It can be observed that different processing conditions such as clayey formulation, compacting pressure, drying, and firing step (heating rates and firing temperatures) have been used. Soil blocks have been produced in that the sandy soil was replaced with 10 wt.% of SCBA waste. Ceramic tiles have been produced in that the feldspar (flux material) was replaced with up to 40 wt.% of SCBA waste. Several physical and mechanical tests have been conducted on the clay-based ceramics bearing SCBA waste. However, the main technological properties determined were linear shrinkage, water absorption, apparent density, and mechanical strength. In addition, special attention is given on the water absorption and mechanical strength values. According to the technical standards, these technical properties define the class to which any clay-based ceramics suits.

Table 4 summarizes the technical properties of clay-based ceramics bearing SCBA waste. It can be seen that the incorporation of SCBA waste has led to obtain different values of technical properties. In addition, a direct comparison between the different results is very complex. The reasons for this are: i) the different typologies of clay-based ceramics; ii) the use of SCBA waste with different chemical and mineralogical compositions; and iii) the use of different ceramic processing conditions.

According to the literature the possibility of reuse of SCBA waste into clay-based ceramic has been evaluated by comparing the technological properties of the ceramic pieces with the specified values in the technical standards. Some of the maim results achieved on the incorporation of SCBA waste in clay-based ceramics are: i) SCBA waste from northeastern Brazil – ceramic tiles containing up to 10 wt.% of SCBA waste fired between 1,150 and 1,250 ºC met all the requirements of the Brazilian technical standard; ii)

SCBA waste from southeastern Brazil – clay bricks containing up to 10 wt.% of SCBA waste fired at 1,000 °C met all the requirements of the Brazilian technical standard; iii) SCBA waste from Veracruz (Mexico) – compacted soil blocks containing up to 10 wt.% of SCBA waste cured in a curing room (90 % relative humidity) met all the requirements of mechanical strength of the Mexican building technical standard; and iv) SCBA waste from Thailand – clay bricks containing up to 30 wt.% of SCBA waste fired at 1,050 °C meet all the requirements of brick quality of Thai technical standard.

Conclusion

This chapter presented a brief review on the possibility of reuse of SCB waste to produce clay-based ceramics. The review suggests that the field of clay-based ceramic is highly promising to absorb moderate amounts of SCBA waste produced worldwide. Different types of clay-based ceramics could be produced using SCBA waste as a low-cost alternative raw material. A broad spectrum of ceramic processing conditions (origin and SCBA waste amount, preparation of the clay body, shaping, drying, and firing/curing) were used to produce clay-based ceramics. It was found that the clay-based ceramics produced with different SCBA wastes meets the specifications in terms of technological properties as described in the technical standards. Thus, the valorization of SCBA waste as a partial replacement of natural raw materials to produce clay-based ceramics has proven to be a safe and sustainable way for final disposal of SCBA waste produced in the sugarcane industry.

References

[1] Stanmore, B. R. *Waste Biomass Valor.* 2010, 1, 77–89.

[2] Faria, K. C. P. *Master Thesis*, UENF-PPGECM, Campos dos Goytacazes-RJ, Brazil 2011.

[3] Dondi, M.; Marsigli, M.; Fabbri, B. *Tile Brick Int.* 1997, 13, 218-225.

[4] Ferreira, C.; Ribeiro, A.; Ottosen, L. *J. Hazard. Mater.* 2002, 1–16.

[5] Weng, C. H.; Lin, D. F.; Chiang, P. C. *Adv. Environ. Res.* 2003, 7, 679-685.

[6] Morsy, M. S. *Silic. Ind.* 2006, 71, 75-78.

[7] Rawlings, R. D.; Wu, J. P.; Boccaccini, A. R. *J. Mater. Sci.* 2006, 41, 733-761.

[8] Oliveira, E. M. S.; Sampaio, V. G.; Holanda, J. N. F. *Ind. Ceram.* 2007, 27, 191-196.

[9] Chiang, K. Y.; Chien, K. L.; Hwang, S. J. *J. Hazard. Mater.* 2008, 159, 499–504.

[10] Souza, A. J.; Pinheiro, B. C. A.; Holanda, J. N. F. *J. Environ. Manag.* 2010, 91, 685-689.

[11] Raut, S. P.; Ralegaonkar, R. V.; Mandavgane, S. A. *Const. Build. Mater.* 2011, 25, 4037-4042.

[12] Santos, P. S. *Ciência e Tecnologia de Argilas*, 2nd ed. Edgard Blücher: São Paulo, SP, 1989, Vol. 1.

[13] Faria, K. C. P.; Gurgel, R. F.; Holanda, J. N. F. *J. Environ. Manag.* 2012, 101, 7-12.

[14] Martirena, J. F.; Middendorf, H. B.; Gehrke, M; Buldelmann, H. *Cem. Concr. Res.* 1998, 28, 1525-1536.

[15] Payá, J.; Monzó, J.; Borrachero, M. V.; Diaz-Pinzón, L.; Ordónez, L. M. *J. Chem. Technol. Biotechnol.* 2002, 77, 321–325.

[16] Cordeiro, G. C.; Toledo, F. R. D.; Fairbair, E. M. R. *Quim. Nova,* 2009, 32, 82-86.

[17] Alavéz-Ramírez, R.; García, P. M.; Reys, J. M.; Juarez, D. C. A.; Ponce, Y. G. *Const. Build. Mater.* 2012, 34, 296-305.

[18] Tonnayopas, D. *Proc. of ASEAN Forum Clean Coal Technology*, Chiang Mai, Thailand, 2013.

[19] Aigbodion, V. S.; Hassan, S. B.; Ause, T.; Nyior, G. B. *J. Miner. Mater. Charact. Eng.* 2010, 9, 67-77.

[20] Paranhos, R. J. S. *Master Thesis*, UFRN-PPGEM, Natal-RN, Brazil 2010.

[21] Govindarajan, D.; Jayalakshmi, G. *Adv. Appl. Sci. Res.* 2011, 2, 544-549.

[22] Borlini, M. C. *PhD Thesis*, UENF-PPGECM, Campos dos Goytacazes-RJ, Brazil 2006.

[23] Teixeira, S. R.; Souza, A. E.; Santos, G. T. A.; Pena, A. F. V. *J. Am. Ceram. Soc.* 2008, 91, 1883-1887.

In: Sugarcane
Editor: Eleanore Webb

ISBN: 978-1-61122-356-9

Chapter 6

NON-LINEAR ANISOTROPIC DIFFUSION FOR SUGARCANE CONTOUR EXTRACTION ON LANDSAT - TM

Edinéia Aparecida dos Santos Galvanin,[1*] Paulo Henrique Hack de Jesus[2] and Jéssica Cocco[3]

[1]Department of Mathematics, University of Mato Grosso State, Barra do Bugres, MT, Brazil
[2]Department of Computer Science, University of Mato Grosso State, Barra do Bugres, MT, Brazil
[3]Master Science student of Graduate Program in Environment and Agricultural Production Systems, University of Mato Grosso State, Tangará da Serra, MT, Brazil

ABSTRACT

This paper proposes a method for the extraction of sugarcane region contour from a LANDSAT - TM image using image enhancement, recursive splitting technique by quadtree structure, region merging and non-linear anisotropic diffusion via Partial Differential Equation. In this context, the proposed methodology comprises preprocessing steps:

[*] Email: galvanin@gmail.com

Initially is produced a LANDSAT TM false-color image. After this, is applied an enhancement technique. This technique is based on the spatial domain. This enhancement operation use an adaptive average of the pixel value, based on a specifically function which adjusts the intensity of each pixel based on its relative magnitude with respect to the neighboring pixels. The recursive splitting technique using the quad tree structure consists of splitting the image into four homogeneous subregions of identical size. Each subregion is checked for homogeneity using a predefined threshold based on prior knowledge of objects presented in the scene. The splitting process proceeds recursively until no regions can be subdivided. In the end, the result is the input image organized according to the quad tree structure, where all homogeneous regions are explicitly represented. In order to meet the goal, the resulting regions are firstly structured by using the neighborhood structure. Next, the resulting regions are classified using similarity criteria, in this case regions presenting high probability of similarity are merged. The algorithm for contour filling is applied to the regions. The sugarcane contours are segmented using techniques such as, anisotropic diffusion detector that is used to previously focus the edge structure due to its notable characteristic in selectively smoothing the image, leaving the homogeneous regions strongly smoothed and mainly preserving the physical edges, i.e., those that are really related to objects presented on the image. The resulting regions are extracted by using techniques well-known, such as, vectorization, and polygonization. TM-LANDSAT images from 2008, bands 3, 4 and 5 were used. The results showed that the proposed methodology is promising for application involving extraction of cultures, because it has made possible the extraction of regions usually related to sugarcane culture.

Keywords: Sugarcane contour extraction, Quadtree structure, anisotropic diffusion

1. INTRODUCTION

Contour extraction methods are of fundamental importance in the context of mapping and updating for GIS (Geographic Information Systems) applications. Thus, to detect and to extract information from the objects, several techniques of the image analysis are used, among them, the edge detection. Depending on the goal, the edge detection can be considered as an end or as a pre-processing for subsequent processes of image analysis.

Anyway, in order to obtain the wished result, it is necessary that the strategy of edge detection be efficient and reliable.

The non-linear anisotropic diffusion via Partial Differential Equation (PDE) is an edge detection methodology which is in principle promising to solve the problem of the duality. The main idea of this methodology is to carry out a selective smooth of the image in a previous stage (Barcelos et al. 2002, Santos, 2002). This process smoothes homogeneous regions of the image more intensely, removing the information of smaller contrast. This information is related to noise and texture elements. Consequently, the PDE detector preserves, in terms of completeness and localization, the edges of better contrast, making possible to detect mainly the contour of the objects (road, buildings, limit of cultures, etc.).

Methods based on remote sensing image have been proposed to evaluate several aspects related to the sugarcane cultivation. For example, Rudorff et al. (2010) have proposed a study to establish the expansion of sugarcane for ethanol production in São Paulo State (Brazil) using LANDSAT data. Mello et al. (2010) focused on the automatic classification of sugarcane harvest based on spectral Linear Mixing. In this paper, a method by PDE for sugarcane culture contour extraction from LANDSAT TM is proposed. Our method is based on the concept of first produced a LANDSAT TM false-color image 5R4G3B. After this, is applied an enhancement technique. This enhancement operation use an adaptive average of the pixel value, based on a specifically function which adjusts the intensity of each pixel based on its relative magnitude with respect to the neighboring pixels.

The motivation of this research are that the area test used in this paper is situated in the Upper Paraguay Basin, Mato Grosso State, Brazil, this basin encompasses the floodplain region known as Pantanal, the largest inundated area on Earth. This area presents an expansion of sugarcane cultivation. In the scope of our research, contour extraction of sugarcane culture, due to scene complexity, requires the development of specific methods in the Remote Sensing image that permit to obtain the interest object.

The applications involving the Image classification and extraction have been discussed widely. The ISPRS (International Society for Photogrammetry and Remote Sensing) Commission VII/4 included Image classification techniques and new algorithms for the image extraction as a reference term.

2. THEORETICAL ASPECTS

Section 2 describes the theoretical framework for the sugarcane culture contour extraction.

2.1. Recursive Splitting Using the Quad Tree Structure

In the recursive splitting technique using the quad tree structure a region is split into subregions of identical size if the values of height variation in this region do not exceed a specified threshold. Each subregion is analyzed in relation to its homogeneity using a threshold based on knowledge of the objects present in the scene. The splitting process is repeated until there are no regions in the tree to be subdivided. The result is an organized image according to the quad tree structure, where all homogeneous regions are represented explicitly. Figure 1 shows an example of this technique, and more details can be found in Jain et al. (1995).

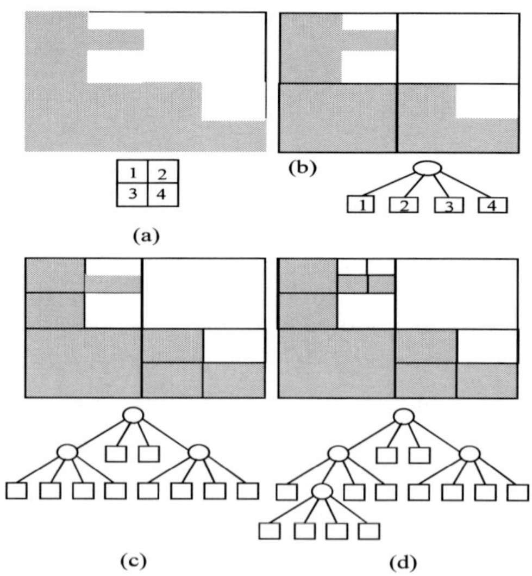

Figure 1. The building of a quad tree. (a) Original Image. (b) Original splitting into four subregions (the left node in the tree corresponds to the top left region in the image). (c) Splitting the regions from (b) into four subregions. One of these regions is still a gray region. (d) Splitting of the last gray region and the final quad tree (Source: Adapted from Jain et al., 1995).

In this application, R represents the image, R_i is each image subregion and P a property (for example, the variance in heights of the points). This splitting is based on a hypothesis testing $H_0 : P(R_i) \leq \sigma_0^2 = \lambda$ (where λ is a pre-established value in agreement with the values of height of the scene objects) against the hypothesis $H_1 : P(R_i) > \sigma_0^2$. If $H_1 : P(R_i) > \sigma_0^2$, H_1 is accepted and H_0 is rejected. The segmentation of R is performed from successive subdivisions. Thus, if the H_1 hypothesis is accepted, then the image is split into smaller subregions. This technique generates a quad tree data structure, i.e., a tree in which each node is either a leaf node or has four children. This approach can be summarized in the following stages:

1) Split the image into four regions.
2) For each region compute the variance of the height values.
3) If $P(R_i) > \sigma_0^2$, split the region into four subregions.

Whenever the H_1 hypothesis is accepted the second and the third stage should be recursively performed for all image regions. The process is concluded when the H_0 hypothesis is accepted for all regions. That means that the strategy should be performed recursively until there are no regions in the tree to be subdivided. Thus, the algorithm is concluded and a structure is generated. That structure corresponds to a segmented image, where each R_i is labeled with the mean height level of the corresponding region.

2.2. Anisotropic Diffusion Detector via PDE

The mathematical model proposed by Barcelos et al. (2002) is based on the non-linear anisotropic diffusion equation. This equation follows the idea formulated by Perona and Malik (1990), mathematically expressed as:

$$\frac{\partial u}{\partial t} = \bar{g} \,|\nabla u|\, div\left(\frac{\nabla u}{|\nabla u|}\right) - \lambda(1-\bar{g})(u-I),$$

$$u(x,y,0) = I(x,y), \ (x,y) \in \Re^2, \tag{1}$$

where $\bar{g} = \dfrac{1}{1+k|\nabla(G_T * u)|^2}$, with $0 \leq \bar{g} \leq 1$; ∇ is the gradient; *div* denotes the divergent operator; λ is a parameter that acts as weight for the term $(1-\bar{g})$; k is a constant in the function $\bar{g}$; I represents the original image; u is the smoothed image from I at the scale t; T represents the smoothing optimum level which is necessary to obtain an adequate level of smoothing and G_T is the Gaussian function.

The term $|\nabla u|\, div\left(\dfrac{\nabla u}{|\nabla u|}\right)$ in equation 1 diffuses the image u along the orthogonal direction to its gradient ∇u. Then, the image u is smoothed on both sides of an edge with minimal smoothing along the own edge.

The mathematical model given by equation 1 has the purpose of selectively smoothing the image. The function $\bar{g}$ in equation 1 is used to control the diffusion speed, that is, the selective smoothing is carried out at speeds which are lower in the surroundings of a point where the term $|\nabla(G_T * u)|$ is small. Consequently, the second term in the denominator of function $\bar{g}$ will be very small. In this conditions $\bar{g} \sim 1$, thus the term $(1-\bar{g}) \sim 0$ in equation 1. Therefore, the term $(u-I)$ that preserves the edge does not act in the model. Consequently, the diffusion accomplished by the first term of equation 1 will be higher within homogeneous regions. On the other hand, if the term $|\nabla(G_T * u)|$ were high, then the analyzed point would be considered an edge point. If this occurs, the second term in the denominator of function $\bar{g}$ will be high, so $(1-\bar{g}) \sim 1$ in equation 1 when $\bar{g} \sim 0$. In this case, the term $(u-I)$ will act strongly in the image, keeping the original characteristic of edges. Thus, the diffusion process carried out by the first term of equation 1 will have an inexpressive effect along the edge regions.

The Gaussian function used in equation 1 was slightly modified by substituting the scale parameter σ by $(aT)^{1/2}$, i.e., $\sigma^2 = aT$ (a is a real constant). This relation was suggested by Teixeira (2001). It indicates that the smoothing optimum level (T) depends on the parameter σ, which controls the smoothing intensity of Gaussian kernel. The modified Gaussian function is given by:

$$G_T(x,y) = \frac{1}{2a\pi T} e^{-(x^2+y^2)/2aT} \qquad (x,y) \in \Re^2 \tag{2}$$

The smoothing optimum level concept introduced by Barcelos et al. (2002) is then given by:

$$T = \frac{\sigma^2}{a} \tag{3}$$

According to Santos (2002) the temporal evolution (t) in the model of anisotropic diffusion is directly related to the smoothing optimal level (T):

$$t = \frac{T}{\Delta t} \tag{4}$$

where Δt represents the step size of temporal evolution.

The model consists of an iterative process, controlled by the temporal evolution defined in equation 4. The process continues up to a smoothing optimal level (T). The estimation of parameter σ is subjective, meaning that the choice of a suitable threshold is difficult and it involves trial and error.

After the application of the PDE model to the image the result will be a smoothed image. With the smoothed image, the second stage can be carried out consisting in detecting edges in the image. The function used in the second stage is:

$$g(|\nabla u|) = \frac{1}{1 + k_1 |\nabla u|^2} \tag{5}$$

where k_1 is the constant in the function g, with $0 \leq g \leq 1$.

After the application of (5) to the smoothed image *u,* the pixels whose values of g are next to the unitary value are changed to the null gray value, and the pixels with values of g are next to null value are changed to the unitary gray value. The result is a binary image where the edge pixels are white and the background pixels are black.

3. METHODS

We propose a method for sugarcane culture contour extraction that is based on LANDSAT TM image. A enhancement technique, recursive splitting and PDE detector are sequentially used for image segmentation with a minimum fragmentation level. In other words, regions compatible with the objects on the scene are sought. The recursive splitting technique (Jain et al., 1995) using the quad tree structure consists of splitting the image into four homogeneous subregions of identical size. Each subregion is checked for homogeneity using a predefined threshold based on prior knowledge of objects presented in the scene. The splitting process proceeds recursively until no regions can be subdivided. In the end, the result is the input image organized according to the quad tree structure, where all homogeneous regions are explicitly represented. In order to meet the goal, the resulting regions are firstly structured by using the neighborhood structure. Next, the resulting regions are classified using similarity criteria, in this case regions presenting high probability of similarity are merged. The algorithm for contour filling is applied to the regions that is, in essence, the same procedure described by Ballard and Brown (1982). This procedure uses three steps: 1) scan the binary grid until a region point is encountered; 3) if the point is a region point, turn left and step; otherwise, turn right and step; and 3) terminate upon returning to the starting pixel. The algorithm stops when all original segments generated by the recursive splitting algorithm have been properly analyzed and grouped. At the end of the segmentation process, all regions that match our concept of culture are categorized accordingly, and the fundamental result is a binary grid where sugarcane culture grid points are assigned a zero value and other objects grid points are assigned a value of one.

The sugarcane contours are segmented using techniques such as, anisotropic diffusion detector that is used to previously focus the edge structure due to its notable characteristic in selectively smoothing the image, leaving the homogeneous regions strongly smoothed and mainly preserving the physical edges, i.e., those that are really related to objects presented on the image. The mathematical model proposed by Barcelos et al. (2002) is based on the non-linear anisotropic diffusion equation. This model follows the idea formulated by Perona and Malik (1990). Finally, we generate polyline representations for the ordered lists of contour points obtained using the three-step contour following algorithm.

4. Experimental Results

LANDSAT-TM images from 2008, bands 3, 4 and 5 over upper Paraguay Basin (Mato Grosso State, Brazil) is used to evaluate the proposed approach (Figure 2).

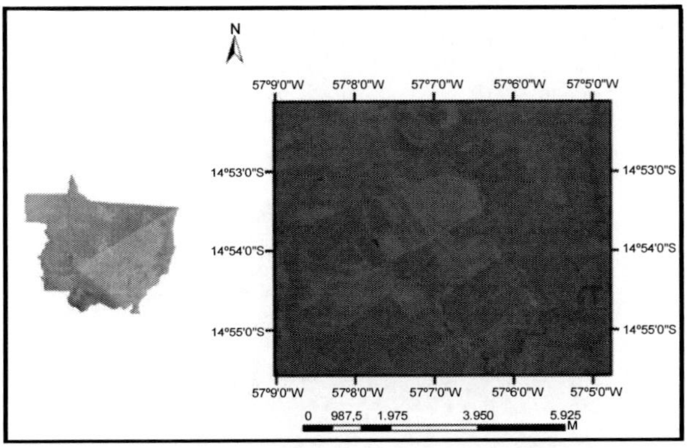

Figure 2. Image of area test over upper Paraguay Basin (Mato Grosso State, Brazil).

Below, we present and analyze the results obtained (Figure 3).

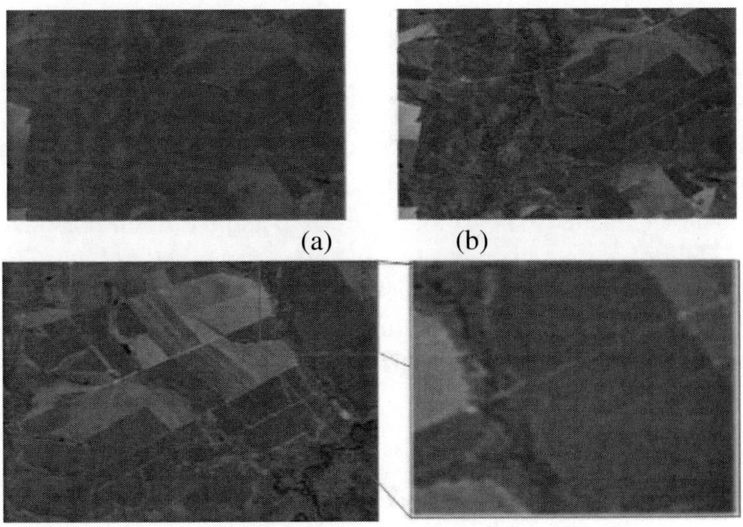

(a) (b)

(c)

Figure 3. (Continued)

(d)

(e)

(f)

Figure 3. (a) LANDSAT-TM images. (b) Result of enhancement technique. (c) Recursive splitting. (d) Sugarcane regions. (e) Contours of the regions. (f) Sugarcane contours overlaid on the original image.

Figure 2 shows the result obtained for sugarcane culture contour extraction. The LANDSAT-TM false-color image 5R4G3B applied an enhancement technique is showed in figure 2(b) and the result obtained by the recursive splitting process is presented in zoomed window in figure 2(c). The detected sugarcane regions are displayed in figure 2(d) using a binary grid (dark areas). The corresponding polylines are visualized in Figure 2(e). Edges of figure 2(e) overlaid on the original image.

CONCLUSIONS AND FUTURE PLANS

This paper presented the theoretical bases and experimental analysis regarding the process of region segmentation via recursive splitting and PDE detector. The steps in the process were described and an experiment was presented using a LANDSAT TM image.

The results showed that the proposed methodology is promising for application involving extraction of cultures, because it has made possible the extraction of regions usually related to sugarcane culture.

ACKNOWLEDGMENTS

The authors thank the FAPEMAT for the financial support.

REFERENCES

Barcelos, C. A. Z.; Boaventura, M.; Silva Jr., E. C. Well-balanced Flow Equation for Noise Removal and Edge Detection, *IEEE Trans. Image Processing,* 2002.

Ballard, D. H.; Brown, C. M. 1982, Computer vision. *Englewood Cliffs,* Prentice-Hall.

Jain, R.; Kasturi, R.; Schunck, B. G. Machine Vision, MIT Press and McGraw-Hill, Inc New York, 549p, 1995.

Mello, M. P.; Rudorff, B. F. T.; Vieira, C. A. O.; Aguiar, D. A. Automatic Classification of Sugarcane Harvest Using Spectral Linear Mixing Model, Brazilian Journal of Cartography No 62/02, 2010.

Perona, P.; Malik, J. Scale space and edge detection using anisotropic diffusion, In *IEEE Transactions on Pattern Analysis and Machine Intelligence*, 12, 7, p. 629-639, 1990.

Rudorff, B. F. T.; Aguiar, D. A.; Silva, W. F.; Sugawara, L. M.; Adami, M.; Moreira, M. A. Studies of the rapid expansion of sugarcane for ethanol production in São Paulo State (Brazil) using LANDSAT data. *Remote Sensing,* 2010.

Santos E. A. Use of partial differential equations in the treatment of orbital images. 2002. 108 f. Dissertation (Master in Cartography) - Faculty of Science and Technology, São Paulo State University, Presidente Prudente.

Teixeira R. 2001. Introduction to the scale spaces. 23 Brazilian Mathematics Colloquium. IMPA, Rio de Janeiro.

In: Sugarcane
Editor: Eleanore Webb

ISBN: 978-1-61122-356-9

Chapter 7

CARBOHYDRATE OF SUGARCANE BAGASSE: PROMISING BIOMASS OF BIOETHANOL

Junli Ren[1*], Cundian Gao[1], Huiling Li[1], Aojie Deng[1] and Runcang Sun[1,2]
[1]State Key Laboratory of Pulp and Paper Engineering, South China University of Technology, Guangzhou, China
[2]Beijing Key Laboratory of Lignocellulosic Chemistry, Beijing Forestry University, Beijing, China

ABSTRACT

Intensive efforts are being made to use the renewable lignocelluloses biomass for the production of energy and high value-added chemicals due to the global challenge of the face of depleting fossil carbon resources and growing concerns about environmental issues. Sugarcane bagasse is a by-product of the cane sugar industry, which consists of mainly tree polymeric components, namely cellulose (40-45%), hemicelluloses (xylan, 28-30%), and lignin (19-21%). The carbohydrate of sugarcane bagasse can be used to generate valuable products of commercial interest. In the period of the last few decades, carbohydrate resource of sugarcane bagasse as a promising biomass has been explored to produce bioethanol by many technologies mainly including pretreatment, hydrolysis and fermentation. The pretreatment to selectively fractionate components of

* Corresponding author: renjunli@scut.edu.cn.

the cell wall of sugarcane bagasse has the potential application for improving cellulose hydrolysis in the bioethanol production. Physical, chemical and biological treatments have been applied and the most potential pretreatment process is brought forward. The removal of hemicelluloses and lignin during the pretreatment could be as useful raw materials for preparing high value-added products like platform chemicals or biomaterials. The hydrolysis process for carbohydrate (cellulose and hemicelluloses) by biotechnology and non-biotechnology is summarized in view of economy, efficiency and environmental issues. The different fermentation technologies are described. Biotransformation can offer fantastic opportunities for the economic utilization of sugarcane bagasse in the production of ethanol, which displays sustainable, economic, environmental, and strategic advantages.

1. INTRODUCTION

Due to the depletion of fossil resources, the environmental pollution and the ever-increasing demand on fuels, exploring renewable biomass resources to produce biofuels has been considered to be an environmentally friendly option to solve the serious issues mentioned above. [1] Biofuel can reduce the demand for petroleum-based fuel for the society security of energy and is as an alternative for mitigating greenhouse gas emission. It contains energy from carbon fixation which occurs in plants and microalgae biomass. The biofuel can be made by biomass conversion referring to three different ways such as thermal conversion, chemical conversion, and biochemical conversion, which yielded solid fuel, liquid fuel and gas fuel. Among them, bioethanol is the most common liquid biofuel particularly in Brazil and the United State which accounts for about 62% of the world production. The major feedstock for ethanol production in Brazil is sugarcane, whereas corn grain is the main feedstock in the United States. [2] Other feedstocks include wheat, corn, sugar beets, molasses and any sugar or starch from which alcoholic beverages such as whiskey. The conversion of grain-based carbohydrate to bioethanol has been industrially performed, but this process need grain resource, consequently, resulting in an increase of food price. [3] Moreover, the bioethanol production from grain depends on factors such as feedstock, characteristics, production location, and the need amounts of grain for edible food. So it is unfeasible for the sustainable development of bioethanol industry with a long-term sight. To avoid the competition between gain-derived bioethanol and food, using renewable, cheap, and sustainable lignocellulose biomass resources to produce bioethanol is importantly taken into account.

Lignocellulose biomass such as trees and grasses is being regarded as a great potential feedstock for ethanol production because of the abundant, renewable, low-cost property. Lignocellulose biomass is rich in carbohydrate polymers which account for typically 75% of lignocellulosic biomass. [4] The carbohydrate possesses cellulose and hemicelluloses, which can both as feedstock for producing bioethanol. Due to the rapid development of agricultural industries, millions of tons of wastes and byproducts are generated every year. [5-6] These renewable agro-industrial residues generally contain 15-25% of lignin, 25-38% of hemicelluloses and 35-45% of cellulose, which can be as the resource for producing biofuel and chemicals. [7] Sugarcane bagasse is one of these byproducts. About 54 million dry tons of sugarcane bagasse is produced annually throughout world, [8] and it can be used in the production of industrial enzymes, ethanol, xylitol, organic acids, etc. [7,9]. Sugarcane bagasse is a residue obtained from sugarcane after it is crushed to obtain the juice used for sugar and ethanol production. It is rich in carbohydrates (chains of sugar molecules) in the form of cellulose and hemicelluloses and has been considered as the promising feedstock for production of bioethnoal. [10,11]

Cellulose and hemicelluloses in sugarcane bagasse are the major components to generate sugar for further producing bioethnoal. Cellulose is composed of glucose units, which are connected by β-1,4 glycosidic bonds to form a straight chain polymer, which is insoluble in water. Hemicelluloses represent a type of hetero-polysaccharides with complex structures containing glucose, xylose, mannose, galactose, arabinose, fucose, glucuronic acid, and galacturonic acid in various amounts depending on the source. [12] These hemicellulosic polymers from sugarcane bagasse have a classical structure, with a backbone of a backbone of β-(1→4)-linked xylosyl residues substituted with arabinose and 4-*O*-methyl-D-glucuronic acid at C-2 and/or C-3 of the main chain. The hemicelluloses from sugarcane bagasse are proved to be composed mainly of L-arabino-(4-*O*-methyl-D-glucurono) xylan. [8]

The cell wall structure of lignocellulose is complex. In the cell wall, hemicelluloses form hydrogen bonds with cellulose, and form covalent bonds with lignin and ester linkages with acetyl units and hydroxycinnamic acids. [13] Lignocellulose biomass is highly resistant to enzymatic degradation in its native state because of the complex linkage among celluloses, hemicelluloses and lignin. So there are the main three steps to produce ethanol which contain the pretreatment process, the hydrolysis process, and the fermentation process, as seen in Figure 1. The objective of pretreatment is to breakdown the cell wall, liberate hemicelluloses and lignin, and destroy the crystalline structure of cellulose and increase its surface area for improving the accessibility of carbohydrate materials. [14] The pretreatment technology has physical, chemical and physical-chemical and biological pretreatments. The objective of the hydrolysis process is to make the polysaccharides to monosaccharides by the chemical hydrolysis or the enzymatic hydrolysis. Fermentation is the process by which sugars (pentose and hexose) are converted to alcohol using different yeasts. Simultaneous saccharification and fermentation (SSF) or simultaneous saccharification and co-fermentation (SSCF) are the desirable options for production of ethanol from lignocellulosic biomass.

Extensive research and development programs have been initiated to convert sugarcane bagasse to bioethanol. This chapter presents a comprehensive overview of the technology and economic status for the bioethanol production from sugarcane bagasse including the pretreatment method, hydrolysis methods and fermentation methods.

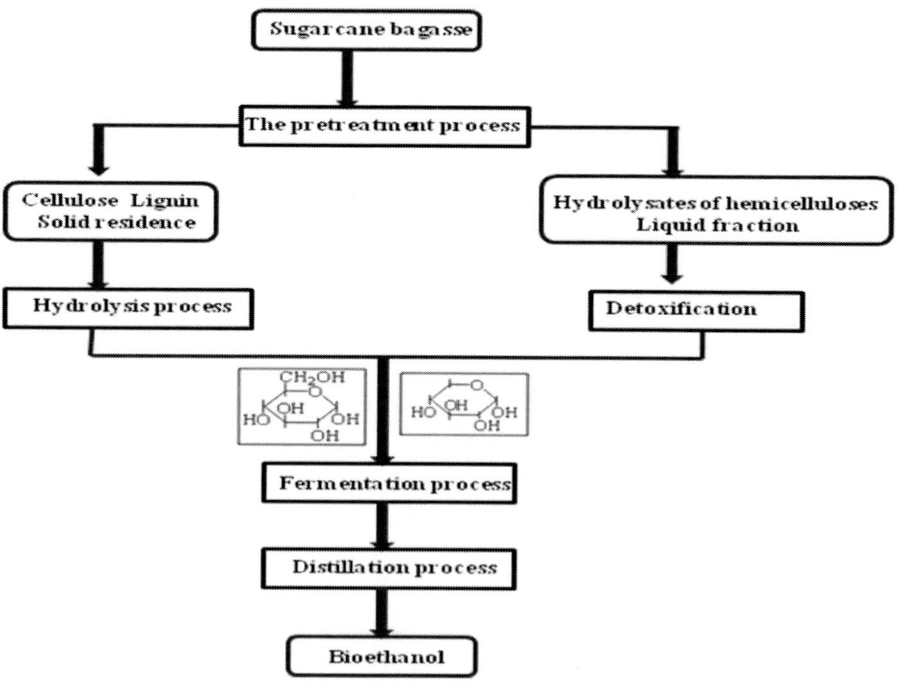

Figure 1. The procedure for the bioethanol production from sugarcane bagasse.

2. Pretreatment Process

The pretreatment process is one of the important processes in the production of bioethanol from lignocellulosic biomass. The pretreatment technology of lignocellulosic materials includes physical-chemical pretreatment, chemical pretreatment, physical pretreatment and biological pretreatment. Because the different lignocellulosic biomasses have various properties, it is significance to select a suitable pretreatment method to alter the structure of biomass, and make the cellulose more accessible to enzymes or chemicals. Physical pretreatment possesses milling pretreatment, microwave-assisted pretreatment and ultrasonic pretreatment. Chemical pretreatment was applied by the addition of chemicals such as acid agents, alkaline agents, organic solvents, and ionic liquids. Physical-chemical pretreatment includes hydrothermal pretreatment, steam explosion pretreatment, supercritical CO_2 explosion pretreatment and ammonia fiber explosion pretreatment. Here, we described the definite pretreatment technology for different lignocellulosic biomass and these methods mentioned below could be applied in the sugarcane bagasse pretreatment.

2.1. Physico-Chemical Pretreatment

2.1.1. Hydrothermal Pretreatment

Hydrothermal treatment (HTP), which is also called liquid hot water (LHW) pretreatment, hot compressed water (HCW), hydrothermolysis, aqueous processing and pressure-cooking in water, has been considered a cost-effective approach to destroy the recalcitrant nature of lignocellulose materials (LCMs) and convert them into cellulose, hemicelluloses and lignin (Fig. 2). [15-20] Three major advantages of HTP are: (1) the process does not require the addition and recovery of chemicals different from water, (2) limited equipment corrosion problems, (3) simple and economical operation. [21, 22] Therefore, hydrothermal processing can be considered as an environmentally friendly fractionation process for bioethanol production, papermaking production and a technology for converting agro-food by-products into useful food ingredients. [23-26]

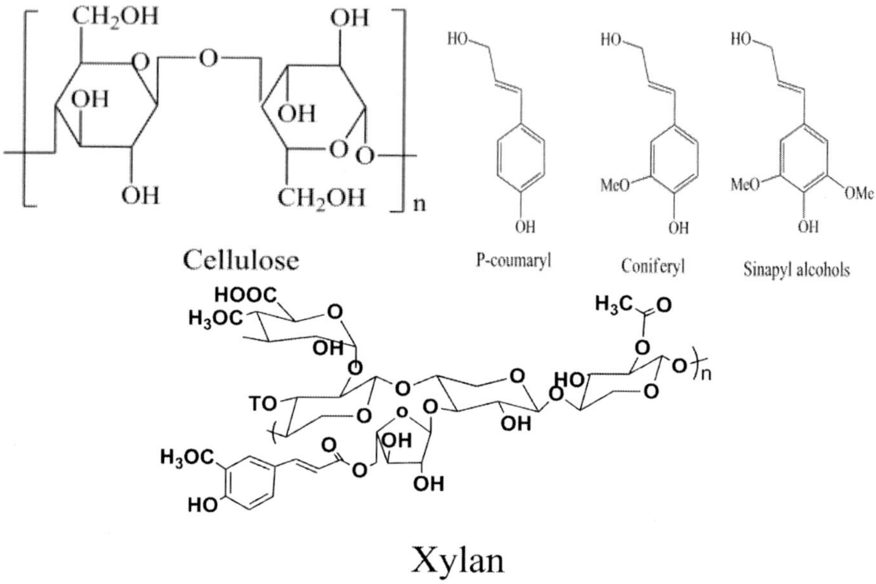

Figure 2. Structure of three main components in lignocelluloses materials.

HTP employs high temperatures (160-220 °C) and pressure to keep water in liquid state and in contact with biomass without addition of any chemicals or catalysts. [27-29] Rapid decompression or expansion is not required and utilization of pressure is only for maintaining water and preventing evaporation during HTP process. During the hydrothermal process, acidic hydronium ions (H_3O^+) that act as catalysts are generated from the autoionization of water at the high temperature and the organic acids from acetyl groups in hemicelluloses. [30-34] Hemicelluloses in biomass can be completely solubilized into aqueous compounds, the lignin seal is broken, and cellulose is almost entirely preserved in the solid product (Fig. 3). [35] Up to 80% of hemicelluloses can be removed by hydrothermal process, leading to the increasing cellulose digestibility. [36,37] In view of the effect removal of reactive hemicelluloses from biomass, HTP will be a potential pretreatment method to improve the products quality through decreasing the formation of water, acids, and other reactive compounds in the following hydrolysis process. [38]

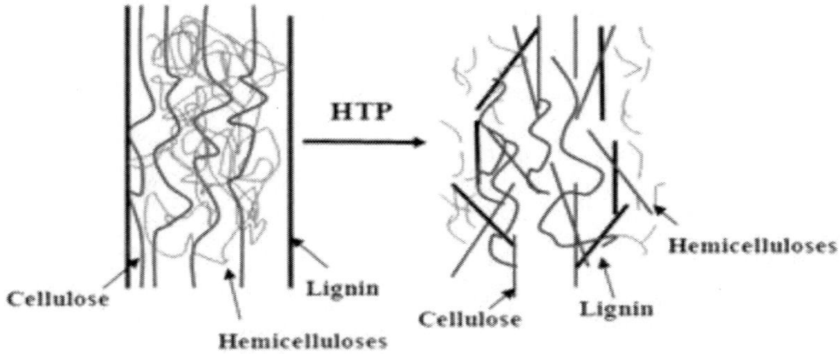

Figure 3. Scheme of hydrothermal pretreatment.

Cellulose is the most abundant biopolymer that can be obtained from LCMs (Table 1). Since the solid fraction from hydrothermal processing are enriched into cellulose and lignin, a variety of applications can be visualized for this phase from hydrothermal process of LCMs. One of the most promising approaches is enzymatic hydrolysis of the cellulose content after pretreatment for second generation bioethanol production. [39, 40] Lignin on the surface of LCMs delocated during the hydrothermal processing, thus favored the accessibility of enzyme to the LCMs structure in the pretreated material, increasing the potential of cellulose saccharification. [41] Moreover, the increase of pore size and accessible area can enhance enzyme penetration and saccharification using HTP process as pretreatment. [42-46] Goh et al. [47] evaluated the effect of HTP for oil palm fronds to enhance glucose recovery for the production of second generation bioethanol. 92.78% saccharification yield was obtained when the feedstock was treated at 178 °C for 11.1 min. Lee et al. [48] studied the autohydrolysis pretreatment of costal *Bermuda grass* for increased enzyme hydrolysis, about 67.4% saccharification yield was achieved after pretreated at 150 °C for 60 min. Xiao et al. [49] demonstrated the influences of hot compressed water pretreatment on the structural changes of woody biomass for bioethanol production. After the treatment of *Tamarix ramosissima* at 200 °C for 180 min, up to 88% saccharification yield was obtained. Thomsen et al. [50] investigated hydrothermal pretreated wheat straw at pilot plant scale using a three-step reactor system at 195 °C for 3 min. High hemicelluloses recovery, high cellulose digestibility and low lignin hydrolysis was observed with 72% saccharification yield. Yu et al. [51] investigated the two-step liquid hot water pretreatment of *Eucalyptus grandis* to enhance sugar recovery and enzymatic digestibility of cellulose. Saccharification yield of 96.6% was obtained when the pretreatment temperature and time were 200 °C and 20 min, respectively. Considering the great fractionation ability and little degradation of cellulose after pretreatment, HTP is a promising pretreatment in the development of biorefineries. [52-54]

Hemicelluloses are the second most abundant polysaccharide in nature and are made up of 14-50% of the raw LCMs dry weight (Table 1). Under optimized conditions, hemicelluloses can be almost totally removed from LCMs during HTP process, being decomposed into valuable soluble products such as oligosaccharides, monosaccharides, sugar-decomposition products and acetic acid. [57-60] Under harsh operational conditions, oligosaccharides can be break down to form monosaccharides and xylo-oligosaccharides, which are the bioactive molecules with great potential as ingredients for functional foods. [61-63] Moreover, monosaccharides, such as xylose and glucose, can be dehydrated to furfural and hydroxymethyl furfural, which are the important molecules for the production of high-added value chemicals. [64-67] Garrote et al. [68] reported that 23.2 g of oligosaccharides/100 g of oven-dried corncobs was obtained at 202 °C through hydrothermal process in non-isothermal reaction conditions.

Table 1. Main components and content of lignocellulosic materials (%, dry materials) [55, 56]

Raw material	Cellulose (%)	Hemicelluloses (%)	Lignin (%)	Water soluble component (%)	Wax (%)	Ash (%)
Wheat straw	38.6	32.6	14.1	4.7	1.7	5.9
Rice straw	36.5	27.7	12.3	6.1	3.8	13.3
Ryegrass	37.9	32.8	17.6	4.1	2.0	3.0
Barley straw	34.8	27.9	14.6	6.8	1.9	5.7
Oat grass	38.5	31.7	16.8	4.6	2.2	6.1
Corn stalks	38.5	28	15	5.6	3.6	4.2
Corn cob	43.2	31.8	14.6	4.2	3.9	2.2
Sugar beet pulp	18.4	14.8	5.9	5.9	1.4	3.7
Sugarcane bagasse	39.2	28.7	19.4	4.0	1.6	5.1
Oil palm fiber	40.2	32.1	18.7	5.0	0.5	3.4

Boussarsar et al. [69] also reported that an acceptable xylose extraction yield and low degradation of sugar monomers were obtained at 170 °C for 2 h. Vegas et al. [70] demonstrated that about 90% of the xylo-oligosaccharides present in hydrolysis liquors when using ultra- and nano-filtration for the purification of oligosaccharides from rice husk hydrothermal processing liquors. Garrote et al. [71,72] investigated hydrothermal pretreated corncobs and *Eucalyptus globules* and favorable features in terms of substrate conversion, reaction, selectivity and low inhibitor concentration were shown when xylose solution was used as fermentation media through sequential stages of hydrothermal processing-posthydrolysis. Li et al. [73] studied the catalytic hydrothermal conversion of corncob into xylose and furfural, highest furfural yield (6.18 g/100 g) could be obtained at 180 °C for 120 min with 6.80 g/100 g xylose yield when then corncob/water ratio was 10:100. Other important application of hemicellulosic liquid phase after hydrothermal pretreatment is the production of xylanases and xylitol. [74-77]

Lignin is the most abundant aromatic heterogeneous polymer formed by phenolic compounds and are bonded together with over two-third being ether bonds (C-O-C) and the rest being C-C bonds between three aromatic alcohols including *p*-coumaryl, coniferyl and sinapyl alcohols. [28] During the HTP process, lignin and lignin-hemicellulose linkages can undergo degradation, partial depolymerizatin and profound re-localization. [78, 79] Phenolics obtained from the HTP process are considered as an attractive source for natural antioxidants and might have potential applications as food additives. Tsubaki et al. [80] found that vanillin, vanillic acid, dihydroconiferyl alcohol and guaiacol are degradation compounds originated from guaiacyl units of lignin under microwave irradiation. Pourali et al. [81] reported the production of eleven phenolic compounds in HTP conditions, and found that the content of phenolic compounds increased with the temperature. Besides, many researchers had focus on the antioxidant activity of the hydrothermal products of lignin. [15, 16, 82-86]

2.1.2. Steam Explosion Pretreatment

Steam explosion is one of the most widely implemented pretreatment methods that can degrade hemicelluloses and transform lignin to increase the crystallinity of cellulose by promoting crystallization of the amorphous portions prior to bioethanol production. [27, 87] During the steam explosion pretreatment process, high-pressure saturated steam (0.69-4.83 Mbar, 160-260 °C) is introduced to heat biomass particles for several seconds to a few minutes followed by the release of pressure to atmospheric, and then condensed moisture evaporates and desegregation of lignocellulosic matrix takes places. [88, 89] Compared to mechanical comminution, lower energy requirement for steam explosion pretreatment is present and no recycling or environmental costs. Water soluble compounds, such as acetic acid, formic acid, phenols, levulinic acid, and lignin degradation products, are generated during steam explosion pretreatment. [90, 91]

There are many factors that affect steam explosion pretreatment, such as residence time, temperature, particle size and moisture content. [34, 92, 93] During the steam explosion pretreatment, hemicelluloses hydrolyzed accomplished by the generation of organic acids such as acetic acids from hydrolysis of acetyl groups in hemicelluloses and formic and levulinic acids derived from other functional groups. [94-96] Similar to HTP process, acidic hydronium ions (H_3O^+) also form in steam explosion pretreatment at high temperature and pressure and act as acid catalyst for the removal of hemicelluloses from the cellulose microfibrils surface, thus increasing the enzyme accessibility and enzymatic hydrolysis rate of cellulose by exposing the cellulose surface. [97, 98]

Ballestros et al. [99] studied the effect of particle size on the steam explosion pretreatment of chipped *B. carinata* biomass (5% moisture). Results showed that higher cellulose recoveries were observed at larger particle sized. After the pretreatment, the maximum yield of sugar in the water insoluble fiber was determined by enzymatic hydrolysis at 50 °C on a rotary shaker at 150 rpm for 72 h and at 2% (w/v) substrate concentration. Up to 99% yield of enzymatic hydrolysis was obtained for samples pretreated at 210 °C. Reczey and co-workers [100] evaluated that the enzymatic conversion (50 °C, 24 h) of cellulose to glucose was enhanced about four times after the steam explosion pretreatment at 200 °C for 5 min with 2% sulfuric acid, and about 90% ethanol yield was achieved. Cara et al. [101] extended the production of fuel ethanol from olive-tree pruning which was subjected to steam explosion at 190-240 °C with and without previous impregnation by water or sulfuric acid solution. Maximum yield of ethanol (7.2 g/100 g) was obtained from water-impregnated residue pretreated at 240 °C.

Since the destruction of a portion of the xylan fraction, incomplete disruption of the lignin-carbohydrate matrix, and generation of inhibitors for enzymatic hydrolysis, steam explosion pretreatment is among the very limited number of cost-effective pretreatment technologies for pilot scale demonstration and commercialized applications. [93, 102,103]

2.1.3. Super Critical CO_2 Explosion Pretreatment

Supercritical fluid refers to a fluid that is in a gaseous form but is compressed at temperatures above its critical point to a liquid like density. Super critical CO_2, whose critical temperature was 31 °C and critical pressure was 7.4 MPa, exhibits gas-like transport properties of diffusivity and viscosity like liquid density. [104] Since CO_2 molecules have a similar size property to those of water, super critical CO_2 can penetrate into small pores of lignocellulosic material with lower temperature and energy consumption than steam explosion, which makes it as an ideal choice among the explosion-type methods. [105] On the other hand, when CO_2 is dissolved in water, carbonic acid forms as an acid catalyst, which will increase the hydrolysis rate. Furthermore, no waste product and no further recovery requirements are acquired due to the easy removal of CO_2 by depressurization. [106] The properties of supercritical CO_2 have provided the impetus for applying CO_2 explosion technology to a amount of separations problems experienced in many segments of the industry. [107]

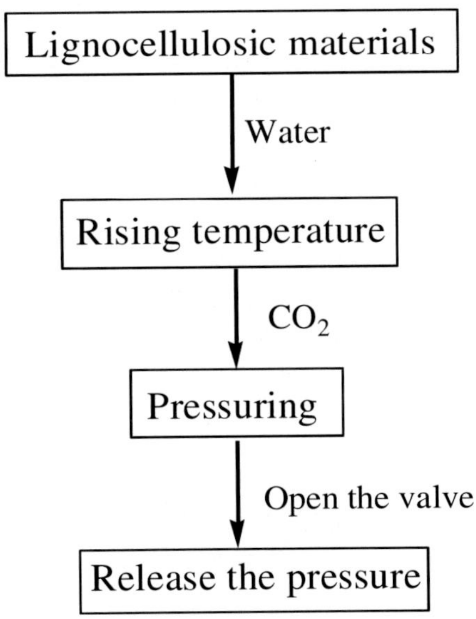

Figure 4. Scheme for the super critical CO_2 explosion pretreatment.

There are four operation factors may affect the performance of supercritical pretreatment: temperature, pressure, moisture content of biomass, and the pretreatment duration. [108] Normal super critical CO_2 explosion pretreatment involves the following steps: adding a small amount of water to the lignocellulosic materials loaded in the reactor, raising the system temperature, pressuring using supercritical CO_2, holding the system for a period of time, and exploding the lignocellulosic materials by rapidly opening the valve to release the pressure (Fig. 4). [109] During the pretreatment process, the low temperature prevents any appreciable decomposition of monosaccharides by the acid, and the disruption of the cellulosic structure increases the accessible surface area of the substrate to hydrolysis. [110, 111]

Narayanaswamy et al. [112] used supercritical CO_2 as a green solvent to pre-treat corn stover and switchgrass at various temperatures and pressures for lignocellulosic ethanol production. The biomass was hydrolyzed after pretreatment using cellulase combined with β-glucosidase. Results showed that the glucose yields from corn stover samples pretreated with supercritical CO_2 were higher than the untreated ones, and the highest yield of glucose of 30% was achieved with supercritical CO_2 pretreatment at 3500 psi and 150 °C for 60 min. However, very limited improvement was observed after pretreatment for switchgrass. Alinia et al. [90] investigated the effect of pretreatment of dry and wet wheat straw using supercritical CO_2 alone and the combination of supercritical CO_2 and steam by varying the temperature (160-200 °C) and the residence time (10, 30, 60 or 70 min). Highest yield of sugar (149.1g/1000 g of wheat straw) was obtained at 190 °C for 30 min by supercritical CO_2 alone. However, the best overall yield for sugar was 208.4 g/1000 g of wheat straw when impregnated with water by supercritical CO_2 at 185 °C for 30 min. Moreover, up to 234.6 g/1000 g of wheat straw for sugar was obtained when wheat straw was pretreated by steam explosion and supercritical CO_2 at steam temperature and retention time of 200 °C and 15 min and supercritical CO_2 conditions of 12 MPa, 190 °C and 60 min. Due to the low temperature in CO_2 explosion, few degradation of sugar was observed, which is good for the following enzymatic hydrolysis process.

2.1.4. Ammonia Fiber Explosion (AFEX) Pretreatment

Ammonia fiber expansion (AFEX) is a pretreatment method that lignocellulosic biomass is exposed to liquid ammonia at high pressure for a period of time to overcome the recalcitrance of plant biomass and thereby renders such materials reactive for hydrolytic activities. [113, 114] Important advantages of AFEX treatment include low moisture content, lower formation of sugar degradation products due to lower temperature requirements, and complete recovery of solid material. [113, 115] This pretreatment method has been shown effective on multiple lignocellulosic feedstocks such as corn stover, switchgrass, rice straw, miscanthus, alfalfa, wheat straw and wheat chaff. [116-120]

There are four parameters that influence the AFEX pretreatment: ammonia loading, water loading, reaction temperature and residence time. [121, 122] During the pretreatment process, hemicelluloses are degraded to oligomeric sugars and deacetylated, resulting in no removal of hemicelluloses. [123, 124] Furthermore, rapid expansion of the ammonia gas causes cleavage of lignin-carbohydrate complex and consequent physical disruption of biomass fibers leading to increased digestibility of biomass and fermentation rate of various feedstocks. [27, 125]

Various research groups have done a significant amount of researches for AFEX pretreatment of lignocellulosic materials. Bals et al. [126] studied the influences of AFEX pretreatment of eleven different forages, including traditional forages, agricultural residues, and dedicated energy crops, for fiber digestibility in vitro. Compared to untreated samples, AFEX treatment improved 48 h neutral detergent fiber digestion for several moderately indigestible forages, but showed no improvement for highly digestible samples. About 52% and 128% digestibility improvement was observed for untreated corn stover and late-harvest switchgrass, respectively, while 74% and 70% for conventional ammonia treated samples. Harun et al. [127] investigated the influence of particle size on the performance of AFEX pretreated rice straw as the source of fermentable sugars. Results showed that during enzymatic hydrolysis, larger cut rice straw particles (5 cm) significantly demonstrated higher sugar conversion when compared to small particles treated with AFEX conditions. Teymouro et al. [128] evaluated the influences of process conditions and parameters on corn stover for maximum effectiveness of the ammonia fiber explosion process. Optimized pretreated conditions are 90 ^{o}C, ammonia/dry corn stover mass ratio of 1:1, a moisture content of corn stover of 60%, and a residence time of 5 min. However, AFEX process was not effective for high lignin content biomass.

2.2. Physical Pretreatment

2.2.1. Milling Pretreatment

Milling pretreatments, which include ball milling, two-roll milling, hammer milling, colloid milling and disk milling, are mechanical pretreatments that break down the structure of lignocellulosic materials and decrease the cellulose crystallinity. [87, 129] The most common use is ball milling, which is considered as environmentally friendly with no addition of chemicals and generation of inhibitors. During ball milling pretreatment, the contact of biomass with balls inside a cycle machine reduces the particles size. The size of the materials is usually 10-30 mm after chipping and 0.2-2 mm after milling or grinding. [130, 131]

After the ball milling pretreatment, the yields of glucose and xylose are higher than those of other milling methods in some pretreated biomass. Schultz-Jensen et al. [40] compared five pretreatment technologies for the production of bioethanol from *macroalgae Chaetomorpha linum*. Wet oxidation and ball milling showed the highest ethanol yield of 44 g ethanol/100 g glucan, which was close to the theoretical ethanol yield of 57 g ethanol/100 g glucan. Buaban et al. [132] demonstrated a bioethanol production from ball milled bagasse using an on-site produced fungal enzyme cocktail and xylose-fermenting *Pichia stipites*. Results showed that after ball milling pretreated for 2 h, nearly complete cellulose structural in sugarcane bagasse transform to an accessible amorphous form. Up to 84% and 70% saccharification yield for glucose and xylose were achieved after hydrolysis at 45 ^{o}C for 72 h, respectively. Li et al. [133] studied a one-step conversion of wheat straw to sugars by simultaneous ball milling, mild acid, and fungus *Penicillium simplicissimum* treatment. Results showed that the optimized conditions for hydrolysis were ball milling 48 h in citrate solvent (pH=4) with P. simplicissimum H5 at the speed of 500 rpm and the yield of sugar increased with increased milling time.

However, the most important drawback of ball milling is high energy consumption, which always requires hours to obtain the desire particle size. According to the literature, if the final particle size is held to the range of 3-6 mm, the energy input for comminution can be kept below 30 kWh per ton of biomass. [110] So ball milling is not suitable for the widely used considering the economic issue.

2.2.2. Microwave-Assisted Pretreatment

Microwave-assisted system, which displays higher yield and better selectivity for a given reaction time than oil-bath in many types of reactions, is considered as the energy efficiency associated with time saving way for carbohydrate conversion. [134-136] The susceptibility of lignocellulosic materials to enzymatic saccharification of the cellulosic is inhibited because of the presence of the complex structure of lignin and hemicelluloses with the cellulose. Microwave irradiation is a promising alternative to the conventional heating method to alter the ultra structure of cellulose by molecular collision due to dielectric polarization, degrade/partially remove lignin and hemicelluloses, disrupt the silicified waxy surface and finally enhance the enzymatic susceptibility of reducing sugars. [137-139]

Various studies have paid attentions to the pretreatment of microwave irradiation. Kitchaiya et al. [138] developed a technique to improve enzymatic hydrolysis of lignocellulosic wastes (grounded rice straw and sugar cane bagasse) by microwave pretreatment at 240 W for 10 min under atmospheric pressure. Compared with no pretreatment, reducing sugars amount was more than twice produced from enzyme saccharification utilizing the microwave pretreatment. Hashem et al. [140] investigated the use of microwave heating for pretreatment cotton fabrics to reduce the pretreatment time, chemicals and water. Results showed that a complete fabric preparation was obtained in 5 min when using microwave irradiation and the fabric properties were comparable to those obtained in traditional pretreatment process which requires 2.5-3 h for completion. Peng et al. [141] demonstrated a combined pretreatment of ball milling and microwave irradiation for enhancing enzymatic hydrolysis of microcrystalline cellulose. About 54.8% and 77.4% less energy consumption was achieved with ball milling for 1 h and microwave irradiation for 20 min, compared with the only ball milling pretreatment.

2.2.3. Ultrasonic Pretreatment

Ultrasonic pretreatment is a commonly used method for mixing and cavitation bubbles to treat lignocellulosic materials for better removal of hemicelluloses and ligninin a short time. [142] Both enzymatic and acid hydrolyses as well as ethanol yield increased after ultrasonic pretreatment of lignocelluloses biomass. [143] Many factors make sense for the degree of ultrasonic pretreatment, such as biomass loading, particle size, frequency and stirring. [144] During the ultrasonic pretreatment, the acoustic waves can break the cohesion of a liquid and can create micro cavities, thus increase the enzymatic hydrolysis yields and reduce necessary times of pretreatment and saccharification. Moreover, extremely high temperatures in the area of collapsed bubble can be achieved by the implosion of cavitations bubbles. [145]

Sindhu et al. [145] developed a novel surfactant-assisted ultrasound pretreatment of sugarcane tops for improved enzymatic release of sugars as well as to optimize the effect of various operational parameters on pretreatment and hydrolysis. Results showed that hemicelluloses and lignin can be effectively removed during the surfactant-assisted ultrasound pretreatment and the reducing sugar yields from sugarcane tops improve. The maximum sugar yield was 0.661 g/g of pretreated biomass under the optimized hydrolysis conditions. Bussemaker et al. [134] studied the effects of ultrasonic treated parameters on lignocellulosic materials by varying the frequency, particle size, loading and stirring. Via different mechanisms, fractionation was improved at 40 and 995 kHz. Delignification was favored at the ultrasonic treatment frequency of 40 kHz, biomass loading 1/20 (g/ml) with stirring and particle size range of 0.5-1 mm. However, carbohydrate solubilization was favored at 995 kHz. Jadhav et al. [146] investigated the ultrasound-assisted hydrolysis and subsequent esterification of Nagchampa oil under mild operating conditions in order to intensify methyl esters production as shown in Fig. 5. Using ultrasonic irradiation, the extent of esterification reaction increased from 75% to 92.5% in the presence of pretreated enzyme and the time requirement reducing from 20 h to 7.5 h. Goshadrou et al. [147] reported that the ethanol yield from hydrolyzed sorghum bagasse was increased by 4.5% from ultrasonic pretreatment. Therefore, the hydrolysis of lignocellulosic materials can be improved by ultrasonic pretreatment, thus leading the increase of the yields of glucose, xylose, and ethanol in downstream processing as well as enhancing the accessibility and delignification. [143]

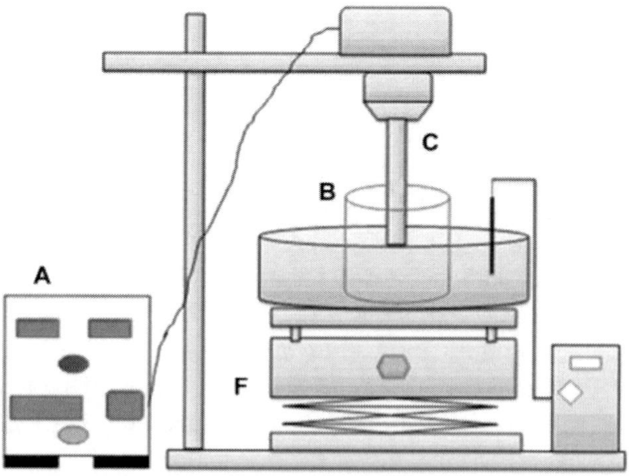

Figure 5. Experimental setup for enzyme sonication A: generator, B: ice bath, C: ultrasound probe, E: temperature indicator, F: stand. [146]

2.3. Chemical Pretreatment

2.3.1. Acid Pretreatment

Mineral acids, such as H_2SO_4, HCl and so on, have been used to treat lignocellulosic materials to improve the enzymatic hydrolysis of lignocellulosic biomass to release fermentable sugars. Due to the drawbacks of toxic, corrosive, hazardous, and thus require reactors that are resistant to corrosion of concentrated acid pretreatment, dilute acid hydrolysis has been successfully developed. Dilute acid can effectively hydrolyze hemicelluloses as well as cellulose to dissolved sugars. The removal of hemicelluloses can greatly enhance the digestibility of cellulose in the residual solids. [147-150]

Dilute-acid pretreatment includes two types: (1) a high temperature (>160 °C), continuous flow process for low solids loadings (weight of substrate/weight of reaction mixture=5-10%); (2) a low temperature (<160 °C), batch process for high solids loadings (10-40%). [110] The common used acids are sulfuric acid, nitric acid, hydrochloric acid, and phosphoric acid. A wide range of feedstocks ranging from hardwoods to grasses and agricultural residues have been tested. Recently, Benjamin et al. [151] investigated the dilute acid pretreatment of bagasse from six varieties of sugarcanes in connection with enzymatic hydrolysis for maximum combined sugar yield. Up to 22.7% differences in combined sugar yield at the optimal conditions were observed between the varieties. Moreover, they found that high ratio of carbohydrates to lignin and low ash content favored the release of sugar from the substrates. Greenwood et al. [152] used the particle scale modeling to study the mild acid pretreatment of sugarcane bagasse. Mild acid pretreatment hydrolyses the hemicellulosic component of biomass, thus allowing enzymes greater access to the cellulosic substrate during saccharification. Zhao et al. [139] developed a novel pseudo-homogenous kinetic model for polysaccharide dissolution during atmospheric acetic acid pretreatment of sugarcane bagasse. Zeng et al. [89] studied the structure of lignin after dilute phosphoric acid plus steam explosion pretreatment process of sugarcane bagasse in a pilot scale as well as the effect of the lignin extracted by ethanol on subsequent cellulose hydrolysis. Results showed that almost 8% of mass weight was extracted by pure ethanol. Compared to pretreated bagasse without extraction, about 22% higher glucose yield was obtained after enzyme hydrolysis. Although acid pretreatment can effectively improve cellulose hydrolysis, its cost is higher than those of other physic-chemical pretreatment processes.

2.3.2. Alkaline Pretreatment

Lignocellulosic materials can be pretreatment by some bases (sodium, potassium, calcium, and ammonium hydroxides) in low temperatures and pressures for hours or days. During the alkaline pretreatment, lignin, acetyl groups and different uronic acid substitution can be removed to enhance the cellulose saccharification rate. Compared with acid pretreatment, alkali pretreatment causes less sugar degradation, less hemicelluloses as well as cellulose solution, and many of the caustic salts can be recovered and/or generated. [110, 154, 155] Moreover, chemical swelling of fibrous cellulose also caused accompanying with the saponification and salvation reactions which led to the disruption of the crosslinks between hemicelluloses and other components. [87, 156] Among various agents, calcium hydroxide has been shown to be an effective pretreatment agent which can be recovered from an aqueous reaction system as insoluble calcium carbonate by neutralizing it with inexpensive carbon dioxide.

Sindhu et al. [157] used the response surface methodology according to Box-Behnken design to study the effect of solid loading, enzyme loading, incubation time and surfactant concentration on enzymatic saccharification for alkali pretreated (3% NaOH) sugarcane tops at 121 ^{o}C for 60 min in a laboratory autoclave. Up to 77.5% sugar was recovered from the pretreated biomass under the optimized conditions, which was seven times higher than that obtained with untreated sugarcane tops. And 90% of lignin was removed by this pretreatment method. Barman et al. [158] investigated the effect of mild alkali pretreatment on structural changes of red straw. Results showed that compared with other pretreatment methods, 2% and 2.5% NaOH pretreated sample exposed more cellulose fibers. About 69.9% and 72.4% lignin removal, and 56.4% and 60.5% hemicelluloses removal were achieved after pretreated by 2% and 2.5% NaOH, respectively. Moreover, reducing sugar yield was increased very marginally when NaOH concentration increased from 2% to 2.5% for reed straw pretreatment. Though less inhibitor was obtained in alkaline pretreatment, long residence time and the need for neutralization of the pretreated samples limit its applications.

2.3.3. Organic Solvent Pretreatment

Many organic solvents, such as ethanol, methanol, and acetone and so on, could be used to treat lignocellulosic materials for the breaking the internal lignin and hemicelluloses bonds and is therefore especially efficient for high lignin lignocellulosic biomass. [159-163] The removal of lignin can increase the surface areas making cellulose more acceptable to enzyme. However, low boiling point of organic solvents, high risk of high pressure operation, flammability and volatility of such solvents pushed them hard for operation. [164-166]

2.3.4. Ionic Liquids (ILs) Pretreatment

Recently, ionic liquids (ILs) have received growing attentions as promising green solvents or catalysts for the dissolving carbohydrates and lignin from lignocellulosic biomass. [167-169] ILs, which are also referred to as room temperature ILs, room temperature molten salts (RTILs), organic ionic liquids, etc., are an ionic system that takes on a liquid state at room temperature or slightly warmer composed of cations and anions with low melting points. [170] The structure of cations, such as the symmetry and the length of alkyl substituents and the presence of hydrophobic groups, and the degree of anion charge delocalization are two major factors that can affect physical chemical and biological properties of ILs. Since their unique properties including high-boiling, negligible vapour pressures, high thermal stabilities and the ability to interact and disrupt the recalcitrant lignocelluloses structure, ILs are very popular nowadays in terms of reducing environmental pollution.

Cellulose could be dissolved by ILs containing chloride, formate, acetate or alkylphosphonate inions by formation of strong hydrogen bonds. [171-173] The most promising ILs in the pretreatment of lignocellulosic materials are imidazolium-based chloride salts: the cations are able to solve the aromatic character of lignin by means of π–π interaction, whereas the chloride counteranion is usually the most effective in disrupting the extensive inter- and intra-molecular hydrogen bonds pattern of cellulose microfibrils. [174] Therefore, utilization of the ILs as green solvents is expected to be an ideal process for the pretreatment of lignocellulosic materials. [175]

1-Ethyl-3-methylimidazole acetate [EMIM]Ac, which presents good solubility properties of lignocellulosic materials and also achieves substrate delignification and the decrystallization of cellulose, is regarded as one of the most suitable ILs for the pretreatment of lignocellulosic biomass. [176, 177] Bian et al. [178] investigated the effect of [EMIM] Ac pretreatment on the structure and enzymatic hydrolysis of sugarcane bagasse cellulose. Results showed that the original cellulose experienced an increase in glucose content from 80-83.3% to 91.6-92.8%, a decrease in the degree of polymerization from 974-1039 to 511-521, a crystal transformation from cellulose I to cellulose II, as well as an increase of surface area during the pretreatment. Lopes et al. [179] studied the pretreatment of wheat straw using various kinds of ILs. Results showed that wheat straw complete dissolute in [BMIM] [HSO_4]. Therefore, the use of ILs as a tool for biomass fractionation exhibits a great potential to maximize the recovery of biomass and simultaneously to add value to the fractionated components within the biorefinery concept.

2.4. Biological Pretreatment

Compared with the physical or chemical pretreatments, no chemicals was acquired for biological pretreatment or microbial pretreatment. Due to its low energy input, cost, reduction in chemical requirement and mild processing conditions, biological pretreatment has advantages in improving enzymatic saccharification of plant biomass. [180] During the biological pretreatment, the plant cell wall can be disrupted by partial breakdown of the lignin-carbohydrate complex through certain microorganisms. [181] Among microorganisms, fungi (white, brown, soft-rot) are the promising lignin degraders and thus have potential for biological breakdown of plant materials. [182] Moreover, the highest efficiency can be obtained by lignin-degrading white-rot fungi for the soft and brown fungi only attacks cellulose. [183, 184]

Many factors affect the lignin degradation and enzymatic hydrolysis yield in biological pretreatment: particle size, moisture content, pretreatment time and temperature. Deswal et al. [185] pretreated the sugarcane bagasse with three white-rot fungi: *Pleurotus florida*, *Coriolopsis caperate* RCK 2011 and *Ganoderma* sp. rckk-01, individually under solid-state fermentation. Results showed that white-rot fungal pretreatment improved the amenability of plant material for enzymatic hydrolysis. Hatakka [186] investigated the influence of white-rot fungi pretreatment of wheat straw for enzymatic saccharification of cellulose. About 35% of the original straw was converted to reducing sugars after 5 weeks pretreatment with Pleurotus ostreatus. Pinto et al. [187] studied the influence of ligninolytic enzymes on straw saccharificatin during fungal pretreatment. Compared with the untreated straw, about 91% sugar yield increased after pretreated by *T. versicolor* (strain Tv2). Moreover, results showed that only the presence of lignin peroxidase during pretreatment could lead to a significant ($P<0.05$) increase in the saccharification yield. Despite its advantages, there are still many drawbacks for biological pretreatment, such as long process time, large space requirement and the need for continuous monitoring of microorganism growth, limited its applications. [188]

3. Hydrolysis Process

The cellulose and hemicelluloses in sugarcane bagasse need to be converted to simple sugars before fermentation, and this process is known as hydrolysis. Complete hydrolysis of cellulose results in glucose, whereas the hemicelluloses give rise to pentose and hexose. In sugarcane bagasse, hemicelluloses are composed mainly of L-arabino-(4-*O*-methyl-D-glucurono) xylan [8], which can be hydrolyzed into pentose. This process could be implemented by chemical hydrolysis and enzymatic hydrolysis. During the chemical hydrolysis procedure, other by-products are produced which are considered as the inhibitory to fermentation such as furfural, hydroxymethyl furfural, acids (acetic, ferulic, glucuronic, vanillic, syringic, and ρ-coumaric), and other chemicals. These resulting products can reduce the yield of bioethanol. So a detoxification process is required to remove these by-products after the hydrolysis to increase the ferment ability of hydrolysates. [189]

3.1. Cellulose and Hemicelluloses Hydrolysis Reactions

Cellulose can be degraded to glucose by acid. The cellulose molecule is characterized by β-1,4-glucosidic linkages between continuous glucose units. There are three reactive hydroxyl groups in each monomer. Acid such as concentrated acid or dilute acid can attack β-1,4-glucosidic linkages in cellulose leading to degradation. [190] The acid or enzymes can promote the reaction of water with glucan molecules to release single glucose by the following reaction:

$$(C_6H_{10}O_5)_n + nH_2O \rightarrow nC_6H_{12}O_6 \qquad (1)$$

Thus, each glucose unit in the long chain combines with a water molecule. [191]

Hemicelluloses can also be hydrolyzed by the addition of water to release individual sugar chains in the hemicelluloses molecules. For sugarcane bagasse, xylan can reacted with water catalyzed by acid or enzymes by the following reaction:

$$(C_5H_8O_4)_n + nH_2O \rightarrow nC_5H_{10}O_5 \qquad (2)$$

Table 2. Comparison between concentrated- and dilute-acid hydrolysis

Hydrolysis methods	Advantages	Disadvantage
Concentrated-acid hydrolysis	-Operated at low temperature	-High acid consumption
		-Equipment corrosion
	-High sugar yield	-High energy consumption for acid recovery
		-Longer reaction time
Dilute-acid hydrolysis	-Low acid consumption	-Operated at high temperature
		-Low sugar yield
	-Short residence time	-Equipment corrosion
		-Formation of undesirable by-product

3.2. Chemical Hydrolysis

Chemical hydrolysis involves the exposure of sugarcane bagasse to chemicals for a period of time at a typical temperature, and results in sugar monomers from cellulose and hemicelluloses polymers. Acids are initially and predominately applied in chemical hydrolysis. Sulfuric acid is the most researched acid, although other acids such as HCl have also been used. [192] Acid hydrolysis can be separated into two groups: (1) concentrated-acid hydrolysis and (2) dilute-acid hydrolysis. A comparison between concentrated-acid hydrolysis and dilute-acid hydrolysis methods is presented in Table 2. [193]

3.2.1. Concentrated-Acid Hydrolysis

Compared with dilute-acid processes, concentrated-acid process gives higher sugar yield and consequently results in higher ethanol yield. Furthermore, the concentrated-acid processes can operate at low temperature (e.g. 40 ^{o}C), it is a significant advantage compared to diluted-acid processes. However, there are some disadvantages about concentrated-acid hydrolysis processes. First, the concentration of acid is very high, consequentially leading to the serious equipment corrosion. Therefore, the process needs the high requirement for equipments made from corrosion-resistant materials, such as ceramic or carbon-brick lining. Moreover, the acid recovery is an energy consumption process. In addition, when sulfuric acid is used, the neutralization process requires large amounts of gypsum. Furthermore, the environmental impact strongly limits the application of hydrochloric acid. The high investment and maintenance costs have greatly reduced the potential commercial interest of this process. [193]

3.2.2. Dilute-Acid Hydrolysis

Among the chemical hydrolysis methods, the most common application is dilute-acid hydrolysis. When heated to high temperatures with dilute sulfuric acid, cellulose break down to shorter groups of molecules that release glucose [194]. But yields are low because dilute acid does not penetrate the crystalline regions of cellulose. Harsh conditions (high temperature) are needed to release glucose from these tightly associated chains. However, other side reactions become dominant above about 220 °C, and the amount of by-products increases as the temperature is raised above these levels [195].

Nowadays, most of dilute-acid hydrolysis processes are performed in a batch mode with a retention time of a few minutes. [196] Batch reactors have been the most widely used reactors for kinetic study of hydrolysis and for laboratory and pilot study of ethanol production from lignocellulosic materials. [197-201] Yields of about 60% of glucose in cellulose can be recovered with temperatures of around 220 °C in typical batch reactors [202]. Other various reactors have been tried over the years to release glucose from cellulose with dilute acid in industrial production. Older reactors tended to be large vessels that could be heated through jackets. More recently, direct steam injection has been applied to rapidly heat sugarcane bagasse to the high temperatures required. Some of these reactors have been tubular flow systems with extruders to promote mixing and control vessel residence times, but such equipment is very expensive and hardly scales to the high throughputs needed for commercial operations. [191]

A main disadvantage of dilute-acid hydrolysis processes, which is degradation of the sugars during hydrolysis reactions to form undesirable by-product. This lowers the yield of sugars, simultaneously the by-products inhibit the formation of ethanol during the fermentation process. In order to avoid degradation of monomers at high temperature and formation of the inhibitors, dilute-acid hydrolysis is carried out in two stages. In the first stage, hemicelluloses are converted to sugar monomers under relatively mild conditions. In the second stage, the residual solid is hydrolyzed under severe conditions, allowing cellulose to be hydrolyzed. In a one-stage pretreatment, a temperature between 140 and 170 °C can be used, but two treatments at about 120 °C for a long time may also be applied. [203-206]

The "two-stage" dilute-acid process is usually preferred to one-stage dilute-acid hydrolysis because there are many advantages. Higher sugar yield could be achieved; there need the minimal energy consumption; more concentrated sugar solution could be obtained; there is less sugar degradation and fewer fermentation-inhibiting components.

3.3. Enzymatic Hydrolysis

Enzymatic hydrolysis is the optimal process for degrading lignocellulosic materials into reducing sugars because of mild reaction conditions (pH between 4.8–5.0 and temperature between 45–50 °C), which does not present corrosion problems in the reactors and results in fewer by-products formation with high sugar yields.

Enzymatic hydrolysis of cellulose is achieved using cellulase, which are usually a mixture of groups of enzymes such as endoglucanases, exoglucanases and β-glucosidases. [207] Endoglucanase attacks the internal structure of cellulose polymers. It cleaves the β-1,4-glucosidic linkages of cellulose structure in random positions to produce oligomeric chain fragments. Exoglucanases attack specific locations on the non-reducing ends in the cellulose structure. They mainly produce cellobiose or glucose units. β-Glucosidase converts cellobiose to glucose. [208] Cellulases are produced by a wide range of bacteria, actinomycetes, fungi, etc. Trichoderma genus appears to produce the greatest quantity of active cellulases.

To breakdown the hemicelluloses, several enzymes such as xylanase, β-xylosidase, glucuronidase, acetylesterase, galactomannanase, and glucomannanase are required. [209] The complexity of hemicellulosic structure requires a high degree of coordination between the enzymes involved in hemicelluloses degradation. Due to their complex nature, enzymatic hydrolysis of xylans is intrinsically more complicated than most of the other plant polysaccharides. Cellulase enzymes when acting together with xylanases on hydrolysis sugarcane bagasse exhibit a better yield due to the synergistic action of enzymes. [210]

However, enzymatic hydrolysis depends on optimized conditions for supreme efficiency (hydrolysis temperature, time, pH, enzyme loading, and substrate concentration) and suffers from end-product inhibition and biomass structural restraints [211, 212]. To overcome the end-product inhibition and reduce the time, hydrolysis and fermentation can be combined, so-called simultaneous saccharification and fermentation (SSF) or simultaneous saccharification and cofermentation (SSCF).

3.4. Detoxification of Hemicellulosic Hydrolysates

The main targets in the pretreatment and hydrolysis of sugarcane bagasse are to minimize the sugars degradation and the formation of by-products, limit the consumption of chemicals, energy and water. [213] The inhibitory compounds could be divided into four groups: (1) substances that are released by the hemicellulosic structure, such as acetic acid, formic acid, levulinic acid; (2) phenolic compounds and other aromatic compounds produced from the degradation of lignin; (3) the furan derivatives, furfural and HMF, resulting from the degradation of pentose and hexose; (4) metals from the equipment. [214,215] These compounds individually as well as together affect the physiology of fermenting microorganisms, therefore, it is essential to eliminate these inhibitory compounds. [191] A number of methods like evaporation, [209,216] neutralization [217], use of membranes, [218] ion exchange resins, [219] and activated charcoal [220,221,222], and enzymatic detoxification using laccases and peroxidases have been attempted to detoxify the hydrolysates. [223-227]

4. Fermentation Processing

After the pretreatment and hydrolysis stages, the hydrolysates were fermented in batch and continuous mode. In order to achieve high qualities and yield of bio-ethanol, researchers have some challenges including the need to convert hydrolysates with high concentrations of sugars and high concentrations of solids [228]. During the last decades, there have been substantial progresses in this field, through a separate hydrolysis and fermentation (SHF), or as a simultaneous saccharification and fermentation (SSF) and a consolidated bioprocess (CBP) [229], a high product yield and cost-efficient recovery of the product were availably achieved [230-232]. The fermentation process is the importance step to obtain ethanol before distillation.

Since ethanol fermentation based on hexose sugars (mainly glucose, mannose) has a long history, the strain and fermentation technology are well established. The yeasts are the most common microorganism in fermentation process. The yeasts such as *Saccharomyces cerevisiae*, and *Brettanomyces custersii*, and bacteria *Zymomonas* mobilis could convert glucose into ethanol in an anaerobic state without an external electron acceptor [233-235]. Among these yeasts, *Saccharomyces cerevisiae* is the optimal choice for ethanol fermentation [236], glucose and the disaccharide sucrose could be cultivated well. Singh et al. used immobilized *Saccharomyces cerevisiae* on various matrices to study on ethanol production from pretreated sugarcane bagasse under solid state fermentation [237]. *Saccharomyces cerevisiae* cell suspension (equivalent to 10%) was added to 4% (w/v) sodium alginate solutions in a 1:1 volume ratio. The results showed that the immobilized cells could be reused for at least for ten cycles retaining its original activity. The repeated batch fermentation of concentrated sugarcane bagasse hydrolyzates showed higher ethanol productivity with sugarcane bagasse immobilized *Saccharomyces cerevisiae* cells. Plessas et al. [238], Dehkhoda et al. [239] Najafpour et al. [240] also researched the utilization of *Saccharomyces cerevisia* in ethanol production in different conditions. However, *Saccharomyces cerevisia* is difficult to assimilate pentose sugars (mainly xylose) due to the absence of genes [241]. Some of other yeasts, like fungi and bacteria are capable of assimilating pentose, but only a few are promising candidates for fermenting xylose into ethanol efficiently [242]. *Pichia stipitis* was found that it was able to ferment both xylose and glucose with high ethanol yields, making it a highly recommendable option for fermenting lignocellulosic hydrolysates [243]. Hande et al. [244] found a kind of yeast *Pichia* strain BY2 affiliated to *Pichia*. In their research, they used this yeast to ferment sugarcane bagasse hemicellulosic hydrolysate, producing 15 $g \cdot L^{-1}$ ethanol from 30 $g \cdot L^{-1}$ xylose under shaking conditions at 28°C. This new yeast isolated showed ethanol yield of 0.45 $g \cdot g^{-1}$ and volumetric productivity of 0.33 $g \cdot L^{-1} \cdot h^{-1}$ from sugarcane bagasse hemicellulosic hydrolysate. *Pachysolen tannophilus* [245] was used to ferment the pretreated sugarcane bagasse hemicellulosic hydrolysate and showed good fermentability (volumetric productivity of $0.59 g \cdot L^{-1} \cdot h^{-1}$). Yeast could affect the fermentability of sugarcane bagasse hydrolyzate, several toxic compounds such as acetic acid, formic acid, furfural, 5-hydroxymethylfurfural (5-HMF), which are usually formed during pretreatment as a result of lignin and sugar degradation [246] may also affect the downstream hydrolysis and fermentation steps [247]. The use of reducing agents in hydrolysates of lignocellulose biomass was beneficial to reduce or avoid the side-effects of toxic compounds in hydrolysis and fermentation steps. Alriksson et al. [248] improved the fermentability of enzymatic hydrolysates of lignocellulose through chemical in-situ detoxification with reducing agents. They found that dithionite as a reducing agent could overcome the obstacles connected with bioconversion of lignocellulose hydrolysates, and a dramatic improvement in fermentability can be achieved with a relatively small addition of dithionite.

After pretreatment, the combination of saccharification and fermentation has been applied to produce ethanol, including separate enzymatic hydrolysis and fermentation (SHF), simultaneous saccharification and fermentation (SSF), non-isothermal saccharification and fermentation (NSSF), simultaneous saccharification and cofermentation (SSCF) [249-251]. Simultaneous saccharification and fermentation (SSF) is one process option for production of ethanol from cellulosic biomass (Fig. 6) [252]. It is a promising method to simultaneously carry out the hydrolysis and fermentation steps [253, 254]. The principal benefits of performing the enzymatic hydrolysis together with the fermentation, instead of in a separate step after the hydrolysis, are the reduced end-product inhibition of the enzymatic hydrolysis, and the reduced investment costs [255-257]. In the SSCF process, enzymes/microbes not only hydrolyze cellulose and hemicelluloses into different sugars but also ferment simultaneously xylose and glucose into ethanol in the same reactor, while SSF ferments glucose (from cellulose hydrolyzed) and pentose (from hemicelluloses) in a separate reactor [258]. This difference conduces that SSCF is better than the SSF technology in aspects of cost effectiveness, yields, and processing time [259]. Nevertheless, the SSF and SSCF process have a drawback of efficient ethanol production due to the differences in optimum temperature between saccharification and fermentation [260-262]. The application of thermotolerant yeast strains would overcome the shortcoming. Nonklang et al. [263] introduced *Kluyveromyces marxianus* which belongs to one of *Saccharomyces cerevisiae* combined with thermostable cellulase genes to produced 43.4 g/L ethanol from 100 g/L cellobiose. Voronovsky et al. [264] and Sakamoto et al. [265] studied hemicellulosic (mainly xylose) thermotolerant yeasts on ethanol production from hemicelluloses. They found that the thermotolerant yeasts could acclimatize the different temperatures between saccharification and fermentation.

Recent years, solid-state fermentation has aroused ever-increasing attentions due to its high efficiency of fermenting agricultural solid wastes like sugarcane bagasse [266-269]. Solid-state fermentation provides potential benefits for the microbial cultivation for bioprocesses and products development, which consists of a three-phase heterogeneous process, solid, liquid and gaseous phases. This method offers more advantages than the conventional submerged fermentation method because it is carried out in the absence of free flowing water and thus, a lower amount of wastewater is generated [270]. Hence, this method could obtain higher concentration of end products and higher fermentation productivity and then low the production cost of bioproducts and has emerged as an attractive alternative to submerged fermentation [271, 272]. In addition, *filamentous fungi* has better growth environment under solid-state fermentation [273]. Solid-state fermentation has been applied to cellulose, actic acid, xylanase and pectinase production [274-276]. Rodríguez-Zúñiga et al. [274] investigated sugarcane bagasse as a substrate for cellulase production using an instrumented bioreactor under solid-state fermentation and results indicated that this method built the foundation for producing complete cellulase and hydrolyzing the cellulosics into fermentable sugars for ethanol production.

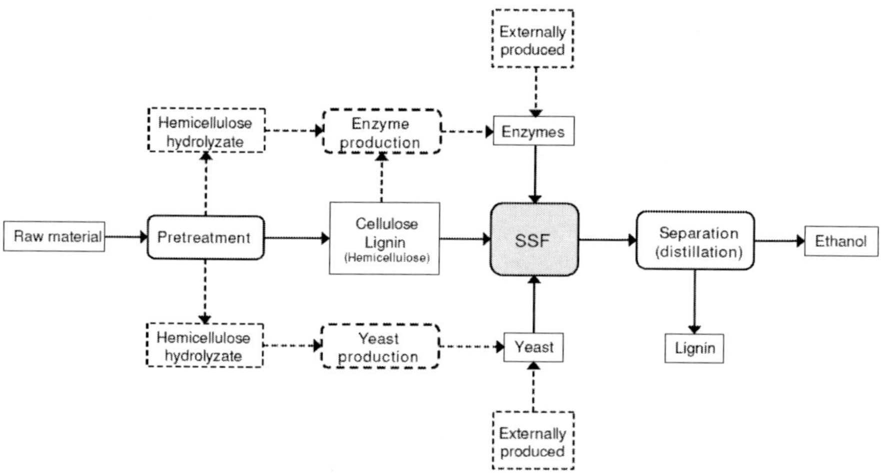

Figure 6. Schematic representation of the SSF process [252].

Conclusion

In summary, as the renewable raw material, sugarcane bagasse has the rich carbohydrate resource which could be as the ideal feedstocks for the ethanol production. Different pretreatment, hydrolysis and fermentation methods can be combined with the advanced technical supports to produce ethanol efficiently. Sugarcane bagasse can be as raw material for ethanol production and also for the production of industrial enzymes, xylitol, organics acids and high-added value chemicals, such as furfural and hydroxymethyl furfural.

However, the efficient bioconversion from sugarcane bagasse into bioethanol is still a challenge to large scale commercialization. The efficient and practical pretreatment and hydrolysis and fermentation should be further developed based on the current advanced technologies for the considerable promise in the future.

ACKNOWLEDGMENTS

This work was supported by the grants from Guangdong Natural Science Funds for Distinguished Young Scholar (S20120011250), the Author of National Excellent Doctoral Dissertation of China (201169), and the Fundamental Research Funds for the Central Universities (2014ZG0003), SCUT.

REFERENCES

[1] A. Romaní, G. Garrote, J. L. Alonso, J. C. Parajó, Bioethanol production from hydrothermally pretreated Eucalyptus globulus wood. *Bioresource technology,* 2010, 101, 8706–8712.

[2] S. Kim, B. E. Dale, Global potential bioethanol production from wasted crops and crop residues. *Biomass and bioenergy,* 2004, 26, 361–375.

[3] B. D. Solomon, J. R. Barnes, K. E. Halvorsen, Grain and cellulosic ethanol: History, economics, and energy policy. *Biomass and Bioenergy,* 2007, 31, 4164–25.

[4] O. Bobleter, Hydrothermal degradation of polymers derived from plants. *Progress in Polymer Science*, 1994, 19, 797–841.

[5] R. Hatti-Kaul, U. Törnvall, L. Gustafsson, & P. Börjesson, Industrial biotechnology for the production of bio-based chemicals–a cradle-to-grave perspective. *Trends in biotechnology,* 2007, 25, 119–124.

[6] C. YSomerville, H. Oungs, C. Taylor, S. C. Davis and S. P. Long, Feedstocks for lignocellulosic biofuels. *Science* (Washington), 2010, 329, 7907–92.

[7] A. Pandey, C. R. Soccol, P. Nigam and V. T. Soccol, Biotechnological potential of agro-industrial residues. I: sugarcane bagasse. *Bioresource Technology,* 2000, 74, 69–80.

[8] J. L. Ren, Z. C. Geng, C. F. Liu, F. Xu, J. X. Sun, R. C. Sun, Fractional isolation and structural characterisation of hemicellulosic polymers from delignified and ultrasonic irradiated sugarcange bgasse. *E-Polymers,* 2006, 067: 1–12.

[9] B. Parameswaran, Sugarcane bagasse. Biotechnology for Agro-Industrial Residues Utilisation. *Springer Netherlands,* 2009: 2392–52.

[10] C. R. Soccol, L. P. S. Vandenberghe, A. B. P. Medeiros, S. G. Karp, M. S. Buckeridge, L. P. Ramos, A. P. Pitarelo, V. Ferreira-Leit˜ao, L. M. F. Gottschalk, M. A. Ferrara, E. P. S. Bon, L. M. P. Moraes, J. A. Ara´ujo and F. A. G. Torres, Bioethanol from lignocelluloses: status and perspectives in Brazil. *Bioresource Technology,* 2010, 101, 4820–4825.

[11] R. C. Cerqueira-Leite, Leal. MRLVL, Cortez. L, Griffin WM and Scandiffio MIG, Can Brazil replace 5% of the 2025 gasoline world demand with ethanol? *Energy,* 2009, 34, 655–661.

[12] J. L. Ren, R.C. Sun, Hemicelluloses. In Cereal Straw as a Resource for Sustainable Biomaterials and biofuels. Ed. By R.C. Sun, *Elsevier,* Oxford, Uk, 2010.

[13] J. X. Sun, F.C. Mao, X.F. Sun and R.C. Sun, Comparative study of hemicelluloses isolated with alkaline peroxide from lignocellulosic materials. *Journal of Wood Chemistry Technology.* 2005, 24, 2392–62.

[14] M. Balat, H. Balat, Recent trends in global production and utilization of bioethanol fuel. *Applied Energy.* 2009, 86, 22732–282.

[15] H. A. Ruiz, D. S. Ruzene, D. P. Silva, F. F. M. da Silva, A. A. Vicente and J. A. Teixeira, Development and characterization of an environmentally friendly process sequence (autohydrolysis and organosolv) for wheat straw delignification. *Applied biochemistry and biotechnology,* 2011, 164, 6296–41.

[16] A. Romaní, G. Garrote, F. López and J. C. Parajó, Eucalyptus globulus wood fractionation by autohydrolysis and organosolv delignification. *Bioresource technology,* 2011, 102, 58965–904.

[17] A. Romaní, G. Garrote and J. C. Parajó, Bioethanol production from autohydrolyzed Eucalyptus globulus by Simultaneous Saccharification and Fermentation operating at high solids loading. *Fuel,* 2012, 94, 305–312.

[18] H. A. Ruiz, D. S. Ruzene, D. P. Silva, M. A. Quintas, A. A. Vicente and J. A. Teixeira, Evaluation of a hydrothermal process for pretreatment of wheat straw—effect of particle size and process conditions. Journal of Chemical Technology and Biotechnology, 2011, 86, 88–94.

[19] G. Yu, S. Yano, H. Inoue, S. Inoue, T. Endo and S. Sawayama, Pretreatment of rice straw by a hot-compressed water process for enzymatic hydrolysis. *Applied biochemistry and biotechnology,* 2010, 160, 539–551.

[20] T. Persson, J. L. Ren, E. Joelsson and A.-S. Jönsson, Fractionation of wheat and barley straw to access high-molecular-mass hemicelluloses prior to ethanol production. *Bioresource technology,* 2009, 100, 3906–3913.

[21] A. Romaní, G. Garrote, J. L. Alonso and J. C. Parajó, Bioethanol production from hydrothermally pretreated Eucalyptus globulus wood. *Bioresource technology,* 2010, 101, 8706–8712.

[22] I. Cybulska, H. Lei and J. Julson, Hydrothermal pretreatment and enzymatic hydrolysis of prairie cord grass. *Energy & Fuels*, 2009, 24, 718–727.

[23] P. Gullón, E. Conde, A. Moure, H. Domínguez and J. C. Parajó, Selected process alternatives for biomass refining: A review. *Open Agriculture Journal,* 2010, 4, 135–144.

[24] F. Gírio, C. Fonseca, F. Carvalheiro, L. Duarte, S. Marques and R. Bogel-Łukasik, Hemicelluloses for fuel ethanol: a review. *Bioresource technology,* 2010, 101, 4775–4800.

[25] C. Vila, J. Romero, J. Francisco, G. Garrote and J. Parajó, Extracting value from Eucalyptus wood before kraft pulping: Effects of hemicelluloses solubilization on pulp properties. *Bioresource technology,* 2011, 102, 5251–5254.

[26] H. A. Ruiz, D. P. Silva, D. S. Ruzene, L. F. Lima, A. A. Vicente and J. A. Teixeira, Bioethanol production from hydrothermal pretreated wheat straw by a flocculating Saccharomyces cerevisiae strain–Effect of process conditions. *Fuel,* 2012, 95, 528–536.

[27] S. Haghighi Mood, A. Hossein Golfeshan, M. Tabatabaei, G. Salehi Jouzani, G. H. Najafi, M. Gholami and M. Ardjmand, Lignocellulosic biomass to bioethanol, a comprehensive review with a focus on pretreatment. Renewable and Sustainable *Energy Reviews,* 2013, 27, 77–93.

[28] H. A. Ruiz, R. M. Rodríguez-Jasso, B. D. Fernandes, A. A. Vicente and J. A. Teixeira, Hydrothermal processing, as an alternative for upgrading agriculture residues and marine biomass according to the biorefinery concept: A review. *Renewable and Sustainable Energy Reviews,* 2013, 21, 35–51.

[29] H. M. Liu, X. A. Xie, J. L. Ren and R. C. Sun, 8-Lump reaction pathways of cornstalk liquefaction in sub-and super-critical ethanol. *Industrial crops and products,* 2012, 35, 250–256.
[30] G. Garrote, H. Dominguez and J. Parajo, Hydrothermal processing of lignocellulosic materials. *European Journal of Wood and Wood Products,* 1999, 57, 191–202.
[31] W. He, G. Li, L. Kong, H. Wang, J. Huang and J. Xu, Application of hydrothermal reaction in resource recovery of organic wastes. *Resources, Conservation and Recycling,* 2008, 52, 691–699.
[32] A. Loppinet-Serani, C. Aymonier and F. Cansell, Supercritical water for environmental technologies. *Journal of chemical technology and biotechnology,* 2010, 85, 583–589.
[33] M. Möller, P. Nilges, F. Harnisch and U. Schröder, Subcritical water as reaction environment: fundamentals of hydrothermal biomass transformation. *ChemSusChem,* 2011, 4, 566–579.
[34] S. J. Duff and W. D. Murray, Bioconversion of forest products industry waste cellulosics to fuel ethanol: a review. *Bioresource Technology,* 1996, 55, 1–33.
[35] M. Sasaki, T. Adschiri and K. Arai, Fractionation of sugarcane bagasse by hydrothermal treatment. *Bioresource technology,* 2003, 86, 301–304.
[36] M. Laser, D. Schulman, S. G. Allen, J. Lichwa, M. J. Antal Jr and L. R. Lynd, A comparison of liquid hot water and steam pretreatments of sugar cane bagasse for bioconversion to ethanol. *Bioresource technology,* 2002, 81, 33–44.
[37] G. P. Van Walsum, S. G. Allen, M. J. Spencer, M. S. Laser, M. J. Antal Jr and L. R. Lynd, Conversion of lignocellulosics pretreated with liquid hot water to ethanol. *Seventeenth Symposium on Biotechnology for Fuels and Chemicals,* 1996, 157–170.
[38] S. Chang, Z. Zhao, A. Zheng, X. Li, X. Wang, Z. Huang, F. He and H. Li, Effect of hydrothermal pretreatment on properties of bio-oil produced from fast pyrolysis of eucalyptus wood in a fluidized bed reactor. *Bioresource technology,* 2013, 138, 321–328.
[39] H. Y. Mou, E. Orblin, K. Kruus and P. Fardim, Topochemical pretreatment of wood biomass to enhance enzymatic hydrolysis of polysaccharides to sugars. *Bioresource technology,* 2013, 142, 540–545.
[40] N. Schultz-Jensen, A. Thygesen, F. Leipold, S. T. Thomsen, C. Roslander, H. Lilholt and A. B. Bjerre, Pretreatment of the macroalgae Chaetomorpha linum for the production of bioethanol–Comparison of

five pretreatment technologies. *Bioresource technology,* 2013, 140, 36–42.

[41] J. B. Kristensen, L. G. Thygesen, C. Felby, H. Jørgensen and T. Elder, Cell-wall structural changes in wheat straw pretreated for bioethanol production. *Biotechnol Biofuels,* 2008, 1, 1–9.

[42] T. Zhang, C. E. Wyman, K. Jakob and B. Yang, Rapid selection and identification of Miscanthus genotypes with enhanced glucan and xylan yields from hydrothermal pretreatment followed by enzymatic hydrolysis. *Biotechnol. Biofuels,* 2012, 5, 56.

[43] D. Cannella, C.-W. Hsieh, C. Felby and H. Jørgensen, Production and effect of aldonic acids during enzymatic hydrolysis of lignocellulose at high dry matter content. *Biotechnol Biofuels*, 2012, 5, 26.

[44] A. Hideno, H. Inoue, T. Yanagida, K. Tsukahara, T. Endo and S. Sawayama, Combination of hot compressed water treatment and wet disk milling for high sugar recovery yield in enzymatic hydrolysis of rice straw. *Bioresource technology,* 2012, 104, 743–748.

[45] D. Deswal, Y. P. Khasa and R. C. Kuhad, Optimization of cellulase production by a brown rot fungus Fomitopsis sp. RCK2010 under solid state fermentation. *Bioresource technology,* 2011, 102, 6065–6072.

[46] V. Menon and M. Rao, Trends in bioconversion of lignocellulose: biofuels, platform chemicals & biorefinery concept. *Progress in Energy and Combustion Science,* 2012, 38, 5225–50.

[47] C. S. Goh, K. T. Lee and S. Bhatia, Hot compressed water pretreatment of oil palm fronds to enhance glucose recovery for production of second generation bio-ethanol. *Bioresource technology,* 2010, 101, 7362–7367.

[48] J. M. Lee, J. Shi, R. A. Venditti and H. Jameel, Autohydrolysis pretreatment of Coastal Bermuda grass for increased enzyme hydrolysis. *Bioresource technology,* 2009, 100, 6434–6441.

[49] L. P. Xiao, Z. J. Sun, Z. J. Shi, F. Xu and R. C. Sun, Impact of hot compressed water pretreatment on the structural changes of woody biomass for bioethanol production. *Bioresources,* 2011, 6, 1576–1598.

[50] M. H. Thomsen, A. Thygesen and A. B. Thomsen, Hydrothermal treatment of wheat straw at pilot plant scale using a three-step reactor system aiming at high hemicellulose recovery, high cellulose digestibility and low lignin hydrolysis. *Bioresource technology,* 2008, 99, 4221–4228.

[51] Q. Yu, X. Zhuang, Z. Yuan, Q. Wang, W. Qi, W. Wang, Y. Zhang, J. Xu and H. Xu, Two-step liquid hot water pretreatment of Eucalyptus grandis

to enhance sugar recovery and enzymatic digestibility of cellulose. *Bioresource technology,* 2010, 101, 4895–4899.

[52] S. Caparrós, J. Ariza, F. López, J. Nacimiento, G. Garrote and L. Jiménez, Hydrothermal treatment and ethanol pulping of sunflower stalks. *Bioresource technology,* 2008, 99, 1368–1372.

[53] A. Alfaro, F. López, A. Pérez, J. García and A. Rodríguez, Integral valorization of tagasaste (Chamaecytisus proliferus) under hydrothermal and pulp processing. *Bioresource technology,* 2010, 101, 7635–7640.

[54] T. Treasure, R. Gonzalez, R. Venditti, Y. Pu, H. Jameel, S. Kelley and J. Prestemon, Co-production of electricity and ethanol, process economics of value prior combustion. *Energy Conversion and Management,* 2012, 62, 141–153.

[55] P. Feng, R. JunLi, X. Feng, S. RunCang, J. Zhu, X. Zhang and X. Pan, Chemicals from hemicelluloses: a review. *Sustainable production of fuels, chemicals, and fibers from forest biomass,* 2011, 219–259.

[56] R. Sun, Cereal straw as a resource for sustainable biomaterials and biofuels: chemistry, extractives, lignins, hemicelluloses and cellulose. 2010.

[57] R. Vegas, M. Kabel, H. A. Schols, J. Alonso and J. Parajó, Hydrothermal processing of rice husks: effects of severity on product distribution. *Journal of Chemical Technology and Biotechnology,* 2008, 83, 965–972.

[58] H. L. Li, S. Y. Wang, W. J. Wang, J. L. Ren, F. Peng, R. C. Sun and L. Liang, One-Step Heterogeneous Catalytic Process for the Dehydration of Xylan into Furfural. *Bioresources,* 2013, 8, 3200–3211.

[59] X. W. Peng, L. X. Zhong, J. L. Ren and R. C. Sun, Highly effective adsorption of heavy metal ions from aqueous solutions by macroporous xylan-rich hemicelluloses-based hydrogel. *Journal of agricultural and food chemistry,* 2012, 60, 3909–3916.

[60] H. Li, A. Deng, J. Ren, C. Liu, W. Wang, F. Peng and R. Sun, A modified biphasic system for the dehydration of d-xylose into furfural using $SO_4^{2-}/TiO_2/-ZrO_2/La^{3+}$ as a solid catalyst. *Catalysis Today,* 2014,

[61] G. Garrote, J. M. Cruz, H. Domínguez and J. C. Parajó, Non-isothermal autohydrolysis of barley husks: Product distribution and antioxidant activity of ethyl acetate soluble fractions. *Journal of food engineering,* 2008, 84, 544–552.

[62] B. Gullón, R. Yáñez, J. Alonso and J. Parajó, Production of oligosaccharides and sugars from rye straw: A kinetic approach. *Bioresource technology,* 2010, 101, 6676–6684.

[63] F. Carvalheiro, T. Silva-Fernandes, L. C. Duarte and F. M. Gírio, Wheat straw autohydrolysis: process optimization and products characterization. *Applied biochemistry and biotechnology,* 2009, 153, 84–93.

[64] P. Gullón, P. Moura, M. a. P. Esteves, F. M. Girio, H. Domínguez and J. C. Parajó, Assessment on the fermentability of xylooligosaccharides from rice husks by probiotic bacteria. *Journal of agricultural and food chemistry,* 2008, 56, 7482–7487.

[65] P. Gullón, G. Pereiro, J. L. Alonso and J. C. Parajó, Aqueous pretreatment of agricultural wastes: characterization of soluble reaction products. *Bioresource technology,* 2009, 100, 5840–5845.

[66] F. Peng, J. Bian, P. Peng, H. Xiao, J. L. Ren, F. Xu and R. C. Sun, Separation and characterization of acetyl and non-acetyl hemicelluloses of Arundo donax by ammonium sulfate precipitation. *Journal of agricultural and food chemistry,* 2012, 60, 4039–4047.

[67] H. Li, W. Wang, J. Ren, F. Peng and R. Sun, Preparation and Characterization of SO_4^{2-}/TiO 2–ZrO_2/La $^{3+}$ and their photocatalytic performance for the dehydration of xylose to furfural. *Journal of Biobased Materials and Bioenergy,* 2014, 8, 50–58.

[68] G. Garrote, R. Yáñez, J. L. Alonso and J. C. Parajó, Coproduction of oligosaccharides and glucose from corncobs by hydrothermal processing and enzymatic hydrolysis. *Industrial & Engineering Chemistry Research,* 2008, 47, 1336–1345.

[69] H. Boussarsar, B. Rogé and M. Mathlouthi, Optimization of sugarcane bagasse conversion by hydrothermal treatment for the recovery of xylose. *Bioresource technology,* 2009, 100, 6537–6542.

[70] R. Vegas, A. Moure, H. Domínguez, J. C. Parajó, J. R. Álvarez and S. Luque, Purification of oligosaccharides from rice husk autohydrolysis liquors by ultra-and nano-filtration. *Desalination,* 2006, 199, 541–543.

[71] G. Garrote, H. Domínguez and J. C. Parajó, Manufacture of xylose-based fermentation media from corncobs by posthydrolysis of autohydrolysis liquors. *Applied biochemistry and biotechnology,* 2001, 95, 195–207.

[72] G. Garrote, H. Domínguez and J. C. Parajó, Generation of xylose solutions from Eucalyptus globulus wood by autohydrolysis–posthydrolysis processes: posthydrolysis kinetics. *Bioresource technology,* 2001, 79, 155–164.

[73] H. Li, A. Deng, J. Ren, C. Liu, Q. Lu, L. Zhong, F. Peng and R. Sun, Catalytic hydrothermal pretreatment of corncob into xylose and furfural via solid acid catalyst. *Bioresource technology,* 2014, 158, 313–320.

[74] M. Michelin, T. Maria de Lourdes, D. S. Ruzene, D. P. Silva, H. A. Ruiz, A. A. Vicente, J. A. Jorge, H. F. Terenzi and J. A. Teixeira, Production of xylanase and β-xylosidase from autohydrolysis liquor of corncob using two fungal strains. *Bioprocess and biosystems engineering,* 2012, 35, 1185–1192.

[75] G. Garrote, H. Dominguez and J. Parajo, Study on the deacetylation of hemicelluloses during the hydrothermal processing of Eucalyptus wood. *Holz als Roh-und Werkstoff,* 2001, 59, 53–59.

[76] B. Rivas, J. Domınguez, H. Domınguez and J. Parajó, Bioconversion of posthydrolysed autohydrolysis liquors: an alternative for xylitol production from corn cobs. *Enzyme and Microbial Technology,* 2002, 31, 431–438.

[77] L. C. Duarte, F. Carvalheiro, S. Lopes, S. Marques, J. C. Parajo and F. M. Gírio, Comparison of two posthydrolysis processes of brewery's spent grain autohydrolysis liquor to produce a pentose-containing culture medium. *Proceedings of the Twenty-Fifth Symposium on Biotechnology for Fuels and Chemicals Held May 4–7, 2003, in Breckenridge, CO,* 2004, 1041–1058.

[78] C. Chi, Z. Zhang, H.-m. Chang and H. Jameel, Determination of furfural and hydroxymethylfurfural formed from biomass under acidic conditions. *Journal of wood chemistry and technology,* 2009, 29, 265–276.

[79] C. Heitner, D. Dimmel and J. Schmidt, Lignin and Lignans: Advances in chemistry. 2010,

[80] S. Tsubaki, M. Sakamoto and J.-i. Azuma, Microwave-assisted extraction of phenolic compounds from tea residues under autohydrolytic conditions. *Food chemistry,* 2010, 123, 1255–1258.

[81] O. Pourali, F. S. Asghari and H. Yoshida, Production of phenolic compounds from rice bran biomass under subcritical water conditions. *Chemical Engineering Journal,* 2010, 160, 259-266.

[82] E. Conde, A. Moure, H. Domínguez and J. C. Parajó, Production of antioxidants by non-isothermal autohydrolysis of lignocellulosic wastes. *LWT-Food Science and Technology*, 2011, 44, 436–442.

[83] E. Conde, P. Gullón, A. Moure, H. Domínguez and J. C. Parajó, Fractionation of industrial solids containing barley husks in aqueous media. *Food and Bioproducts Processing,* 2009, 87, 208–214.

[84] O. Akpinar, K. Gunay, Y. Yilmaz, O. Levent and S. Bostanci, enzymatic processing and antioxidant activity of agricultural waste autohydrolysis liquors. *Bioresources,* 2010, 5, 699–711.

[85] S. Tsubaki, Y. Ozaki and J. Azuma, Microwave-Assisted Autohydrolysis of Prunus mume Stone for Extraction of Polysaccharides and Phenolic Compounds. *Journal of food science,* 2010, 75, C152–C159.

[86] J. Li and G. Gellerstedt, Improved lignin properties and reactivity by modifications in the autohydrolysis process of aspen wood. *Industrial crops and products,* 2008, 27, 175–181.

[87] Y. Sun and J. Cheng, Hydrolysis of lignocellulosic materials for ethanol production: a review. *Bioresource technology,* 2002, 83, 1–11.

[88] W. E. Mabee, D. J. Gregg, C. Arato, A. Berlin, R. Bura, N. Gilkes, O. Mirochnik, X. Pan, E. K. Pye and J. N. Saddler, Updates on softwood-to-ethanol process development. Applied biochemistry and biotechnology, 2006, 129, 55–70.

[89] J. Zeng, Z. Tong, L. Wang, J. Zhu and L. Ingram, Isolation and structural characterization of sugarcane bagasse lignin after dilute phosphoric acid plus steam explosion pretreatment and its effect on cellulose hydrolysis. *Bioresource technology,* 2014, 154, 274–281.

[90] E. Palmqvist, B. Hahn-Hägerdal, M. Galbe and G. Zacchi, The effect of water-soluble inhibitors from steam-pretreated willow on enzymatic hydrolysis and ethanol fermentation. *Enzyme and Microbial Technology,* 1996, 19, 470–476.

[91] H. B. Klinke, A. Thomsen and B. K. Ahring, Inhibition of ethanol-producing yeast and bacteria by degradation products produced during pre-treatment of biomass. *Applied microbiology and biotechnology,* 2004, 66, 10–26.

[92] J. D. Wright, Ethanol from biomass by enzymatic hydrolysis. *Chemical Engineering Progress,* 1988, 84, 62–79.

[93] E. Tomas-Pejo, J. Oliva and M. Ballesteros, Realistic approach for full-scale bioethanol production from lignocellulose: a review. *Journal of Scientific and Industrial Research,* 2008, 67, 874.

[94] J. Li, K. Wu, W. Xiao, J. Zhang, J. Lin, Y. Gong and Z. Liu, Effect of antioxidant extraction on the enzymatic hydrolysis and bioethanol production of the extracted steam-exploded sugarcane bagasse. *Biochemical Engineering Journal,* 2014, 82, 91–96.

[95] B. Esteves, L. Nunes and H. Pereira, Properties of furfurylated wood (Pinus pinaster). *European Journal of Wood and Wood Products,* 2011, 69, 521–525.
[96] T. Ding, L. Gu and T. Li, Influence of steam pressure on physical and mechanical properties of heat-treated Mongolian pine lumber. *European Journal of Wood and Wood Products,* 2011, 69, 121–126.
[97] M. A. Kabel, G. Bos, J. Zeevalking, A. G. Voragen and H. A. Schols, Effect of pretreatment severity on xylan solubility and enzymatic breakdown of the remaining cellulose from wheat straw. *Bioresource Technology,* 2007, 98, 2034–2042.
[98] J. Li, G. Henriksson and G. Gellerstedt, Lignin depolymerization/repolymerization and its critical role for delignification of aspen wood by steam explosion. *Bioresource technology,* 2007, 98, 3061–3068.
[99] I. Ballesteros, J. Oliva, M. Negro, P. Manzanares and M. Ballesteros, Enzymic hydrolysis of steam exploded herbaceous agricultural waste (Brassica carinata) at different particule sizes. *Process Biochemistry,* 2002, 38, 187–192.
[100] E. Varga, K. Réczey and G. Zacchi, Optimization of steam pretreatment of corn stover to enhance enzymatic digestibility. *Applied biochemistry and biotechnology,* 2004, 114, 509–523.
[101] C. Cara, E. Ruiz, M. Ballesteros, P. Manzanares, M. J. Negro and E. Castro, Production of fuel ethanol from steam-explosion pretreated olive tree pruning. *Fuel,* 2008, 87, 692–700.
[102] K. Mackie, H. Brownell, K. West and J. Saddler, Effect of sulphur dioxide and sulphuric acid on steam explosion of aspenwood. *Journal of Wood Chemistry and Technology,* 1985, 5, 405–425.
[103] J. D. McMillan, Pretreatment of lignocellulosic biomass. *ACS symposium series,* 1994, 566, 292–324.
[104] R. Alinia, S. Zabihi, F. Esmaeilzadeh and J. F. Kalajahi, Pretreatment of wheat straw by supercritical CO_2 and its enzymatic hydrolysis for sugar production. *Biosystems engineering,* 2010, 107, 61–66.
[105] Y. Zheng, H.-M. Lin, J. Wen, N. Cao, X. Yu and G. T. Tsao, Supercritical carbon dioxide explosion as a pretreatment for cellulose hydrolysis. *Biotechnology Letters,* 1995, 17, 845–850.
[106] C. Schacht, C. Zetzl and G. Brunner, From plant materials to ethanol by means of supercritical fluid technology. *The Journal of Supercritical Fluids,* 2008, 46, 299–321.

[107] S. P. Sawan, Supercritical fluid cleaning: fundamentals, technology and applications. 1998.
[108] N. Srinivasan and L. K. Ju, Statistical optimization of operating conditions for supercritical carbon dioxide-based pretreatment of guayule bagasse. *Biomass and Bioenergy,* 2012, 47, 451–458.
[109] N. Srinivasan and L. K. Ju, Pretreatment of guayule biomass using supercritical carbon dioxide-based method. *Bioresource technology,* 2010, 101, 9785–9791.
[110] P. Kumar, D. M. Barrett, M. J. Delwiche and P. Stroeve, Methods for pretreatment of lignocellulosic biomass for efficient hydrolysis and biofuel production. *Industrial & Engineering Chemistry Research,* 2009, 48, 3713–3729.
[111] K. H. Kim and J. Hong, Supercritical CO_2 pretreatment of lignocellulose enhances enzymatic cellulose hydrolysis. *Bioresource technology,* 2001, 77, 139–144.
[112] N. Narayanaswamy, A. Faik, D. J. Goetz and T. Gu, Supercritical carbon dioxide pretreatment of corn stover and switchgrass for lignocellulosic ethanol production. *Bioresource technology,* 2011, 102, 6995–7000.
[113] L. da Costa Sousa, S. P. Chundawat, V. Balan and B. E. Dale, 'Cradle-to-grave'assessment of existing lignocellulose pretreatment technologies. *Current Opinion in Biotechnology,* 2009, 20, 339–347.
[114] M. J. Lau, M. W. Lau, C. Gunawan and B. E. Dale, Ammonia fiber expansion (AFEX) pretreatment, enzymatic hydrolysis, and fermentation on empty palm fruit bunch fiber (EPFBF) for cellulosic ethanol production. *Applied Biochemistry and Biotechnology,* 2010, 162, 1847–1857.
[115] J. Kudakasseril Kurian, G. Raveendran Nair, A. Hussain and G. Vijaya Raghavan, Feedstocks, logistics and pre-treatment processes for sustainable lignocellulosic biorefineries: A comprehensive review. *Renewable and Sustainable Energy Reviews,* 2013, 25, 205–219.
[116] M. W. Lau and B. E. Dale, Cellulosic ethanol production from AFEX-treated corn stover using Saccharomyces cerevisiae 424A (LNH-ST). *Proceedings of the National Academy of Sciences,* 2009, 106, 1368–1373.
[117] H. Alizadeh, F. Teymouri, T. I. Gilbert and B. E. Dale, Pretreatment of switchgrass by ammonia fiber explosion (AFEX). *Applied biochemistry and biotechnology,* 2005, 124, 1133–1141.
[118] C. Zhong, M. W. Lau, V. Balan, B. E. Dale and Y.-J. Yuan, Optimization of enzymatic hydrolysis and ethanol fermentation from

AFEX-treated rice straw. *Applied Microbiology and Biotechnology,* 2009, 84, 667–676.

[119] H. K. Murnen, V. Balan, S. P. Chundawat, B. Bals, L. d. C. Sousa and B. E. Dale, Optimization of ammonia fiber expansion (AFEX) pretreatment and enzymatic hydrolysis of Miscanthus x giganteus to fermentable sugars. *Biotechnology Progress,* 2007, 23, 846–850.

[120] M. Mes-Hartree, B. Dale and W. Craig, Comparison of steam and ammonia pretreatment for enzymatic hydrolysis of cellulose. *Applied Microbiology and Biotechnology*, 1988, 29, 462–468.

[121] B. Bals, C. Wedding, V. Balan, E. Sendich and B. Dale, Evaluating the impact of ammonia fiber expansion (AFEX) pretreatment conditions on the cost of ethanol production. *Bioresource Technology*, 2011, 102, 1277–1283.

[122] V. Bitra, A. Womac, W. Hart and G. Melnichenko, Liquid AFEX pretreatment and enzymatic hydrolysis of switchgrass from different harvest and storage conditions. *Transactions of the ASABE,* 2013, 56, 1511–1520.

[123] L. E. Gollapalli, B. E. Dale and D. M. Rivers, Predicting digestibility of ammonia fiber explosion (AFEX)-treated rice straw. 2002.

[124] D. Kumar and G. S. Murthy, Life cycle assessment of energy and GHG emissions during ethanol production from grass straws using various pretreatment processes. *The International Journal of Life Cycle Assessment,* 2012, 17, 388–401.

[125] C. Zhang, F. Pang, B. Li, S. Xue and Y. Kang, Recycled aqueous ammonia expansion (RAAE) pretreatment to improve enzymatic digestibility of corn stalks. *Bioresource Technology,* 2013, 138, 314–320.

[126] B. Bals, H. Murnen, M. Allen and B. Dale, Ammonia fiber expansion (AFEX) treatment of eleven different forages: Improvements to fiber digestibility in vitro. *Animal Feed Science and Technology,* 2010, 155, 147–155.

[127] S. Harun, V. Balan, M. S. Takriff, O. Hassan, J. Jahim and B. E. Dale, Performance of AFEXTM pretreated rice straw as source of fermentable sugars: the influence of particle size. *Biotechnology for Biofuels*, 2013, 6, 40.

[128] F. Teymouri, L. Laureano-Pérez, H. Alizadeh and B. E. Dale, Ammonia fiber explosion treatment of corn stover. *Proceedings of the Twenty-Fifth Symposium on Biotechnology for Fuels and Chemicals Held May 4–7, 2003, in Breckenridge, CO,* 2004, 951–963.

[129] L. Canilha, A. K. Chandel, T. Suzane dos Santos Milessi, F. A. F. Antunes, W. Luiz da Costa Freitas, M. das Graças Almeida Felipe and S. S. da Silva, Bioconversion of sugarcane biomass into ethanol: an overview about composition, pretreatment methods, detoxification of hydrolysates, enzymatic saccharification, and ethanol fermentation. *Journal of Biomedicine & Biotechnology,* 2012, 2012, 989572.

[130] H. Inoue, S. Yano, T. Endo, T. Sakaki and S. Sawayama, Combining hot-compressed water and ball milling pretreatments to improve the efficiency of the enzymatic hydrolysis of eucalyptus. *Biotechnol Biofuels,* 2008, 1, 1–9.

[131] A. Hideno, H. Inoue, K. Tsukahara, S. Fujimoto, T. Minowa, S. Inoue, T. Endo and S. Sawayama, Wet disk milling pretreatment without sulfuric acid for enzymatic hydrolysis of rice straw. *Bioresource technology,* 2009, 100, 2706–2711.

[132] B. Buaban, H. Inoue, S. Yano, S. Tanapongpipat, V. Ruanglek, V. Champreda, R. Pichyangkura, S. Rengpipat and L. Eurwilaichitr, Bioethanol production from ball milled bagasse using an on-site produced fungal enzyme cocktail and xylose-fermenting Pichia stipitis. *Journal of Bioscience and Bioengineering,* 2010, 110, 18–25.

[133] L. Yuan, Z. Chen, Y. Zhu, X. Liu, H. Liao and D. Chen, One Step Conversion of Wheat Straw to Sugars by Simultaneous Ball Milling, Mild Acid, and Fungus Penicillium simplicissimum Treatment. *Applied Biochemistry and Biotechnology,* 2012, 167, 39–51.

[134] P. Verma, T. Watanabe, Y. Honda and T. Watanabe, Microwave-assisted pretreatment of woody biomass with ammonium molybdate activated by H_2O_2. *Bioresource Technology,* 2011, 102, 3941–3945.

[135] C. Z. Liu and X. Y. Cheng, Improved hydrogen production via thermophilic fermentation of corn stover by microwave-assisted acid pretreatment. International Journal of Hydrogen Energy, 2010, 35, 8945–8952.

[136] X. W. Peng, J. L. Ren, L. X. Zhong and R. C. Sun, Synthesis and characterization of amphoteric xylan-type hemicelluloses by microwave irradiation. *Journal of Agricultural and Food Chemistry,* 2012, 60, 1695–1702.

[137] H. Ma, W. W. Liu, X. Chen, Y. J. Wu and Z. L. Yu, Enhanced enzymatic saccharification of rice straw by microwave pretreatment. *Bioresource Technology,* 2009, 100, 1279–1284.

[138] P. Intanakul, M. Krairiksh and P. Kitchaiya, Enhancement of enzymatic hydrolysis of lignocellulosic wastes by microwave pretreatment under

atmospheric pressure. Journal of Wood Chemistry and Technology, 2003, 23, 217–225.

[139] X. W. Peng, J. L. Ren, L. X. Zhong, X. F. Cao and R. C. Sun, Microwave-induced synthesis of carboxymethyl hemicelluloses and their rheological properties. *Journal of Agricultural and Food Chemistry,* 2010, 59, 570–576.

[140] M. Hashem, M. A. Taleb, F. El-Shall and K. Haggag, New prospects in pretreatment of cotton fabrics using microwave heating. *Carbohydrate Polymers,* 2014, 103, 385–391.

[141] H. Peng, H. Li, H. Luo and J. Xu, A novel combined pretreatment of ball milling and microwave irradiation for enhancing enzymatic hydrolysis of microcrystalline cellulose. *Bioresource Technology,* 2013, 130, 81–87.

[142] K. Suslick, W. McNamara III and Y. Didenko, Hot spot conditions during multi-bubble cavitation. *Sonochemistry and Sonoluminescence,* 1999, 191–204.

[143] M. J. Bussemaker and D. Zhang, Effect of ultrasound on lignocellulosic biomass as a pretreatment for biorefinery and biofuel applications. *Industrial & Engineering Chemistry Research,* 2013, 52, 3563–3580.

[144] M. J. Bussemaker, F. Xu and D. Zhang, Manipulation of ultrasonic effects on lignocellulose by varying the frequency, particle size, loading and stirring. *Bioresource Technology*, 2013, 148, 15–23.

[145] R. Sindhu, M. Kuttiraja, V. Elizabeth Preeti, S. Vani, R. K. Sukumaran and P. Binod, A novel surfactant-assisted ultrasound pretreatment of sugarcane tops for improved enzymatic release of sugars. *Bioresource Technology,* 2013, 135, 67–72.

[146] S. H. Jadhav and P. R. Gogate, Ultrasound assisted enzymatic conversion of non edible oil to methyl esters. *Ultrasonics Sonochemistry,* 2014, 21, 1374–1381.

[147] A. Goshadrou, K. Karimi and M. J. Taherzadeh, Bioethanol production from sweet sorghum bagasse by Mucor hiemalis. *Industrial Crops and Products,* 2011, 34, 1219–1225.

[148] N. Mosier, C. Wyman, B. Dale, R. Elander, Y. Lee, M. Holtzapple and M. Ladisch, Features of promising technologies for pretreatment of lignocellulosic biomass. *Bioresource Technology,* 2005, 96, 673–686.

[149] K. J. Zeitsch, The chemistry and technology of furfural and its many by-products. 2000, 13.

[150] D. Diedericks, E. van Rensburg and J. F. Görgens, Enhancing sugar recovery from sugarcane bagasse by kinetic analysis of a two-step dilute acid pretreatment process. *Biomass and Bioenergy,* 2013, 57, 149–160.

[151] Y. Benjamin, H. Cheng and J. Görgens, Optimization of Dilute Sulfuric Acid Pretreatment to Maximize Combined Sugar Yield from Sugarcane Bagasse for Ethanol Production. *Applied Biochemistry and Tiotechnology,* 2013, 1–21.

[152] A. A. Greenwood, T. W. Farrell and I. M. O'Hara, Understanding mild acid pretreatment of sugarcane bagasse through particle scale modeling. *Biotechnology and Bioengineering,* 2013, 110, 3114–3125.

[153] X. Zhao, Y. Morikawa, F. Qi, J. Zeng and D. Liu, A novel kinetic model for polysaccharide dissolution during atmospheric acetic acid pretreatment of sugarcane bagasse. *Bioresource Technology,* 2014, 151, 128–136.

[154] F. Peng, J. Bian, J. L. Ren, P. Peng, F. Xu and R. C. Sun, Fractionation and characterization of alkali-extracted hemicelluloses from peashrub. *Biomass and Bioenergy,* 2012, 39, 20–30.

[155] K. Wang, H. Yang, Q. Chen and R. C. Sun, Influence of delignification efficiency with alkaline peroxide on the digestibility of furfural residues for bioethanol production. *Bioresource Technology,* 2013, 146, 208–214.

[156] F. Carvalheiro, L. C. Duarte and F. M. Gírio, Hemicellulose biorefineries: a review on biomass pretreatments. *Journal of Scientific & Industrial Research,* 2008, 67, 849–846.

[157] R. Sindhu, M. Kuttiraja, P. Binod, R. K. Sukumaran and A. Pandey, Physicochemical characterization of alkali pretreated sugarcane tops and optimization of enzymatic saccharification using response surface methodology. *Renewable Energy,* 2014, 62, 362–368.

[158] D. Nath Barman, M. A. Haque, T. H. Kang, G. H. Kim, T. Y. Kim, M. K. Kim and H. D. Yun, Effect of mild alkali pretreatment on structural changes of reed (Phragmites communis Trinius) straw. *Environmental Technology,* 2014, 35, 232–241.

[159] X. Zhao, K. Cheng and D. Liu, Organosolv pretreatment of lignocellulosic biomass for enzymatic hydrolysis. *Applied Microbiology and Biotechnology,* 2009, 82, 815–827.

[160] L. Mesa, E. González, C. Cara, M. González, E. Castro and S. I. Mussatto, The effect of organosolv pretreatment variables on enzymatic hydrolysis of sugarcane bagasse. *Chemical Engineering Journal,* 2011, 168, 1157–1162.

[161] X. W. Peng, J. L. Ren, L. X. Zhong, F. Peng and R. C. Sun, Xylan-rich hemicelluloses-graft-acrylic acid ionic hydrogels with rapid responses to pH, salt, and organic solvents. *Journal of Agricultural and Food Chemistry,* 2011, 59, 8208–8215.
[162] X. Q. Zhang, M. J. Chen, C. F. Liu and R. Sun, Dual-component system DMSO/LiCl as solvent and catalyst for homogeneous ring opening grafted polymerization of ε-caprolactone onto xylan. *Journal of Agricultural and Food Chemistry,* 2014, 62, 682–690.
[163] J. L. Wen, B. L. Xue, S. L. Sun and R. C. Sun, Quantitative structural characterization and thermal properties of birch lignins after auto-catalyzed organosolv pretreatment and enzymatic hydrolysis. *Journal of Chemical Technology and Biotechnology,* 2013, 88, 1663–1671.
[164] B. W. Koo, H. Y. Kim, N. Park, S. M. Lee, H. Yeo and I. G. Choi, Organosolv pretreatment of Liriodendron tulipifera and simultaneous saccharification and fermentation for bioethanol production. *Biomass and Bioenergy,* 2011, 35, 1833–1840.
[165] F. Sun and H. Chen, Organosolv pretreatment by crude glycerol from oleochemicals industry for enzymatic hydrolysis of wheat straw. *Bioresource Technology,* 2008, 99, 5474–5479.
[166] J. L. Wen, S. L. Sun, T. Q. Yuan, F. Xu and R. C. Sun, Structural elucidation of lignin polymers of eucalyptus chips during organosolv pretreatment and extended delignification. *Journal of Agricultural and Food Chemistry,* 2013, 61, 11067–11075.
[167] D. Ray, N. Mittal and W. J. Chung, Phosphorous pentoxide mediated synthesis of 5-HMF in ionic liquid at low temperature. *Carbohydrate Research,* 2011, 346, 2145–2148.
[168] T. Ståhlberg, M. G. Sørensen and A. Riisager, Direct conversion of glucose to 5-(hydroxymethyl) furfural in ionic liquids with lanthanide catalysts. *Green Chemistry,* 2010, 12, 321–325.
[169] X. W. Peng, J. L. Ren and R. C. Sun, Homogeneous esterification of xylan-rich hemicelluloses with maleic anhydride in ionic liquid. *Biomacromolecules,* 2010, 11, 3519–3524.
[170] S. N. Sun, M. F. Li, T. Q. Yuan, F. Xu and R. C. Sun, Effect of ionic liquid pretreatment on the structure of hemicelluloses from Corncob. *Journal of Agricultural and Food Chemistry,* 2012, 60, 11120–11127.
[171] H. Zhao, C. L. Jones, G. A. Baker, S. Xia, O. Olubajo and V. N. Person, Regenerating cellulose from ionic liquids for an accelerated enzymatic hydrolysis. *Journal of Biotechnology,* 2009, 139, 47–54.

[172] J. L. Wen, T. Q. Yuan, S. L. Sun, F. Xu and R. C. Sun, Understanding the chemical transformations of lignin during ionic liquid pretreatment. *Green Chemistry,* 2014, 16, 181–190.

[173] Y. C. Sun, J. K. Xu, F. Xu, R. C. Sun and G. L. Jones, Dissolution, regeneration and characterisation of formic acid and Alcell lignin in ionic liquid-based systems. *RSC Advances,* 2014, 4, 2743–2755.

[174] R. P. Swatloski, S. K. Spear, J. D. Holbrey and R. D. Rogers, Dissolution of cellose with ionic liquids. Journal of the American *Chemical Society,* 2002, 124, 4974–4975.

[175] T. Q. Yuan, W. Wang, F. Xu and R. C. Sun, Synergistic benefits of ionic liquid and alkaline pretreatments of poplar wood. Part 1: Effect of integrated pretreatment on enzymatic hydrolysis. *Bioresource Technology,* 2013, 144, 429–434.

[176] S. H. Lee, T. V. Doherty, R. J. Linhardt and J. S. Dordick, Ionic liquid-mediated selective extraction of lignin from wood leading to enhanced enzymatic cellulose hydrolysis. *Biotechnology and Bioengineering,* 2009, 102, 1368–1376.

[177] A. Sant'Ana da Silva, S. H. Lee, T. Endo and E. PS Bon, Major improvement in the rate and yield of enzymatic saccharification of sugarcane bagasse via pretreatment with the ionic liquid 1-ethyl-3-methylimidazolium acetate ([Emim] [Ac]). *Bioresource Technology,* 2011, 102, 10505–10509.

[178] J. Bian, F. Peng, X. P. Peng, X. Xiao, P. Peng, F. Xu and R. C. Sun, Effect of [Emim] Ac pretreatment on the structure and enzymatic hydrolysis of sugarcane bagasse cellulose. *Carbohydrate Polymers,* 2014, 100, 211–217.

[179] A. M. da Costa Lopes, K. G. João, D. F. Rubik, E. Bogel-Łukasik, L. C. Duarte, J. Andreaus and R. Bogel-Łukasik, Pre-treatment of lignocellulosic biomass using ionic liquids: Wheat straw fractionation. *Bioresource Technology,* 2013, 142, 198–208.

[180] M. Saritha, A. Arora and L. Nain, Pretreatment of paddy straw with Trametes hirsuta for improved enzymatic saccharification. *Bioresource Technology,* 2012, 104, 459–465.

[181] C. Sánchez, Lignocellulosic residues: biodegradation and bioconversion by fungi. *Biotechnology Advances,* 2009, 27, 185–194.

[182] A. A. Dias, G. S. Freitas, G. S. Marques, A. Sampaio, I. S. Fraga, M. A. Rodrigues, D. V. Evtuguin and R. M. Bezerra, Enzymatic saccharification of biologically pre-treated wheat straw with white-rot fungi. *Bioresource Technology,* 2010, 101, 6045–6050.

[183] J. Chen, S. L. Fales, G. A. Varga and D. J. Royse, Biodegradation of cell wall components of maize stover colonized by white-rot fungi and resulting impact on in-vitro digestibility. *Journal of the Science of Food and Agriculture,* 1995, 68, 91–98.

[184] X. L. Ping, S. Z. Jun, B. Y. Yuan, W. Wei, Z. X. Ming and R. C. Sun, Biodegradation of lignocellulose by white-rot fungi: structural characterization of water-soluble hemicelluloses. *BioEnergy Research,* 2013, 6, 1154–1164.

[185] D. Deswal, R. Gupta, P. Nandal and R. C. Kuhad, Fungal pretreatment improves amenability of lignocellulosic material for its saccharification to sugars. *Carbohydrate Polymers,* 2014, 99, 264–269.

[186] A. I. Hatakka, Pretreatment of wheat straw by white-rot fungi for enzymic saccharification of cellulose. *European Journal of Applied Microbiology and Biotechnology,* 1983, 18, 350–357.

[187] P. A. Pinto, A. A. Dias, I. Fraga, G. Marques, M. A. Rodrigues, J. Colaço, A. Sampaio and R. M. Bezerra, Influence of ligninolytic enzymes on straw saccharification during fungal pretreatment. *Bioresource Technology,* 2012, 111, 261–267.

[188] C. E. Wyman, B. E. Dale, R. T. Elander, M. Holtzapple, M. R. Ladisch and Y. Lee, Coordinated development of leading biomass pretreatment technologies. *Bioresource Technology,* 2005, 96, 1959–1966.

[189] E. Palmqvist and B. Hahn-Hägerdal, Fermentation of lignocellulosic hydrolysates. II: inhibitors and mechanisms of inhibition. *Bioresource Technology,* 2000, 74(1), 25–33.

[190] Y. J. Lee, Oxidation of sugarcane bagasse using a combination of hypochlorite and peroxide (Doctoral dissertation, Faculty of the Louisiana State University and Agricultural and Mechanical College in partial fulfillment of the requirements for the degree of Master of Science in The Department of Food Science by Yong-Jae Lee B. Sc., Chonnam National University). 2005.

[191] C. E. Wyman, S. R. Decker, M. E. Himmel, J. W. Brady, C. E. Skopec and L. Viikari, Hydrolysis of cellulose and hemicellulose. *Polysaccharides: Structural diversity and functional versatility,* 2005, 1, 1023–1062.

[192] A. M. Hashem and M. M. Rashad, Production of ethanol by yeasts grown on hydrolysate of Egyptian sweet potato. *Egyptian Journal of Food Science,* 1993, 21, 171–180.

[193] M. J. Taherzadeh and Karimi, Acid-based hydrolysis processes for ethanol from lignocellulosic materials: a review. *BioResources,* 2007, 2, 472–499.

[194] R. Katzen and D. F. Othmer, Wood Hydrolysis. A Continuous Process. *Industrial & Engineering Chemistry,* 1942, 34, 314–322.

[195] R. W. Torget, J. S. Kim and Y. Y. Lee, Fundamental aspects of dilute acid hydrolysis/fractionation kinetics of hardwood carbohydrates. 1. Cellulose hydrolysis. *Industrial & Engineering Chemistry Research,* 2000, 39, 2817–2825.

[196] K. Karimi, S. Kheradmandinia and M. J. Taherzadeh, Conversion of rice straw to sugars by dilute-acid hydrolysis. *Biomass and Bioenergy,* 2006, 30, 247–253.

[197] T. Brandberg, N. Sanandaji, L. Gustafsson and C. Johan Franzén, Continuous Fermentation of Undetoxified Dilute Acid Lignocellulose Hydrolysate by Saccharomycescerevisiae ATCC 96581 Using Cell Recirculation. *Biotechnology Progress,* 2005, 21, 1093–1101.

[198] R. Purwadi, C. Niklasson and M. J. Taherzadeh, Kinetic study of detoxification of dilute-acid hydrolyzates by Ca(OH)2. *Journal of Biotechnology,* 2004, 114, 187–198.

[199] G. Sanchez, L. Pilcher, C. Roslander, T. Modig, M. Galbe and G. Liden, Diluted-acid hydrolysis for fermentation of the Bolivian straw material Paja Brava. *Bioresource Technology,* 2004, 93, 249–256.

[200] A. Sues, R. Millati, L. Edebo and M. J. Taherzadeh, Ethanol production from hexoses, pentoses, and dilute-acid hydrolyzate by Mucor indicus. F*EMS Yeast Research,* 2005, 5, 669–676.

[201] M. J. Taherzadeh, C. Niklasson and G. Lidén, On-line control of fed-batch fermentation of dilute-acid hydrolyzates. *Biotechnology and Bioengineering,* 2000, 69, 330–338.

[202] L. Amaya-Delgado, M. E. Hidalgo-Lara and M. C. Montes-Horcasitas, Hydrolysis of sucrose by invertase immobilized on nylon-6 microbeads. *Food Chemistry,* 2006, 99, 299–304.

[203] T. S. El-Tayeb, A. A. Abdelhafez, S. H. Ali and E. M. Ramadan, Effect of acid hydrolysis and fungal biotreatment on agro-industrial wastes for obtainment of free sugars for bioethanol production. *Brazilian Journal of Microbiology,* 2012, 43, 1523–1535.

[204] S. Zu, W. Z. Li, M. Zhang, Z. Li, Z. Wang, H. Jameel and H. M. Chang, Pretreatment of corn stover for sugar production using dilute hydrochloric acid followed by lime. *Bioresource Technology,* 2014, 152, 364–370.

[205] J. W. Kim, K. S. Kim, J. S. Lee, S. M. Park, H. Y. Cho, J. C. Park and J. S. Kim, Two-stage pretreatment of rice straw using aqueous ammonia and dilute acid. *Bioresource Technology,* 2011, 102, 8992–8999.

[206] X. Wang, X. Liu and G. Wang, Two-stage hydrolysis of invasive algal feedstock for ethanol fermentation. *Journal of Integrative Plant Biology,* 2011, 53, 246–252.

[207] S. J. Duff and W. D. Murray, Bioconversion of forest products industry waste cellulosics to fuel ethanol: a review. *Bioresource Technology,* 1996, 55, 1–33.

[208] A. M. Milagres, W. Carvalho and A. Ferraz, Topochemistry, porosity and chemical composition affecting enzymatic hydrolysis of lignocellulosic materials. In Routes to Cellulosic Ethanol, *Springer* New York. 2011, 53–72.

[209] Y. J. Lee, Chung, C. H., & Day, D. F. Sugarcane bagasse oxidation using a combination of hypochlorite and peroxide. *Bioresource Tchnology,* 2009, 100, 935–941.

[210] M. N. de Almeida, V. M. Guimarães, D. L. Falkoski, G. B. Paes, J. I. Ribeiro Jr, E. M. Visser and S. T. de Rezende, Optimization of endoglucanase and xylanase activities from fusarium verticillioides for simultaneous saccharification and fermentation of sugarcane bagasse. *Applied Biochemistry and Biotechnology,* 2014, 172, 1332–1346.

[211] L. Canilha, A. K. Chandel, T. Suzane dos Santos Milessi, F. A. F. Antunes, W. Luiz da Costa Freitas, M. das Graças Almeida Felipe and S. S. da Silva, Bioconversion of sugarcane biomass into ethanol: An overview about composition, pretreatment methods, detoxification of hydrolysates, enzymatic saccharification, and ethanol fermentation. *Journal of Biomedicine and Biotechnology,* 2012.

[212] A. Teramoto and M. Fuchigami, Changes in Temperature, Texture, and Structure of Konnyaku (Konjac Glucomannan Gel) During High-pressure-freezing. *Journal of Food Science,* 2000, 65, 491–497.

[213] W. Zhang, X. Zhang, L. Bao and G. Wei, Bio-flocculation Detoxification Treatment with Acid Hydrolyzate of Corn Straw for Ethanol Production. *Chemical and Biochemical Engineering Quarterly,* 2012, 26, 127–136.

[214] A. Hande, S. Mahajan and A. Prabhune, Evaluation of ethanol production by a new isolate of yeast during fermentation in synthetic medium and sugarcane bagasse hemicellulosic hydrolysate. *Annals of Microbiology,* 2013, 63, 63–70.

[215] A. Kadam, J. D. Kamatkar, R. V. Khandare, J. P. Jadhav and S. P. Govindwar, Low-cost $CaCl_2$pretreatment of sugarcane bagasse for enhancement of textile dyeadsorption and subsequent biodegradation of adsorbed dye under solid-statefermentation. *Bioresource Technology,* 2013, 132, 276–284.

[216] S. I. Mussatto and I. C. Roberto, Alternatives for detoxification of diluted-acid lignocellulosic hydrolyzates for use in fermentative processes: a review. *Bioresource Technology,* 2004, 93, 1–10.

[217] E. Palmqvist and B. Hahn-Hägerdal, Fermentation of lignocellulosic hydrolysates. I: inhibition and detoxification. *Bioresource Technology,* 2000, 74, 17–24.

[218] D. L. Grzenia, D. J. Schell and S. Ranil Wickramasinghe, Membrane extraction for detoxification of biomass hydrolysates. *Bioresource Technology,* 2012, 111, 248–254.

[219] W. Carvalho, M. A. Batista, L. Canilha, J. C. Santos, A. Converti and S. S. Silva, Sugarcane bagasse hydrolysis with phosphoric and sulfuric acids and hydrolysate detoxification for xylitol production. *Journal of Chemical Technology and Biotechnology,* 2004, 79, 1308–1312.

[220] L. Canilha, W. Carvalho, M. Giulietti, M. D. G. A. Felipe, A. E. Silva and J. Batista, Clarification of a wheat straw-derived medium with ion-exchange resins for xylitol crystallization. *Journal of Chemical Technology and Biotechnology,* 2008, 83, 715–721.

[221] N. O. Nilvebrant, A. Reimann, S. Larsson and L. J. Jönsson, Detoxification of lignocellulose hydrolysates with ion-exchange resins. *Applied Biochemistry and Biotechnology,* 2001, 91, 35–49.

[222] A. Converti, J. M. Dominguez, P. Perego, S. S. Da Silva and M. Zilli, Wood hydrolysis and hydrolyzate detoxification for subsequent xylitol production. *Chemical Engineering and Technology,* 2000, 23, 1013–1020.

[223] A. K. Chandel, G. Chandrasekhar, K. Radhika, R. Ravinder and P. Ravindra, Bioconversion of pentose sugars into ethanol: a review and future directions. *Biotechnology and Molecular Biology Review,* 2011, 6, 008–020.

[224] B.Yang and C. E. Wyman, Pretreatment: the key to unlocking low-cost cellulosic ethanol. *Biofuels, Bioproducts and Biorefining,* 2008, 2, 26–40.

[225] A. D. Moreno, D. Ibarra, J. L. Fernández, M. Ballesteros, Different laccase detoxification strategies for ethanol production from

lignocellulosic biomass by the thermotolerant yeast Kluyveromyces marxianus CECT 10875. *Bioresource technology,* 2012, 106, 101–109.
[226] W. Parawira and M. Tekere, Biotechnological strategies to overcome inhibitors in lignocellulose hydrolysates for ethanol production: review. *Critical Reviews in Biotechnology,* 2011, 31, 20–31.
[227] B. G. Fonseca, R. de Oliveira Moutta, F. de Oliveira Ferraz, E. R. Vieira, A. S. Nogueira, B. F. Baratella and S. S. da Silva, Biological detoxification of different hemicellulosic hydrolysates using Issatchenkia occidentalis CCTCC M 206097 yeast. *Journal of Industrial Microbiology & Biotechnology,* 2011, 38, 199–207.
[228] M. Galbe, P. Sassner, A. Wingren and G. Zacchi, Process engineering economics of bioethanol production. Advances in Biochemical Engineering-Biotechnology, 2007, 108, 303–327.
[229] M. Galbe and G. Zacchi, Pretreatment of lignocellulosic materials for efficient bioethanol production. *Advances in Biochemical Engineering-Biotechnology,* 2007, 108, 41–65.
[230] I. Park, I. Kim, K. Kang, H. Sohn, I. Rhee and I. Jin, Cellulose ethanol production from waste newsprint by simultaneous saccharification and fermentation using Saccharomyces cerevisiae KNU5377. *Process Biochemistry.* 2010, 45, 487–492.
[231] L.R. Lynd, W.H. Van Zyl, J.E. McBride, M. Laser, Consolidated bioprocessing of cellulosic biomass: an update. *Current Opinion in Biotechnology,* 2005, 16, 577–583.
[232] Y. Lin, W. Lee, K. Duan, Y. Lin, Ethanol production by simultaneous saccharification and fermentation in rotary drum reactor using thermotolerant Kluveromyces marxianus. *Applied Energy,* 2013, 105, 389–394.
[233] H. Q. Gu, J. Zhang and J. Bao, Inhibitor analysis and adaptive evolution of Saccharomyces cerevisiae for simultaneous saccharification and ethanol fermentation from industrial waste corncob residues. *Bioresource Technology.* 2014, 157, 6–13.
[234] K. K. Cheng, W. Wang, J. A. Zhang, Q. Zhao, J. P. Li and J. W. Xue Statistical optimization of sulfite pretreatment of corncob residues for high concentration ethanol production. *Bioresource Technology,* 2011, 102, 3014–3019.
[235] Y. Tang, D. Zhao, C. Cristhian and J. Jiang, Simultaneous saccharification and cofermentation of lignocellulosic residues from commercial furfural production and corn kernels using different nutrient media, *Biotechnol Biofuels,* 2011, 4, 22.

[236] Y. Lin and S. Tanaka, Ethanol fermentation from biomass resources: current state and prospects. *Applied Microbiology and Biotechnology,* 2006, 69, 627–642.
[237] A. Singh, P. Singh and A. K. Saran, Comparative study on ethanol production from pretreated sugarcane bagasse using immobilized Saccharomyces cerevisiae on various matrices. *Renewable Energy,* 2013, 50, 488–493.
[238] S. Plessas, A. A. Bekatorou, M. Koutinas, M. SoupioniI and B. R. Marchant, Use of Saccharomyces cerevisiae cells immobilized on orange peel as biocatalyst for alcoholic fermentation. *Bioresource Technology,* 2007, 98, 58–60.
[239] A. Dehkhoda, T. Brandberg and J. Taherzadeh Mohammad, Comparison of vacuum and high pressure evaporated wood hydrolysate for ethanol production by repeated fed batch using flocculating *Saccharomyces cerevisiae*. *BioResources* 2008, 4, 20–309.
[240] G. Najafpour, H. Younesi, S. Younesi and I. Ku, Ethanol fermentation in an immobilized cell reactor using *Saccharomyces cerevisiae*. *Bioresource Technology,* 2004, 92, 60–251.
[241] R. E. Hector, J. A. Mertens, M. J. Bowman, N. N. Nichols, M. A. Cotta, and S. R. Hughes, Saccharomyces cerevisiae engineered for xylose metabolism requires gluconeogenesis and the oxidative branch of the pentose phosphate pathway for aerobic xylose assimilation. *Yeast,* 2011, 28, 645–660.
[242] L. K. Singh, G. Chaudhary, C. B. Majumder, and S. Ghosh, Utilization of hemicellulosic fraction of lignocellulosic material for bioethanol production. *Advances in Applied Science Research,* 2011, 2, 508–521.
[243] C. Bellido, S. Bolado, M. Coca, S. Lucas, G. González-Benito and M.T. García-Cubero, Effect of inhibitors formed during wheat straw pretreatment on ethanol fermentation by Pichia stipitis. *Bioresource Technology,* 2011, 102, 10868–10874.
[244] A. Hande, S. Mahajan and A. Prabhune, Evaluation of ethanol production by a new isolate of yeast during fermentation in synthetic medium and sugarcane bagasse hemicellulosic hydrolysate. *Annals of Microbiology,* 2013, 63, 63–70.
[245] K. K. Cheng, J. P. Ge, J. A. Zhang, Fermentation of pretreated sugarcane bagasse hemicellulose hydrolysate to ethanol by Pachysolen tannophilus. *Biotechnology letters,* 2007, 29, 1051–1055.

[246] R. Kont, M. Kurašin, H. Teugjas and P. Väljamäe Strong cellulase inhibitors from the hydrothermal pretreatment of wheat straw. *Biotechnology Biofuels,* 2013, 6, 135.

[247] J. R. M. Almeida, T. Modig, A. Petersson, B. Hahn-Hägerdal, G. Lidén and M.F. Gorwa-Grauslund, Increased tolerance and conversion of inhibitors in lignocellulosic hydrolysates by Saccharomyces cerevisiae. *Journal of chemical technology and biotechnology,* 2007, 82, 340–349.

[248] B. Alriksson, A. Cavka, L. J. Jönsson, Improving the fermentability of enzymatic hydrolysates of lignocellulose through chemical in-situ detoxification with reducing agents. *Bioresource Technology,* 2011, 102, 1254–1263.

[249] M. J. Taherzaderh and K. Karimi, Enzyme based hydrolysis processes for ethanol from lignocellulosic materials: a review. *BioResources,* 2007, 2, 707–738.

[250] Y. Sun,and J. Cheng, Hydrolysis of lignocellulosic materials for ethanol production: a review. *Bioresource Technology*, 2002, 83, 1–11.

[251] L. C. Teixeira, J. C. Linden, H. A. Schroeder, Optimizing peracetic acid pretreatment conditions for improved simultaneous saccharification and co-fermentation (SSCF) of sugar cane bagasse to ethanol fuel. *Renewable Energy,* 1999, 16, 1070–1073.

[252] K. Olofsson, M. Bertilsson, G. Lidén, A short review on SSF-an interesting process option for ethanol production from lignocellulosic feedstocks. *Biotechnology Biofuels,* 2008, 1, 1–14.

[253] C. A. Cardona, J. A. Quintero, I.C. Paz, Production of bioethanol from sugarcane bagasse: status and perspectives. *Bioresource Technology,* 2010, 101, 4754–4766.

[254] K. Olofsson, M. Bertilsson, G. Lidén, A short review on SSF- an interesting process option for ethanol production from lignocellulosic feedstocks. *Biotechnology Biofuels,* 2008, 1, 1–14.

[255] Y. S. Lin and W. C. Lee, Simultaneous saccharification and fermentation of alkalipretreatd cogongrass for bioethanol production. *BioResources,* 2011, 6, 2744–2756.

[256] S. Srichuwong, M. Fujiwara, X. H. Wang, T. Seyama, R. Shiroma, M. Arakane, N. Mukojima and K. Tokuyasu, Simultaneous saccharification and fermentation (SSF) of very high gravity (VHG) potato mash for the production of ethanol. *Biomass and bioenergy*, 2009, 33, 890–898.

[257] K. Öhgren, J. Vehmaanperä and M. Siika-Aho, High temperature enzymatic prehydrolysis prior to simultaneous saccharification and

fermentation of steam pretreated corn stover for ethanol production. *Enzyme and Microbial Technology,* 2007, 40, 607–613.

[258] M. J. Taherzsadeh and K. Karimi, Enzymatic-based hydrolysis processes for ethanol from lignocellulosic materials: a review. *Bioresources,* 2007, 2, 707–738.

[259] A. K. Chandel, C. Es, R. Rudravaram, M.L. Narasu, L. V. Rao and P. Ravindra, Economics and environmental impact of bioethanol production technologies: an appraisal. *Biotechnology and Molecular Biology Review*, 2007, 2, 14–32.

[260] S. Hari Krishna, T. G. Janardhan Reddy and V. Chowdary, Simultaneous saccharification and fermentation of lignocellulosic wastes to ethanol using a thermotolerant yeast. *Bioresource Technology,* 2001, 77, 6–193.

[261] B. M.A. Abdel-Banat, H. Hoshida, A. Ano, S. Nonklang and R. Akada, High-temperature fermentation: how can processes for ethanol production at high temperatures become superior to the traditional process using mesophilic yeast? *Applied Microbiology Biotechnology*, 2010, 85, 7–861.

[262] I. M. Banat, P. Nigam, D. Nigam, R. Marchant and A. P. McHale, Ethanol production at elevated temperatures and alcohol concentrations: part I. Yeasts in general. *World Journal of Microbiology & Biotechnology,* 1998, 14, 21–809.

[263] S. Nonklang, B. M. A. Abdel-Banat, K. Cha-aim, N. Moonjai, H. Hoshida, S. Limtong, High-temperature ethanol fermentation and transformation with linear DNA in the thermotolerant yeast Kluyveromyces marixianus DMKU3-1042. *Applied and Environmental Microbiology,* 2008, 74, 217–514.

[264] A.Y. Voronovsky, O. V. ARohuya, C. A. Bbas and A. A. Sibirny, Development of strains of the thermotolerant yeast Hansenula polymorpha capable of alcoholic fermentation of starch and xylan. *Metabolic Engineering*, 2009, 11, 234–42.

[265] T. Sakamoto, T. Hasunuma, Y. Hori, R. Yamada and A. Kondo, Direct ethanol production from hemicellulosic materials of rice straw by use of an engineered yeast strain codisplaying three types of hemicellulolytic enzymes on the surface of xylose-utilizing *Saccharomyces cerevisiae* cells. *Journal of Biotechnology,* 2012, 158, 203–10.

[266] A. Kadam, J. D. Kamatkar, R. V. Khandare, J. P. Jadhav and S. P. Govindwar, Low-cost $CaCl_2$ pretreatment of sugarcane bagasse for enhancement of textile dyeadsorption and subsequent biodegradation of

adsorbed dye under solid-statefermentation. *Bioresource Technology,* 2013, 132, 276–284.

[267] A. Pandey, Aspects of design aspects of fermenters in solid-state fermentation. *Process Biochemistry*, 1991, 26, 355–361.

[268] U. F. Rodríguez-Zúñiga, V. B. Neto, S. Neto and C. S. Farinas, Use of spectroscopic and imaging techniques to evaluate pretreated sugarcane bagasse as a substrate for cellulase production under solid-state fermentation. *Applied Biochemistry and Biotechnology,* 2013, 1–15.

[269] M. Mazutti, J. P. Bender, H. Treichelet and M. D. Luccio, Optimization of inulinase production by solid-state fermentation using sugarcane bagasse as substrate. *Enzyme and Microbial Technology,* 2006, 39, 56–59.

[270] A. Pandey, C. R. Soccol, P. Nigam and V.T. Soccol, Biotechnological potential of agro-industrial residues. I: sugarcane bagasse. *Bioresource Technology* 2000, 74, 69–80.

[271] U. Hölker, M. Höfer and J. Lenz, Biotechnological advantages of laboratory-scale solid state fermentation with fungi. *Applied Microbiology and Biotechnology,* 2004, 64, 175–86.

[272] R. R. Singhania, A. K. Patel, C. R. Soccol and A. Pandey, Recent advances in solid-statefermentation. *Biochemical Engineering Journal,* 2009, 44, 13–18.

[273] P. Shrestha, S. K. Khanal, A. L. Pometto III and J. Leeuwen, Ethanol production via in situ fungal saccharification and fermentation of mild alkali and steam pretreated corn fiber. *Bioresource Technology,* 2010, 101, 8698–705.

[274] U. F, Rodríguez-Zúñiga, S. Couri, V. B. Neto, S. Crestana and C. S. Farinas, Integrated strategies to enhance cellulolytic enzyme production using an instrumented bioreactor for solid-state fermentation of sugarcane bagasse. *BioEnergy Research,* 2013, 6, 142–152.

[275] C. R. Soccol, B. Marin, M. Raimbault and J. M. Lebeault, Potential of solid state fermentation for production of L (+)-lactic acid by Rhizopus oryzae. *Applied Microbiology and Biotechnology,* 1994, 41, 286–290.

[276] C. Botella, A. Diaz, I. De Ory, C. Webb and A. Blandino, Xylanase and pectinase production by *Aspergillus awamori* on grape pomace in solid state fermentation. *Process Biochemistry,* 2007, 42: 98–101.

In: Sugarcane
Editor: Eleanore Webb

ISBN: 978-1-61122-356-9

Chapter 8

MULTIDISCIPLINARY APPROACHES FOR ANALYSIS OF SOCIO-ECONOMIC AND ECOLOGICAL CONSTRAINTS FOR DIVERSIFICATION PROJECTS AND SUGARCANE BIOREFINERIES

Noé Aguilar-Rivera**, *Agustín Herrera-Solano, Vidal Enríquez-Ruvalcaba, Daniel Arturo Rodríguez-Lagunes and Adolfo Castillo-Moran

Universidad Veracruzana, Facultad de Ciencias Biológicas y Agropecuarias, Xalapa Enríquez, Mexico

ABSTRACT

World Sugarcane production has more than 500 years of history and integrates the agricultural activities of growing, harvesting and transportation of sugarcane with industrial production in sugar mills and distilleries. The integration of the territory is needed as an agro-industrial cluster to solve logistical problems, productivity, innovation and new

* Corresponding author: Noé Aguilar-Rivera. Universidad Veracruzana, Facultad de Ciencias Biológicas y Agropecuarias. km 1 carretera Peñuela – Amatlan S/N, Colonia Peñuela, CP 94945, Córdoba Veracruz México. E-mail: naguilar@uv.mx.

productions, as productive diversification and transition from conventional sugar mill and distilleries to biorefineries, to increase competitiveness. However, it has challenges related to low agricultural productivity derived from conventional practices of crop management, climate change and other ecological and socio-economic constraints such as vulnerability (pests, diseases, drought, etc.) and environmental impacts that are a risk to food security and the conversion to biorefineries. Therefore, at the present time, sugar industry and sugarcane crops are a potential source and epicenter of renewable energy, biofuels and bio-materials, as well as a food crop, but they are becoming more widely recognized as a source of rural livelihoods in developing countries and it will require a systematic effort, innovative and multidisciplinary methodologies of analysis to determine critical points that threaten the environmental and economic sustainability to improve the profitability and productivity with a reduction in the cost of production. This paper presents an approach or a conceptual framework as a new methodology for analysis based on existing knowledge of the sugar industry and the state of the art, for evaluating diversification using the multicriteria evaluation by analytical hierarchy process (AHP) as a tool suitable for analyzing complex systems, and for the identification of alternatives to the current situation and their discussion to facilitate decision making in the use of sugarcane as raw material in biorefineries to produce sugar based value-added products and derivatives as ethanol and cogeneration, bioplastic, etc. in cleaner production.

Keywords: Sugarcane chain value, analytic hierarchy process, diversification, constraints

INTRODUCTION

Biomass is considered the renewable energy source with the highest potential in this respect, both in industrialized and developing countries worldwide. More specifically, there is global interest in sugarcane (*Saccharum spp.*) as a traditional food crop which has been used for centuries as a feedstock for sugar production. Its systematic use for this purpose dates back more than 2,000 years in India, and from there, it spread in the tropical and subtropical areas around the world, where it was produced in a wide range of management practices and varied growing conditions. Sugarcane is a semiperennial crop, which is planted once and harvested five to six times before being replanted. After planting, it will take around 18 months to be ready for harvesting (plant cane); afterwards, it is harvested annually (ratoon cane). Therefore, a full sugarcane crop cycle averages between 6 to 7 years.

Sugarcane is a C4 plant with one of the highest efficiency levels in converting solar energy to biomass. It is considered an energy feedstock rather than a food feedstock. However, the total biomass and by-products from sugarcane are not currently used to their full potential. (Gómez-Merino et al., 2014; Cheavegatti-Gianotto, 2011). Waclawosvsky et al. (2010) have estimated the theoretical potential at 381 tonnes of cane (tc)/ha/year, wet basis, corresponding to 177 t/ha/year of dry matter. However, sugarcane needs excellent soils, plentiful rainfall, and large amounts of land and the potential to expand sugarcane production depends on the proportion of arable land that is suitable and available and the potential effect of expansion on food crops, agro-industrial products, and irrigation requirements.

Sugarcane is grown in more than 100 countries and the harvested area is reported by FAOSTAT (2013) to be of 25.78 million hectares; it is divided into 10.93 million hectares in Asia, 10.54 million in South America, Central America 0.135 million, 0.69 in the Caribbean, Africa 1.51 million, 0.40 million in Oceania and 0.37 million in North America. Worldwide major producing countries are Brazil (36.52%), India (19.76%), China (6.97%), Thailand (5.05%), Pakistan (4.06%), Mexico (2.85%), Cuba (1.98%), Indonesia (1.77 %), Philippines (1.68%), United States (1.44%), Argentina (1.36%), Australia (1.31%), South Africa (1.24%), Vietnam (1.15%), Guatemala (0.97%) and Colombia (0.65%) There are different levels of productivity at conventional indices: sugarcane yield (t ha^{-1}), sucrose yield (t ha^{-1}) and factory yield (%) (Figures 1 and 2).

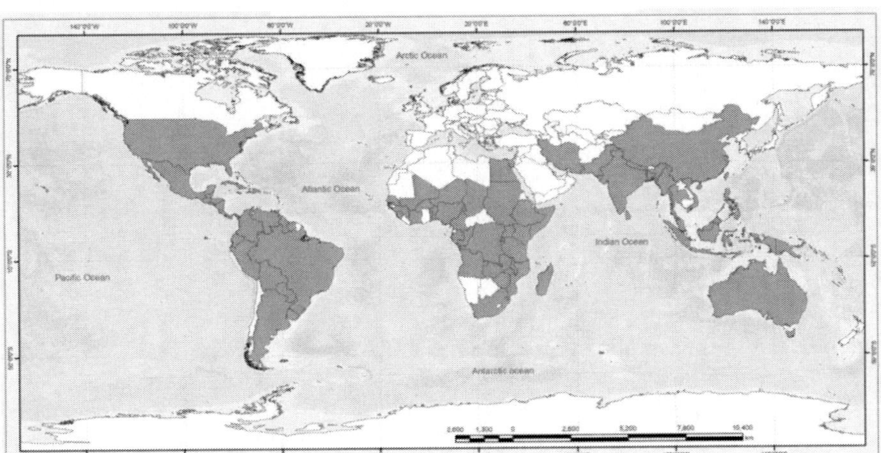

FAOSTAT, 2013.

Figure 1. Sugarcane producer countries.

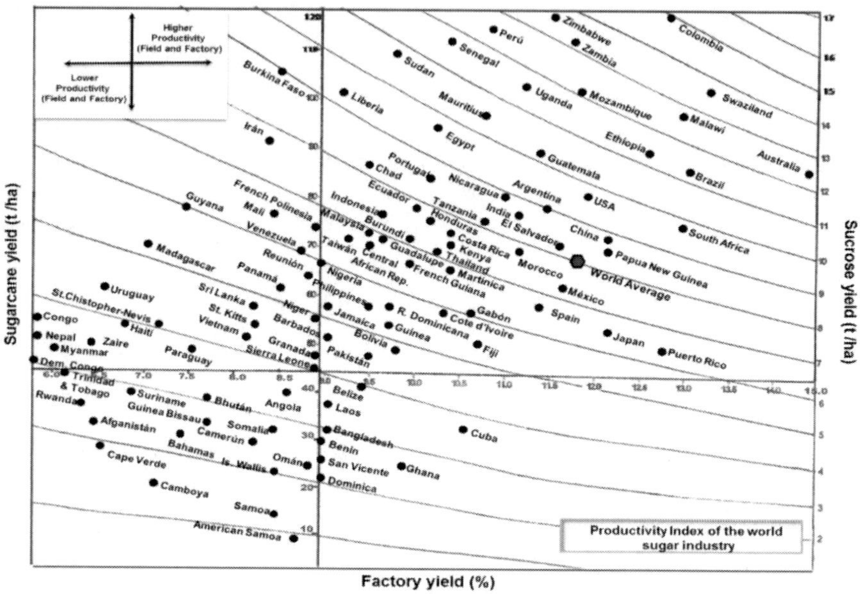

FAOSTAT, 2013.

Figure 2. Productivity Index of Sugar producer countries.

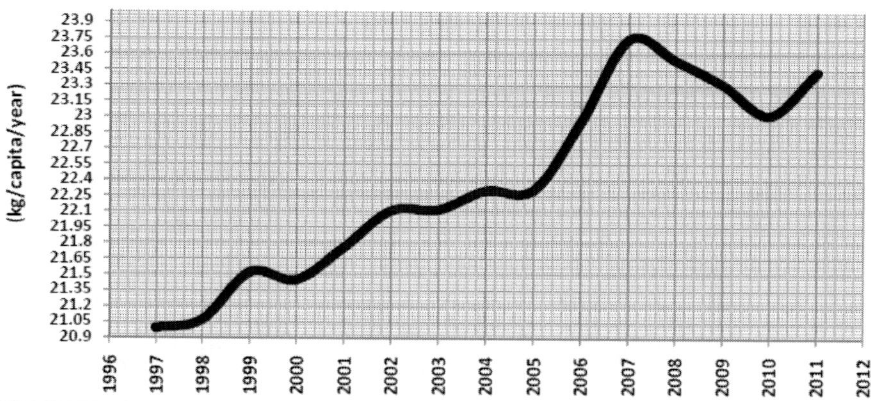

FOLICHTS, 2012.

Figure 3. Sucrose consumption (kg/year/inhabitant).

Sugar as the principal product from sugarcane represents 65–70% of the world's production of sugar and its production is mainly located in developing countries with a decline in global consumption and a sugar market with low growth (Figures 3 and 4).

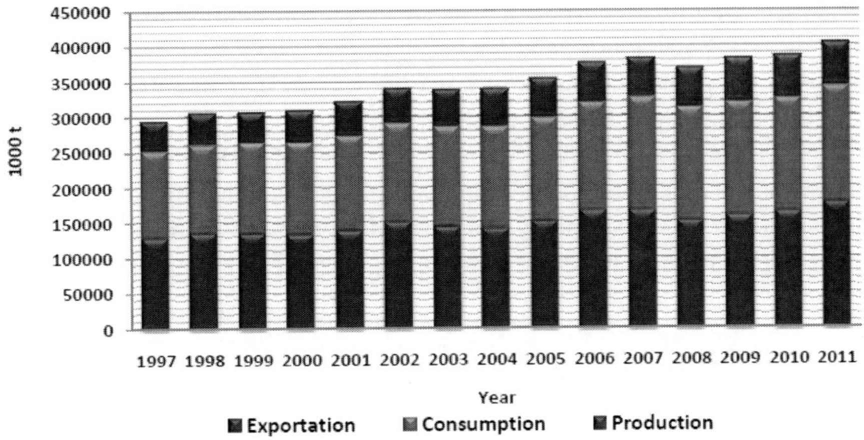

FOLICHTS, 2012.

Figure 4. Global trend in sugar market.

The production process requires high amounts of inputs, steam and electricity at the different stages, and generates important quantities of residues or by-products. The expansion of sugar production by itself may offer some economic growth, but contributes very little to *economic development*; nor does sugar production contribute to enhancing the long-term *sustainability* of the sugarcane regions. Expanded sugar production is merely an extension of economic activity at the margin in a mature industry—it generally does not offer opportunities for enterprise development and innovation. The expansion of renewable energy and other potential co-products of the sugarcane industry offer new economic opportunities with greater potential for sustainable development and long-term competitiveness. Bio-energy expansion, as the new paradigm from sugarcane, will create rural livelihoods, reduce costly energy imports, stimulate economic integration, reduce GHG emissions, and offer new development paths. Bio-ethanol can replace significant shares of petrol in many countries, and export markets could also be developed.

Co-generated electricity from sugar factories can help to diversify the fuel mix in the power sectors of the producers' regions. A variety of non-energy co-products could also be obtained for applications in agricultural and industrial sectors, taking advantage of the fiber resources and/or the sugar resources that can be extracted from sugarcane (Gómez-Merino et al., 2014; Correa do Lago et al., 2012). Harnessing this bio-energy and biomass potential will require significant increases in investment, technology transfer, and international cooperation.

Because of its high efficiency and its concentration in the developing world, the cane resource should be viewed as a global resource for sustainable development and should command much greater focus and concerted policy action through north-south and south-south cooperation.

Sugarcane offers a competitive and environmentally beneficial bio-energy resource for developing countries that can help to facilitate the sustainability transition within the continuing economic development of these producers' countries (Seebaluck et al., 2008; Johnson, et al., 2007).

SUGAR INDUSTRY SUSTAINABILITY

Sugarcane has been grown in different countries since the middle of the 19th century[1], primarily for the production of sugar. Agricultural cultivation of sugarcane includes planting, crop treatments, alternative mechanization for planting, harvesting and hauling, the recovery of the straw, and the agricultural management. From the perspective of the increase of cultivated areas of sugarcane, a lot of alterations need to happen in the agricultural sector.

It was only after the Global Energy crisis of 1973 that the scientists and technologists realized the value of sugarcane, its by-products and co-products.

For several years, plant breeders and agronomists have focused on increasing sucrose yields per hectare and millers have focused on increasing recoverable sucrose per ton of sugarcane in sugar mills. Attempting to exploit the energy potential of sugarcane more fully calls for a more holistic approach focusing on both sucrose and lignocellulosic components of sugarcane biomass, and gaining some insight into the sustainable management practices

[1] During the course of the nineteenth century technologies of manufacture greatly improved and revolutionized the production of sugar worldwide. Sugar was increasingly manufactured to common industrial standards, assured by a combination of steam and steel and their attendant chemistries into marketable forms of sucrose (muscovado, raw, refined and others). Technological convergence in the international sugar economy began in the 1830s and was conditioned by a great number of factors, including a massive increase in demand for the commodity, new tariff regimes in the main consuming countries, a significant fall in international costs of freight – and the increased extent and penetration of the colonial regimes with which cane-sugar production was closely associated throughout the tropical and subtropical world. It was substantially complete by the outbreak of World War I in 1914. By the end of the nineteenth century, the industrialized sugar factory was a global phenomenon like the steamship and the railway engine with the degree of technological convergence that came to characterize their manufacturing sectors, regardless of the type of labor involved, but there was no parallel in the crop field, where there continued to be a striking global divergence between the means and modes by which the industry was supplied with raw material (sugarcane) (Bosma et al., 2004).

required to optimize sugarcane cropping systems in these respects[2]. Such options include genotype selection, harvest date with respect to the crop's growing cycle, crop type (plant crop vs. ratoon crops) and harvesting systems (mechanical vs. manual). The effects of these factors are strongly modulated by climate and soil properties, and these interactions are overall poorly known and sometimes sugarcane is planted and harvested with contrasting management practices and adverse weather conditions (low or high temperature and radiation, water stress and climate change) (Sabatier et al., 2014; Singels et al., 2013, Marin et al., 2013).

The sugar industry biomass which contains cellulose, lignin, pentosans and the effective utilization of all the three components would play a significant role in the economic viability of cellulosic ethanol and can be converted into value-added products through the application of suitable chemical, biochemical and microbial technologies. Sugarcane is a versatile crop, being a rich source of food (sucrose, panela, jaggery and syrups), fiber (cellulose), fodder (green leaves and tops of cane plant, bagasse, molasses and to some extent press mud), fuel and chemicals (bagasse, molasses and ethanol). Besides these main by-products, there are other residues which are produced from sugarcane and have less commercial value such as trash and green tops, wax, boiler or fly ashes and factory and distillery effluents. Usually, sugarcane residues are heterogeneous and have different composition, consistency, and heating values, depending on soil, cane type, climate, region, harvesting method, and mill efficiency, among other factors.

Accordingly, it is very difficult to establish a unique price for biorefinery and diversification projects (Alonso-Pippo et al., 2007). Besides, only in the production of ethanol from sugarcane has the environmental benefits of replacing fossil fuels with ethanol from sugarcane been recognized.

---

[2] The effective achievement of sustainability goals needs the science of engineering to integrate aspects from the social, legal, and economic fields. In this sense, engineering science and conventional indicators (sugarcane yield, sucrose yield and factory yield) only examine the concept of sustainable development by focusing on the industrial and manufacturing sectors by yields. To eliminate industrial impacts on the physical and ecological environment, a range of environmental management practices, frameworks and methodologies have been proposed, such as cleaner production methods, environmental management systems, environmental indicators assessment, life cycle analysis, and environmental performance methodologies with focus only on specific environmental aspects (e.g. water, waste management, and energy); therefore, environmental problems and constraints should be faced at a more global (various production stages) and multiple level (various environmental aspects) and the combination of such practices (multidisciplinary approaches) may provide "deep insight about the environmental sustainability of industrial ecosystems and facilitate the development of the most eco-effective symbiosis for recycling, reuse and resource conservation" (Evangelinos et al., 2014).

It is estimated that ethanol derived from 1 ha of sugarcane avoids the emission of about 14 tons (t) CO_2eq yr^{-1} relative to the use of fossil fuels. Therefore, sugarcane for ethanol requires three steps to successfully and efficiently reduce greenhouse gas (GHG) emissions from agriculture: (i) identification of the most GHG polluting farms; (ii) determination of appropriate mitigation options for these farms; and (iii) selection between these options on the basis of their cost effectiveness (Franks, 2012).

Furthermore, when compared to ethanol derived from other feedstock (such as sugar beet, maize and sorghum), sugarcane is the most effective option in mitigating emissions of greenhouse gases (GHG); but the rapid expansion of the worldwide cultivated sugarcane area, mainly in Brazil, has left many unanswered questions about the true sustainability of biofuels, especially in relation to GHG emissions from the agricultural sector, which accounts for nearly 90% of the GHG footprint of sugarcane ethanol.

Normally, the sugarcane harvest is performed manually with preliminary burning (burned harvest), or mechanically without burning (green harvest). The burning practice aims to facilitate manual harvesting through the removal of leaves and poisonous animals. However, resultant emissions of greenhouse gases cause harm to the environment and human health. GHG emission from pre-harvest burning has been estimated as 941 kg CO_2eq ha^{-1} yr^{-1}, which corresponds to 30% of the total GHG emission in sugarcane production; these emissions are essential in assessing the sustainability of ethanol production and others sugarcane derivatives[3] (De Oliveira et al., 2013).

Sugarcane breeding and better agronomic practices have contributed to a huge increase in sugarcane yield, but worldwide sugarcane yields for sugar, ethanol, diversification projects and biorefineries have been declining over the past 15 years, increasing the acreage and decreasing the recoverable sugar per hectare (Figures 5 and 6).

---

[3] If the use of cane trash is considered for energy production, second generation ethanol and other derivatives, one key issue remains to be settled: the optimal amount of trash to be left in the field, taking into consideration the economic and environmental benefits of using it to enhance the energy efficiency of the ethanol production chain (more ethanol and electricity) and the positive impacts of the trash blanket on the ground such as nutrient recycling, soil protection against erosion, increased soil organic matter and carbon capture, weed control, soil moisture conservation, and increased microbial soil activity; the negative effects should also take into consideration factors such as fire hazard, increased difficulty in some agricultural operations, delay in the ratooning of some sugarcane varieties under the trash blanket, and excess moisture in wet regions. Added to that is the fact that all these effects are highly dependent on the local conditions (soil, climate, and agricultural practices). The solution to this problem is very complex, and there is no consensus about the ideal amount of trash or straw to be recovered or the quantification of the economic and environmental impacts (Leal, 2013).

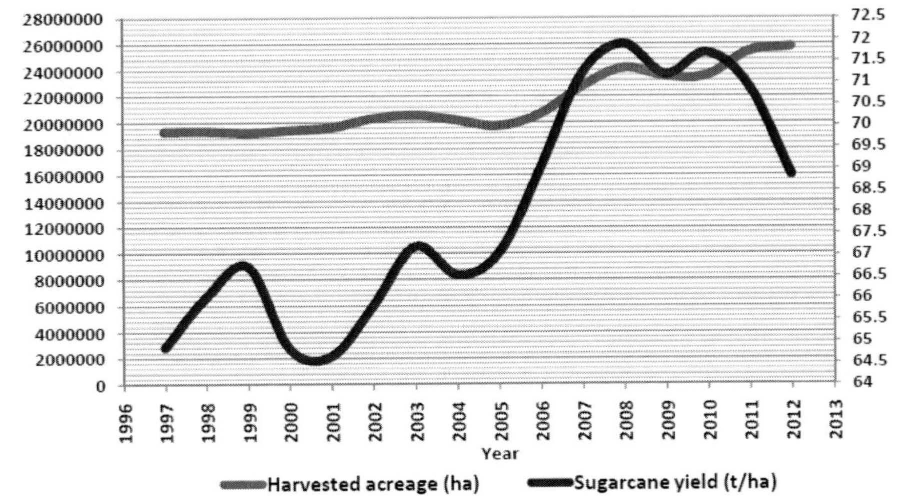

FAOSTAT, 2013.

Figure 5. Acreage and sugarcane yield.

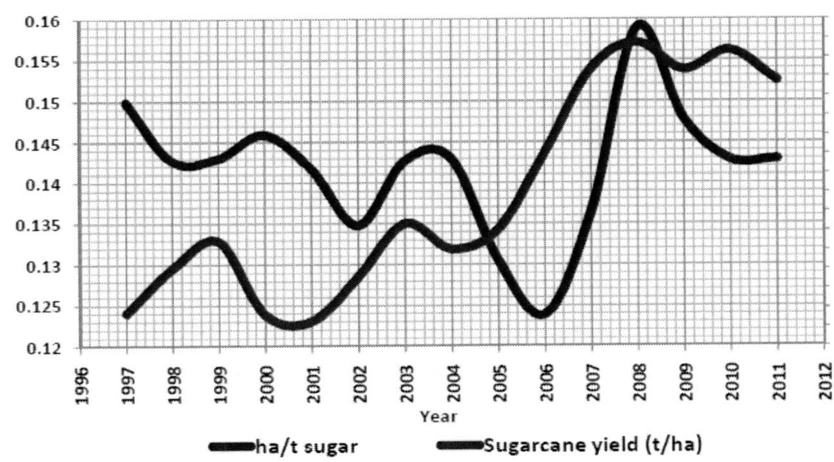

FAOSTAT, 2013.

Figure 6. Required acreage to produce one tonne of sugar and sugarcane yield.

Soils under cane are more compacted, more acid, contain less organic matter and are lower in cation exchange capacity and exchangeable cations. These differences reflect soil degradation caused by intensive cultivation.

Contributing factors to the degradation of soils include soil compaction and structural breakdown occurring during harvest and cultivation operations, losses of organic matter due to burning of crop residues and acidification of soils due to large applications of nitrogen fertilizers; but considering that limitation of suitable land and/or competition with other land uses might occur in all sugarcane countries, additional increases in sugarcane yield are expected to result from the use of biotechnology tools in the near future.

Genetically modified (GM) sugarcane that incorporates genes to increase resistance to biotic and abiotic stresses could play a major role in achieving this goal by developing solutions that would produce more sugarcane with decreased requirements for fertilizers and water.

Besides the assessments of potential crop productivity and management strategies with interactions, the environment and socioeconomics can contribute to an improved planning of land requirements for sugarcane biofuels and derivatives from by-products under high productivity or marginal conditions and the soil management practices should aim to increase soil organic matter levels, provide a more favorable biological environment, reduce physical damage to soils during harvesting and cultivation, reduce soil acidity and improve the effectiveness of fertilizing practices (Cheavegatti-Gianotto, 2011).

The development of sustainable sugarcane cropping systems is a key priority for agronomists and crop scientists.

The first step involves understanding the relationship between cropping system performance and farmers' practices particularly by the selection of indicators for assessing the effects of crop management, soil and weather conditions, and data analysis.

This knowledge of how management affects the quality characteristics of the sugarcane biomass and by-products is needed to aid decision-making and optimize cropping systems for multicommodity production, with a focus on the concomitant exploitation of sugarcane biomass for food and energy purposes in a biorefinery or 'sugar mill–power plant'[4] (Dore et al., 2009 and Wood, 1985).

[4] An assessment of the industry side can be divided into three main components: the set-up of the sugar factory, the configuration of the ethanol distillery, and the cogeneration plant. A cogeneration plant would be installed if the sugar factory processing capacity and in-house energy utilization is efficient enough and if there was a market for surplus heat and electricity. For ethanol production, the Brazilian designs offer the best technical configuration, including the preparation of feedstock (juice and/or molasses). The present tendency for distillery configuration is the so-called 'annexed' distillery, which offers the advantage of flexibility in allowing various combinations of sugar and ethanol production.

SUGAR INDUSTRY DIVERSIFICATION

The economic challenges facing the global sugar industry are volatile sugar prices in world trade and the diversification of the productive structure of sugarcane monoculture, and the sugar mill and distillery by-products (bagasse, molasses, vinasse, filter mud and ashes) should be viewed as a first step in the process of achieving competitiveness and an option to improve productivity in the transition to biorefineries using existing equipment, infrastructure, and increasing specialization resulting from the potential demands for sugarcane derivatives worldwide (Aguilar et al., 2011, 2010, 2009).

Almost all of the countries in the world that produce sugarcane have realized that although the production of sugar from sugarcane is undoubtedly the most costly proposition, it is better to produce many value-added products by diversification and utilize the by-products of the sugar industry, instead of depending on just one product i.e. sugar, but the seasonal characteristics of the sugar agro-industry, sucrose monoculture, and the lack of alternative biomass supply feedstock during the off-season for biorefineries seem to be the main barriers from a technological point of view in developing countries where a feasible alternative supply of biomass is not available. (Solomon, 2011; Alonso-Pippo et al., 2007; Yadav et al., 2006; Gálvez, 2000; Almazan, 1999, Paturau, 1989; ICIDCA, 1986).

Industries and biorefineries based on sugarcane by-products have several advantages. (i) The waste material can be converted into valuable products by adopting suitable technologies; (ii) It creates employment opportunity, especially in rural areas and in turn improves the economic status of people involved in these industries; (iii) The most important aspect of utilizing these by-products is that it helps in minimizing the pollution hazards; (iv) Setting up by-product based industries will help in improving the general economy of the sugar industry *vis-a-vis* their financial status (Solomon, 2011). However, the current complex situation demands that the sugar agro-industry undergoes important changes because the sugar industry diversification and establishment or conversion to biorefineries (Figure 7) is a highly complex project, which has been evaluated since 1970 as a viable techno-economically alternative to offset the volatility from international sugar prices by decreasing production costs and environmental impacts while converting by-products into raw materials for new production cycles with technologies such as combustion, gasification, hydrothermal processing, liquefaction, pyrolysis or biochemical conversion (Menon, 2012; Pellegrini et al., 2011; Contreras et al., 2009;

Renouf, 2007; Johannesburg, 2005; Lodi, 2005 and Galvez, 2000) and actually helping to reduce GHG and carbon footprints with ecological sugarcane agriculture and the integration as cluster or chain value (Table 1).

Co-production concept[5].

Figure 7. Sugar industry by-products and derivatives.

[5] The concept of co-production is based on biophysical, industrial and economic principles; it is a well-established technology in a competitive market. The biophysical basis lies in the special properties of the sugarcane plant, which contains large quantities of biomass along with digestible and fermentable sugars. The industrial facilities that have been designed for co-production have a spatial advantage, because key co-products of one industrial process, such as bagasse, are available onsite for use as raw materials for the additional production processes. A business that wishes to stay competitive in today's global economy must attain flexibility and diversity in adapting to changing markets. The ability to sell three major products such as sugar, ethanol, and electricity is therefore a major asset from an economic perspective. Consequently, a variety of co-products have emerged and found markets during the historical evolution of the sugarcane-based sugar industry (Cornland et al., 2011).

Table 1. Agro-industrial and economic value of sugar industry by-products

Sector of economy	Value-added products from sugarcane and sugar industry residues
Food	Sweeteners (traditional, modern, synthetic), vitamins, acids, beverages, fats and oils, edible proteins, mushrooms)
Health	Chemicals, antibiotics, anti-cholesterol (policosanol), lingo-meds, enzymes, vaccines, juice
Agricultural fertilizers, compost, food, feed, fodder, forages, pesticides	A range of food, feed, fodder, fertilizer and forages
Industry	Solvents, plastics, bioplastic, alcohol-based chemicals, anti-corrosive compounds, tenso-active compounds, biocides
Energy electric power, biogas, bagasse fuel, fuel alcohol	Bagasse as fuel, biogas, cogeneration of power, ethanol from bagasse
Transportation	Ethanol-petrol/diesel blends (gasohol), bio-diesel
Education and culture	Text books, note books, newsprints, writing and printing paper
Housing/construction	Particle boards, hard boards, ac ducts, decorative laminate
Light industry	Textile, polish, bitumen, carbon paper and chemicals
Communication	Insulating materials
Heavy industry	Resins for casting molds
Human resource development	Employment generation in rural areas

Solomon, 2011, 1995; Yadav, 2006.

Diversification has been pursued in many countries as a way to improve the long-term viability of sugarcane agriculture and sugar mills and distilleries by enhancing the profitability and overall stability of the sector.

Diversification has emerged as a central topic of research in strategic management.

This topic has been widely and intensively studied by scholars from other areas such as industrial organization economics, financial economics, organization theory, and marketing (Ramanujan, 1989).

SUGARCANE BIOREFINERIES

The development of biorefineries represents the key for access to an integrated production of food, feed, chemicals, materials, goods, and fuels of the future. Biorefineries combine the necessary technologies of the biogenic raw materials with those of intermediates and final products. The main focus is directed at the precursors carbohydrates, lignin, oils, and proteins and the combination between biotechnological and chemical conversion of substances.

Technology is the key factor allowing the conception and design of biorefineries, where materials that are burned directly for energy, such as firewood, wood chips, pellets, animal waste, forest, bagasse and crop residues, are considered primary biofuels (Hahn-Hägerdal et al., 2006). Logistical challenges of transporting, storing, and maintaining acceptable quality of the biomass will restrict the size of future biorefineries. These biorefineries are also likely to be most economical and energy efficient if they are able to produce a multitude of high value fuels and chemicals (Gibbons and Hughes 2011).

The basic sugarcane biorefinery has three types of product streams (sugarcane resources): sugar/solids, molasses/juice, and crop residues. There are a variety of products that are feasible and marketable within each category.

The sugar/solids category includes various feedstocks and intermediate products in addition to sugar. Molasses and/or cane juice are valued for the fermentable sugars that can be converted into ethanol, as well as being used as industrial and agricultural inputs.

Cane residues, namely bagasse and cane trash, are valued for their fiber content and organic residues as well as their use as fuel in cogeneration plants. The useful cycle of by-products and co-products continues after ethanol production, with stillage serving (vinasses) as input for the production of fertilizers and methane gas (Correa do Lago et al., 2012) (Figure 8).

Sugarcane biorefinery is the industrial unit able to convert integral sugarcane into the desired product and surplus electricity to the grid would include:

- Optimized sugarcane varieties
- Characterization of the production environment
- Total crop mechanization and optimized logistics
- Integral sugarcane production
- Optimal irrigation regime
- Maximum energy generation (steam and energy)
- Maximum EE surplus sales to the local grid

The potential expansion and conversion of land to sugarcane biofuel crop production can potentially alter the water cycle, soil erosion, biomass yield, nutrient loss, C balance, ecosystem biogeochemical impacts and GHG emissions. Quantification of these changes forms the basis of evaluating environmental risks and benefits. In turn, these changes can become feedbacks that impact sugarcane biofuel crop productivity and long-term sustainability.

In order to apply operational research methods to sugarcane agriculture it is necessary to take into account the variation in both space and time of the production conditions as well as the variation in management practices.

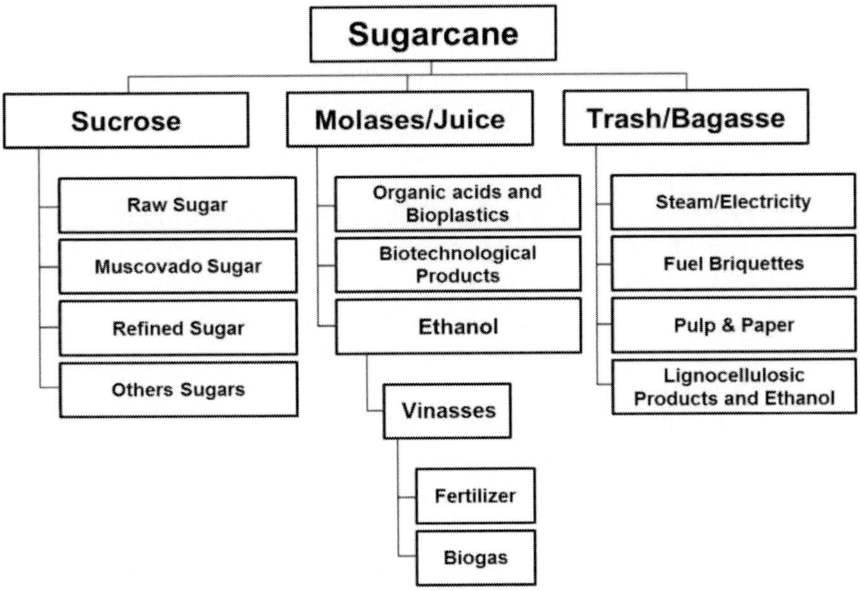

Aguilar et al., 2011.

Figure 8. Simplified sugarcane biorefinery.

Therefore, as the sugarcane cropping systems are exclusively optimized for sucrose production, the design, assessment and optimization of new cropping systems aimed at multicommodity production (e.g. food and energy as diversification projects) are needed but require decision-making tools and support as LCA life cycle assessment, carbon and water footprints, eMergy, remote sensing and GIS, multicriteria evaluation (Tsiropoulos et al., 2014; Amores et al., 2013; Jamali-Zghal et al., 2013; Agostinho, et al., 2013, 2010; Pereira and Ortega, 2012, 2010; Bergquist et al., 2012; Gerbens-Leenes, 2012; Renouf et al., 2011; Chauhan et al., 2011; Cherubini, 2011; Contreras et al., 2009; Luo et al., 2009) and others are recommended to fully understand the requirements regarding the services and environmental impacts of current practices in sugarcane cropping systems from both environmental and economic points of view and may define food and energy production from sugarcane cropping systems according to a wide range of management options (cultivars, environments and irrigation); this is of primary importance.

Therefore, the analysis of diversification projects and the potential for establishing them must be based on scientific knowledge of sugarcane crop fields and socio-economic and ecological constraints, the co-products, energy and material balance in sugar mills, and sugarcane derivatives' trade in order to incorporate and integrate different factors and agro-industrial variables under the biorefinery and sustainability concept and to discuss their possible effects, but mainly to make decisions that will be simultaneously technical, economic, social and ethical (Doré et al., 2009; D'Hont et al. 2008; Brumbley 2007; Birch, 2007; Edye, 2006; Morandini, 2006 and Avram, 2005).

Accordingly, to address the full suite of requirements for future sugarcane biorefinery as well as the inputs (economic and environmental), energy, interconnections, feedbacks, purpose, policy conditions, stakeholder values, location, temporal influences, spatial scale, baselines, reference scenarios and uncertainties that typify complex socio-ecological systems (Figures 9 and 10) in the value chain (with integration across disciplines and scales; it is necessary to identify opportunities for improvements towards sustainability with indicators or novel approaches (Dale et al., 2013; Neves et al., 2011).

The actual question is how bio-energy and derivatives from sugarcane and by-products can support sustainable development and improve global, regional and local competitiveness? There are four fundamental global drivers of renewable energy and sustainability that are specific to sugarcane: (1) global warming and environmental factors; (2) national energy security in terms of a supply and demand imbalance; (3) national balance of payment concerns whereby the importation of energy is often a nation's single biggest cost; and (4) job creation and socio-economic factors (Wynne et al., 2008).

Sugarcane industries face increasingly difficult and complex issues. Management strategies need to take into account:

(1) the need to be sustainable, i.e. options to improve profitability must maintain the resource base and minimize impacts on other ecosystems
(2) diversity in soils and production systems, and highly variable climate from season to season and across locations
(3) variability in sugar price and exchange rate
(4) heterogeneity among farms in terms of size, production systems, cost structures and labor availability (social factors)
(5) the sugar industry being an integrated value chain from farm to mill to markets, with the individual components not being independent, but part of a value system of interdependent components
(6) off-site impacts in terms of expectations of the community and other industries.

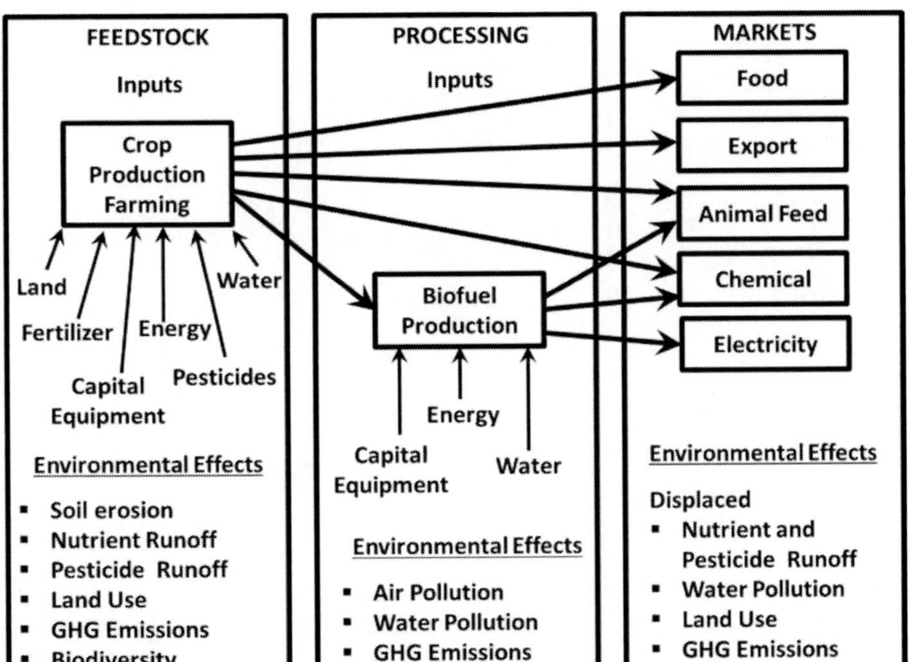

Figure 9. Chain value an impacts of sugarcane biorefineries.

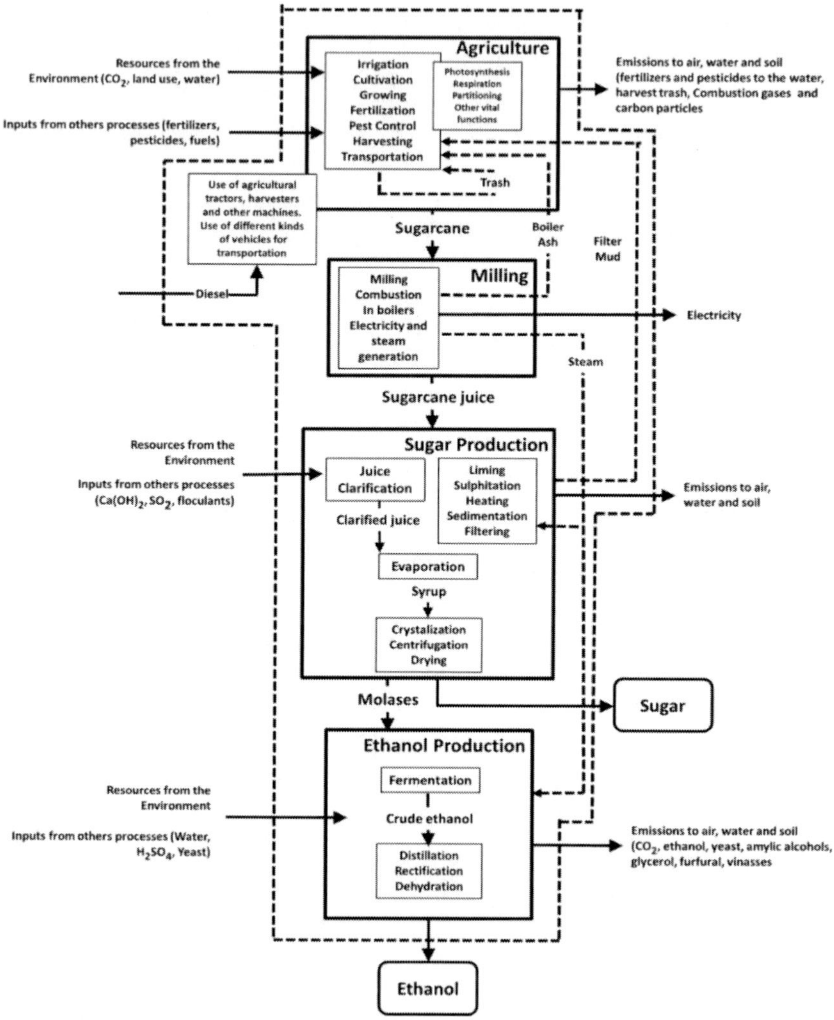

Amores et al., 2013.

Figure 10. Sugarcane biorefinery, inputs, processing, impacts and chain value.

Efforts to improve production efficiency and economic viability in the sugar industry have traditionally focused on maximizing sugarcane yield per hectare of agricultural land and sugar produced per tonne of sugarcane grown. Although some cane co-products (such as bagasse and molasses) are utilized in this process, priority is accorded to sugar production. The traditional focus on sugar has made the industry vulnerable to changing market prices and weather patterns and prone to financial instability. There have been few attempts by sugar companies to consider all sugarcane resources as a bundle of potential products and services whose value could be maximized together. Co-production strategies present attractive options as they offer flexibility in producing varied quantities of sugar, ethanol and electricity depending on prevailing market conditions (Cornland et al., 2011).

Accordingly, sugar industry sustainability requires a redesigning of production, consumption and waste management; thus, Duarte et al., (2013) demonstrated that there are important opportunities for improvement of the sugarcane crop fields sustainability, as the basis (raw material) of diversification and biorefinery projects, under five categories: (1) long-term water availability and water quality maintenance; (2) biodiversity enhancement and reversal of ecological fragmentation; (3) the elimination of sugarcane straw burning in the fields and increasing mechanization; (4) indirect and direct land-use change; and (5) the quality, availability and durability of livelihood opportunities by diversification. The five issues and the sustainable use of economic and environmental inputs all require broader strategic planning, but must also be understood within the local context of the sugar mill and its watershed.

To address these issues requires long-term integrated planning and monitoring, better understanding of cumulative impacts and thresholds, recognition of important tradeoffs, an enforcement of limits, and a credible and collaborative decision-making process that involves and empowers stakeholders to set the agendas and seek common goals or priority issues for a broader national strategy relating to sugarcane ethanol production and biorefineries.

Traditional disciplinary approaches (mainly agronomy, chemical engineering and economy) will be increasingly limited in dealing with these issues, and multidisciplinary approaches using a whole systems analysis framework offer the potential to meet the emerging challenges of the 21st century. A systems approach dealing comprehensively with the financial and biophysical factors involved, and with climate variability in particular, is the only way to improve the profitability and sustainability of sugar production and the sustainability of biorefineries.

Whole Systems Analysis is an essential technology that can cope with complexity and variability in delivering benefits to industry and in ensuring research efficiency in a complex operating environment.

Strategic research is required to enhance the knowledge base (crop physiology and nutrition, soil physics and chemistry, economics, mathematics and information technology) to develop implementation strategies for the use of these tools in a participatory manner, to ensure adoption of new options for enhancing whole industry profitability in harmony with other sustainability considerations to the most effective and practical options for profitable and sustainable sugar production, by-products and biorefineries (Muchow et al., 2001).

The sustainability aspects that can be emphasized in the sugarcane processing are:

1 Productivity,
- Producing more with the same equipment;

2 Efficiency,
- Producing more with the same raw material,
- Reducing losses and emissions – pollution;

3 Energy,
- Producing more with the same energy;

4 Water,
- Producing more with the same water;

5 Chemicals,
- Producing more with the same chemicals,
- Less contamination – pollution (Amaya, 2010).

This multidisciplinary conceptualization[6] in biobased economy should be the starting point for establishing a holistic approach boundary for the goal to simplify, quantify, analyze and communicate the complex and complicated information about the diversification process, biorefinery transition and chain value and identify criteria, constraints and the specific weight of each

[6] Multidisciplinary approaches have an increasingly greater role to play in meeting rising demands for food, fiber, energy, clean environment and good health. Cross-applications of tools and analytical approaches have tremendous potential to fill existing knowledge gaps, clear roadblocks and facilitate translation of basic science discoveries as solutions towards addressing some of the most pressing global issues. Such research should, ideally, be based in the core of each analytical field, following traditions of multidisciplinary research to which well-defined development targets can be added. Thus, development oriented biobased systems research should identify and select promising cropping practices which fit in a farming systems orientation that can integrate biobased production and market potentials. Only in this way can technological and development potentials be realized simultaneously. Such a development requires innovation frameworks not focusing on fixed technological development but having a more process oriented setup. Remnants of farming systems research infrastructure can play a role here, as they can link potentials of technological innovations to the needs of development oriented processes (Langeveld, 2010).

determining factor in sugar mills and sugarcane crop fields. This implies *a priori* the design of new methodologies and use of normalization and aggregation of conventional indicators based on scientific rules, sensitivity analysis and uncertainty analysis and robust statistical methods to create a standardized set of sustainability indicators covering all main aspects of sustainable development for the sugarcane industry and condense the enormous complexity and multidimensionality of dynamic environment (sugarcane production, transformation and markets) with an integral approach taking into account environmental, economic and social aspects to a manageable amount of meaningful information to decision making.

Thus, indices should be constructed within a coherent framework.

Zhang et al. (2010) proposed a modeling framework consisting of three major components (1) a Geographic Information System (GIS)-based geodatabase, (2) a biophysical and biogeochemical process model, and (3) a multiobjective optimization method. Dale and Beyeler (2001) analyzed existing literature on indicator selection for biofuel to identify key criteria:

1 Practical (easy, timely, and cost-effective to measure);
2 Sensitive and responsive to both natural and anthropogenic stresses to the system;
3 Unambiguous with respect to what is measured, how measurements are made, and how response is measured;
4 Anticipatory of impending changes;
5 Predictive of changes that can be averted with management action;
6 Estimable with known variability in response to changes; and
7 Sufficient when considered collectively (i.e. a suite of indicators integrates changes in socio-economic sustainability).

The process of characterizing sustainability by indicators will need to evolve to reflect new information and society's changing priorities. This would help in the selection of suitable parameters or variables which really influence the policy goal and also ensure that the evaluation process involved could change through time according to the interests of the particular stakeholders involved in the construction of the indicator (Efroymson et al., 2013; Mata et al., 2013; Singh et al., 2012, Bojórquez, 2009). The objective of this paper was to develop a methodology for determining the capacity of sugarcane municipalities, sugarcane crops and sugar mills to diversify their basis production (sugarcane and sucrose) to biorefineries through the use of multicriteria method AHP (Analytic Hierarchy Process) with available and conventional data.

MULTI-CRITERIA EVALUATION (MCE) FOR SUGARCANE SUSTAINABILITY

The multi-criteria evaluation (MCE) is based on the weighting and compensation of variables and aptitude determining factors related to decision theory (Gomez, 2006). A multicriteria problem with a discrete number of alternatives, where A is a finite set of n alternatives or feasible actions, G is the set of m evaluation functions gi, i = 1,2, ..., m associated with the evaluation criteria. If A is an alternative, gi (A) is their assessment on the i criterion can be represented in a matrix P of m rows and n columns called evaluation table or impact matrix whose elements pij (i = 1,2, .. ., m, j = 1,2, ..., n) represent the evaluation of alternative j in the i criterion (Falconi, 2004).

The AHP (Analytic Hierarchy Process) is supported by reciprocal judgments and homogeneity of the elements in a complex multivariable problem; these are compared to the same order of magnitude (paired comparisons as ratios) and the construction of hierarchical structure to prioritize their relative importance in different production systems (Freitas, 2006 and Saaty, 1990). It has been used in sugarcane variety selection (Zhong et al., 2009), in the design of strategies to increase competitiveness (Rivera, 2014; Duarte et al., 2013; Akli, 2009; Castellanos, 2007; Nagesha, et al., 2006), management of sugarcane areas (Diogo et al. 2007 and Qureshi, 2003) and bio-ethanol and bio-energy projects (Oliveira et al., 2012; Turcksin, 2010; Hilmola et al. 2010; Lapola et al., 2010; Tienwong, 2009; Lamparelli, 2009; Junqueira et al., 2009; Silva, 2009, Quintero et al. 2008; and Papalexandrou, 2007).

The AHP is a mathematical method of analyzing complex decisions with multiple criteria. The AHP procedure involves six essential steps (Lee et al., 2008):

1 Define the unstructured problem
2 Develop the AHP hierarchy
3 Make a pairwise comparison
4 Estimate the relative weights
5 Check the consistency
6 Obtain the overall rating

Sugar industry diversification or conversion to biorefineries, being a complex problem, is decomposed into a hierarchical structure with decision elements.

This approach takes into account variables or factors; firstly, for sugar mills according to international standards of competitiveness and diversification for the sugar industry: Sugar mill yield (%), sucrose recovery rate (%), total time loss (%), sugarcane quality (% sucrose), external electricity (kWh / t. cane), external fuel consumption (oil) oil L/t.cane and produced sugarcane derivatives (raw, muscovado, refined, ethanol, electricity, compost, fertirigation, rum, spirits and other) corresponding to the International Sugar Organization (ISO, 2005).

Secondly, for sugarcane municipalities and sugarcane crops, the socio-economic and ecological resources and capabilities that have been identified as limiting factors or constraints in diversifying the process of transition to biorefineries are: sugarcane agroclimatic land aptitude, sugarcane yield, sugarcane farm size (acreage), production cycles (plant and ratoon crops), sugarcane pests and diseases (aeneolamia, borer, termites, nematodes, orange rust, smut, mosaic, leaf scald, ratoon stunting, etc.), harvesting systems (green or burned), land tenure (private, communal or public), established sugarcane projects, access to credit, distance to factories and markets, access to irrigation, Human Development Index, other income excluding sugarcane, access to technical assistance and training, access to public services, and access to human labor and mechanization (Eakin et al., 2011; Waclawovsky et al. 2010; Haque et al. 2010; Moore, 2009; Tienwong et al., 2009; Bojórquez, 2009; Pérez, 2007; Kamruzzaman, 2007; Oddershede, 2007; Roebeling et al., 2006; Domacet al., 2005; Nothard et al.2005; Naraine, 2005; Bandaranaike, 2005; Windle, 2005, 2003; Antony, 2005; Mishra, 2004; Escobal, 2004; Gallardo-López et al. 2002; Solomon, 2000; Summer, 2000; Chaplin, 2000; Godoy, 1988; Pope, 1980).

In both cases of the sugar industry value chain, such variables or factors provide *a priori* information about the capacity to diversify the value chain.

Together they can be considered optimal indexes for comparative analysis (benchmarking) at international level, of the dynamics in the sugarcane crop fields, municipalities and the industry, allowing for weighting of the relative importance of each variable or techno-economic indicators on the diversifier process analysis when considering the socio-economic and ecological resources and capacities, the sugar mill capacity, raw material quality and energy balances (Aguilar-Rivera, 2012; ISO, 2005; Banerjee, 2004 and Zimmermann, 2002).

The relative importance of each factor can be evaluated through the building of a hierarchical structure that contains three levels: goal or objective, criteria and alternatives; a matrix pairwise factors comparison (paired) AHP multicriteria evaluation, and the subsequent calculation of the matrix to assign a different value by weight to individual factors or constraints, whereby each possible pair were compared and qualified by applying a continuous hierarchical scale of 17 relative importance factors (Figures 11 to 14).

For the overall set of alternative priorities, AHP evaluates the error or inconsistency of the paired matrix. If it is less than 10% (CI index <0.1), it is considered acceptable and robust (Sipahi 2010; Berumen 2007; Aguarón et al., 2003; Saaty, 1990). The pairwise matrix and the weights can be obtained from Expert Choice 11.5 ® software AHP evaluation module for each factor or AHP Arc Gis Module or similar software (Tables 2 to 4).

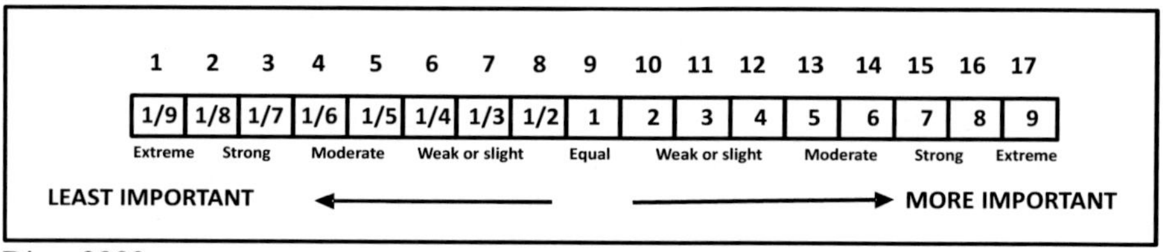

Diaz, 2000.

Figure 11. Hierarchical scale of 17 relative importance indicators for the construction of the comparison matrix between factor pairs or variable decisions.

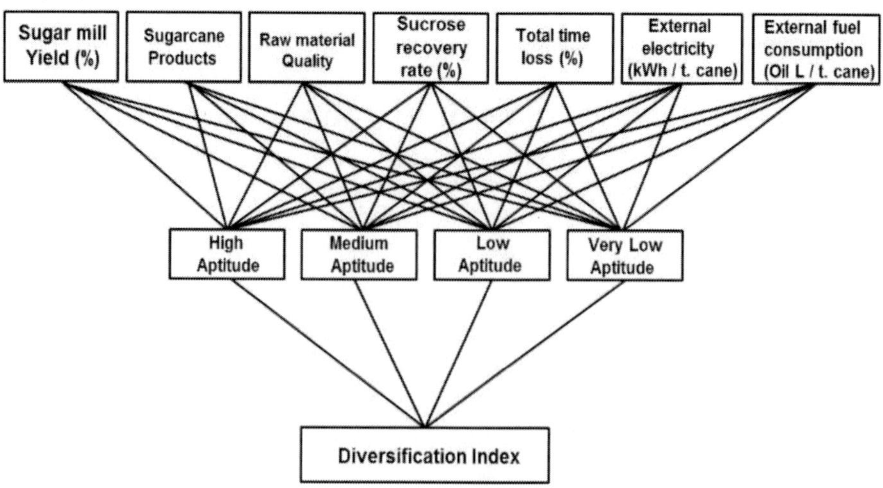

Figure 12. Hierarchical structure for diversification constraints in sugar mills.

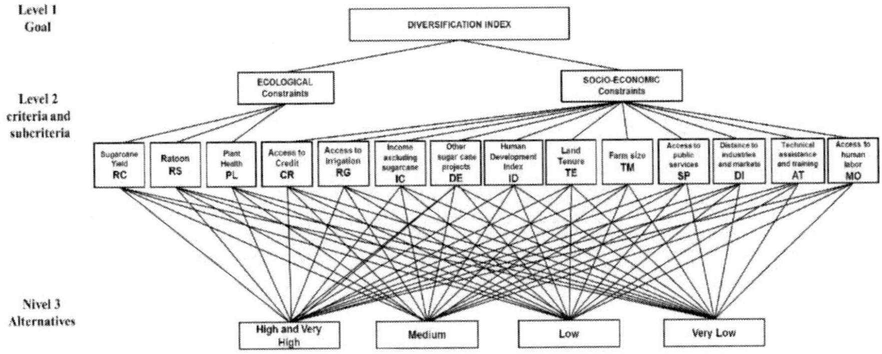

Figure 13. Hierarchical structure for diversification constraints in municipalities.

Table 2. Weights of sugar mill diversification factors

Constraints on sugar mill diversification	Weight
Sugar mill Yield (%)	0.332
Produced sugarcane derivatives (raw, muscovado, refined, ethanol, electricity, compost, fertirigation, rum, spirits and other)	0.327
Raw material quality (% Sucrose, % Fiber)	0.121
Sucrose recovery rate (%)	0.092
Total time loss (%)	0.077
External electricity (kWh / t. cane)	0.027
External fuel consumption (oil L / t. cane)	0.025
Total	1.000

Consistency index = 0.06.

Table 3. Weights of sugarcane municipalities diversification factors

Constraints on sugarcane municipalities			Weight
Other sugarcane projects (%) (Crop diversification)	Socio-economic	DE	0.155
Sugarcane yield (t/ha)	Ecological	RC	0.144
Plant health (sugarcane acreage with affectations by pests and diseases (%)	Ecological	PL	0.123
Access to credit (%)	Socio-economic	CR	0.117
Distance to factories and markets more than 5 km (%)	Socio-economic	DI	0.106
Access to irrigation (%)	Socio-economic	RG	0.077
Human Development Index	Socio-economic	ID	0.075
Other income excluding sugarcane (%)	Socio-economic	IC	0.049
Farm size (ha)	Socio-economic	TM	0.040

Table 3. (Continued)

Constraints on sugarcane municipalities			Weight
Without technical assistance and training (%)	Socio-economic	AT	0.034
Without access to public services (%)	Socio-economic	SP	0.031
Ratoon (%)	Ecological	RS	0.027
Land tenure (% public)	Socio-economic	TE	0.019
Without access to human labor (%)	Socio-economic	MO	0.009
Total			1

Consistency index = 0.05.

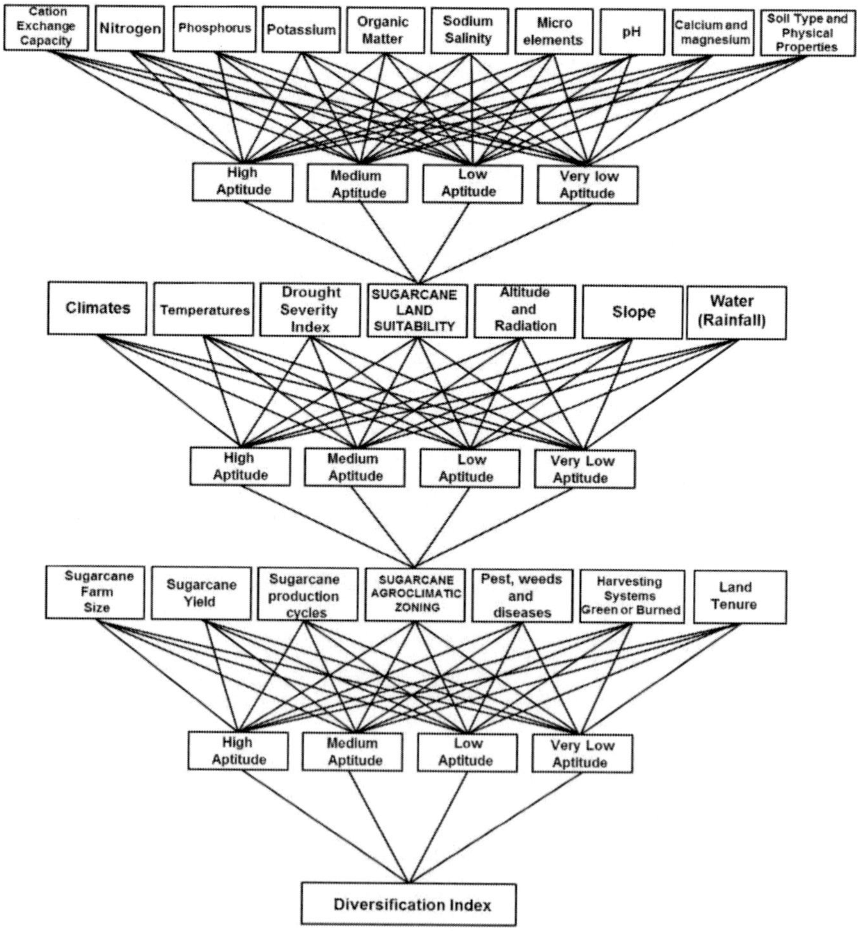

Figure 14. Hierarchical structure for diversification constraints in sugarcane crop fields (rainfed).

Table 4. Weights of sugar crop fields diversification factors

Constraints on sugarcane crop fields	Weight
Sugarcane agroclimatic land aptitude	0.381
Sugarcane yield	0.254
Sugarcane farm size (acreage)	0.144
Sugarcane production cycles (plant and ratoon crops)	0.091
Sugarcane pests and diseases (Aeneolamia, borer, termites, nematodes, orange rust, smut, mosaic, leaf scald, ratoon stunting)	0.061
Harvesting systems (green or burned)	0.042
Land tenure (private, communal or public)	0.028
Total	1.000

Consistency Index = 0.02.

Therefore, for sugar mills the following factors: Sugar yield (%) and number of goods or produced sugarcane derivatives (raw, muscovado, refined, ethanol, electricity, compost, fertirigation, rum, spirits and others) have a greater weight than the rest of the factors, since they imply a greater capacity of the sugar mill to transform raw material into sucrose and generate by-products (molasses, bagasse, filter mud, ashes and vinasses) for energy and other productions (cogeneration, ethanol, compost, refined, muscovado, organic, etc.) along with conventional sugar production (raw or refined), and the existence of facilities and technical knowledge (learning curve or know-how) in the production of sugarcane derivatives and trade. The next most important factors are sugarcane quality (% sucrose, % fiber) and sucrose recovery rate (%) as their importance is directly related to sugarcane production and intrinsic productivity. Secondly, sucrose recovery rate (%), is related to the core technology and/or obsolescence of process equipment for handling raw materials, external electricity and fuel consumption both by cane that are the result of energy diversification in relation to external power (Aguilar-Rivera, 2012).

The sugarcane agroclimatic land aptitude is the most important constraint, followed by sugarcane yield and access to land or size of sugarcane farm, and together account for 77.9 % of the capacity to diversify.

Sugarcane agroclimatic land aptitude, with a weight of 38.1 %, has a major impact because it determines the capacity of the land to produce raw material in quantity and quality, sustain the production cycle and expand the current agricultural frontier. These effects are closely related to soil quality, climatic conditions and pest affectation under rainfed conditions.

Therefore, there is a strong link between agroclimatic land aptitude and farm management processes and economies of scale, irrigation, mechanization (especially harvest), fertilizers and pest, diseases and weed management for the sugarcane production with higher relative advantages (competiveness) or at least minor disadvantages of physical factors (climate, soil, etc.) and biological and economic forces that constrained opportunities for diversification and transition to biorefineries (Weiss, 2002).

For sugarcane crop field and municipalities, the ecological constraints (RC, PL and RS) have a weight of 0.294 and socio-economic (DE, CR, DI, RG, ID, IC, TM, AT, SP, TE and MO) result of 0.712. This result is consistent with the findings of Eakin et al. (2011) and Bojorquez (2009) who established the ecological or environmental constraints for plantation crops as sugarcane accounts for about 30-35% of the total weight and socio-economic concerns represent 65-70% in relation to productivity for future diversification projects.

Besides Waclawovsky et al. (2010), Roebeling et al. (2006) and Vlosky et al. (2005) concluded that diversification projects based on increased productivity of sugarcane, according to the trend of sugar prices, should not lead to a reduction in the supply of cane sugar to risk the viability and profitability of the sugar factory and distillery, and that a sufficient and constant supply of sugarcane is a prerequisite to supporting the development of any new diversification project or biorefinery.

In relation to each weighting of the factors in Tables 5 to 7, a sugar mill municipalities and sugarcane crops evaluation instrument was made according to Ahumada, (2009); Seebaluck et al., (2008); ISO (2005); Trujillo (2002) and Zimmermann et al. (2002) to determine the weight or relative importance by hierarchical level and threshold for making technical decisions to diversify the chain value such as the diversification index (high, medium, low and very low).

Table 5. Evaluation instrument of sugar mills by technical ability level to diversify

Constraints on sugar mill diversification or conversion to biorefineries	Unit	Value				Final weight
		High (1)	Medium (0.75)	Low (0.50)	Very low (0.25)	
Sugar mill yield	%	>12	12-11.5	11.5-10.5	<10.5	
		0.332	0.249	0.166	0.083	
Sucrose recovery rate	%	>85	85-83	83-80	<80	
		0.092	0.069	0.046	0.023	

Constraints on sugar mill diversification or conversion to biorefineries	Unit	Value				Final weight
		High (1)	Medium (0.75)	Low (0.50)	Very low (0.25)	
Total time loss	%	<10	10-15	15-20	>20	
		0.077	0.05775	0.0385	0.01925	
Sugarcane quality (Raw material)	% Sucrose	>14.5	14.5-13.5	13.5-12.5	<12.5	
		0.121	0.09075	0.0605	0.03025	
External electricity	KWH/t. cane	0	0-0.5	0.5-1	>1	
		0.027	0.02025	0.0135	0.00675	
External fuel consumption	Oil L / t. cane	0	0-2	2-5	>5	
		0.025	0.01875	0.0125	0.00625	
Produced sugarcane derivatives (raw, muscovado, refined, ethanol, electricity, compost, fertirigation, rum, spirits and other)	#	>4	3	2	1	
		0.327	0.24525	0.1635	0.08175	
TOTAL						

1 = High capacity, 0 = no capacity.

Table 6. Evaluation instrument of sugarcane municipalities by socio-economic and ecological resources and capability level to diversify

Constraint on supply of raw material to biorefineries or diversification projects		Unit	Value				Weight	Subtotal
			High (1)	Medium (0.75)	Low (0.50)	Very low (0.25)		
Ecological	Sugarcane yield	t/ha	>75	74-65	64-50	<50		
	Production cycles (ratoon crops)	%	<30	30-40	40-50	>50		
	Plant health (sugarcane acreage with affectations by pests and diseases)	%	<5	5-10	10-15	>15		
Socio-economic	Farm size	ha	>10	10-7	7-4	<4		
	Land tenure (public)	%	<30	30-50	50-75	>75		
	Access to irrigation	%	>75	75-50	50-25	<25		
	Access to credit	%	>50	50-30	30-15	<15		
	Distance to factories and markets more than 5 km	%	>50	50-40	40-30	<30		
	Without technical assistance and training	%	<10	10-20	20-30	>30		
	Without access to human labor	%	<30	30-50	50-70	>70		
Socio-economic	Human Development Index	-	> 0.80	0.80-0.75	0.70-0.75	<0.70		

Table 6. (Continued)

Constraint on supply of raw material to biorefineries or diversification projects		Unit	Value				Weight	Subtotal
			High (1)	Medium (0.75)	Low (0.50)	Very low (0.25)		
	Without access to public services	%	<5	5-10	10-20	>20		
	Other income excluding sugarcane (crop diversification)	%	>50	50-40	40-30	<30		
	Other sugarcane projects	%	>20	10-20	10-5	<5		
Total								

1 = High capacity, 0 = No capacity.

Table 7. Evaluation instrument of sugarcane crop fields by socio-economic and ecological resources and capability level to diversify

Constraint on sugarcane productivity	Value				Weight	Subtotal
Sugarcane agroclimatic land aptitude	High (1)	Medium (0.75)	Low (0.50)	Very low (0.25)	0.381	
	0.381	0.28575	0.1905	0.09525		
Sugarcane yield	>85	84-75	74-65	<65	0.254	
	0.254	0.1905	0.127	0.0635		
Sugarcane farm size (acreage)	>15	15-10	10-5	<5	0.144	
	0.144	0.108	0.072	0.036		
Sugarcane production cycles	Plant	First ratoon	Second ratoon	N ratoon	0.091	
	0.091	0.06825	0.0455	0.02275		
Sugarcane pests and diseases	0	1	2	>3	0.061	
(Aeneolamia, borer, rats, rust)	0.061	0.04575	0.0305	0.01525		
Harvesting systems (green or burned)	0.90		0.10			
	green		burned		0.042	
	0.0378		0.0042			
Land tenure (private, communal or public)	0.90		0.10			
	Private		Public		0.028	
	0.0252		0.0028			
TOTAL						

1 = High capacity, 0 = No capacity.

The developed methodology can be used as a tool to support decision making in the planning processes for the diversification of sugarcane municipalities, sugarcane crops and sugar mills, delimiting the variables, factors and constraints. Subsequent intensive studies are needed, and can be replicated in other settings and enriched through the inclusion of new criteria and constraints, avoiding subjective information derived from the statistics of the sector (and the restrictions of the indicators that have traditionally been used to highlight the national and regional importance of the sugar sector as sugarcane, sucrose and factory yield), and the speculative action to positively manage the controllable factors of sugarcane, sugar and by-product production in a systematic way for diversification and biorefineries.

Roebeling et al. (2006) recommended that for the potential diversification analysis a joint spatial approach of economic and environmental factors is required, limiting factors to optimize future agro-industrial methods of processing as biorefineries projects in combination with management models.

Combining AHP with GIS provides an effective means for the study of regional eco-environmental evaluation as an approach for generating and visualizing criteria weighting (spatial analysis) of diversification and potential for biorefineries because of its capacity to integrate a large amount of heterogeneous data and the ease of obtaining the weights of enormous alternatives (criteria), and therefore, it is applied in a wide variety of decision making problems (Marinoni, 2004).

The selection of evaluation criteria is based on the project objective, spatial scale, and in particular, data availability. MCE in a GIS is primarily concerned with how to combine available information from several criteria to form a single index of evaluation (new information); it therefore becomes one of the most useful methods for spatial planning and management and provides a mechanism to explore the problem while learning how changes in criteria weights affect evaluation outcomes spatially (Rivera, 2014; Sahoo et al., 2007). It permits a range of user and stakeholder defined simulations to be performed to quantitatively evaluate model dynamic changes; it measures the stability of results with respect to the variation of different parameter weights, displays spatial change dynamics and creates more realistic and holistic output scenarios to produce a final suitability map. Because the criteria are measured on different scales, they are standardized and converted at GIS into maps of factors that correlate positively; this, according to the objective of establishing spatial relationships among the options, is in order to develop regional strategies for land use. On this case, we are concerned with the potential for diversification projects inside sugarcane crop fields and transition to biorefineries. (Enough sugarcane for actual processing in sugar mills and distilleries and surplus raw material for other projects as a result of ecological and socio-economic resources and capabilities) (Figures 15 and 16).

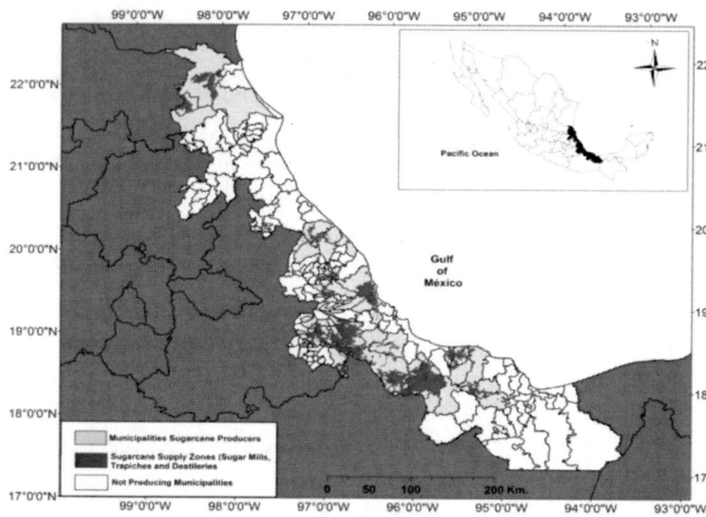

Figure 15. Sugarcane industry in Veracruz, Mexico.

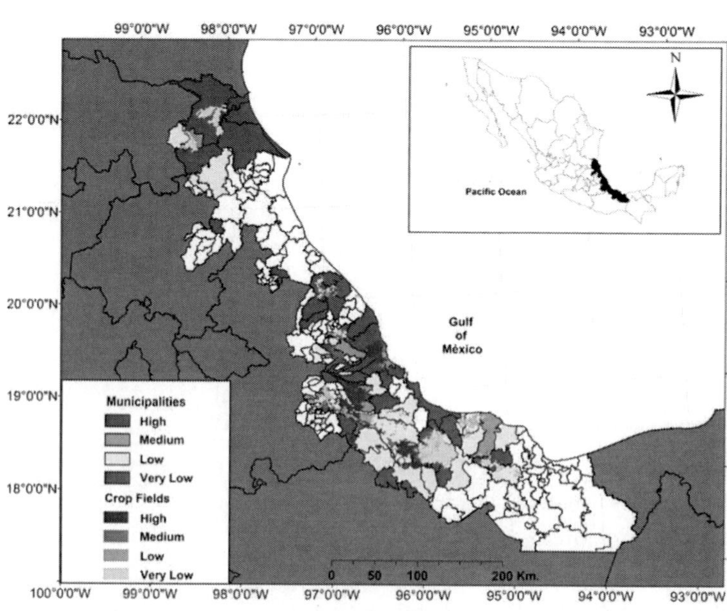

Figure 16. Potential for diversification projects and biorefineries in sugarcane crop fields and municipalities in Veracruz, Mexico.

From the perspective of the suitability map, this management tool is intended to guide decisions for the regional decision-making process of the use of the sugarcane area for the purposes of diversification and transition from sugar mills to biorefineries.

CONCLUSION

The results in IC <0.1 have established that the independent variables used were enough to explain *a priori* the dependent variable "capacity to diversify sugar mills, municipalities and sugarcane crop fields."

This method firstly allowed the generation of a parameter or synthesis through indicators (Diversification Index), an aspect which is an advantage, since it significantly reduces the amount of data to be analyzed. Secondly, the parameter is not simply the aggregation of indicators; therefore, each of them is weighted according to its relative importance and, having real dimensions, it is applicable to any sugar industry worldwide, as an initial process in the transition to biorefineries, first considering the socio-economic and ecological factors limiting the potential to produce raw material in quantity and quality, and secondly considering the technological factors for efficient processing with energetic autonomy.

REFERENCES

Agostinho, F., & Ortega, E. (2013). Energetic-environmental assessment of a scenario for Brazilian cellulosic ethanol. *Journal of Cleaner Production*, *47*, 474-489.

Agostinho, F., Ambrósio, L. A., & Ortega, E. (2010). Assessment of a large watershed in Brazil using Emergy Evaluation and Geographical Information System. *Ecological Modelling*, *221*(8), 1209-1220.

Aguaron, J., & Moreno-Jiménez, J. M. (2003). The geometric consistency index: Approximated thresholds. *European Journal of Operational Research*, *147*(1), 137-145.

Aguilar, N., (2012). A methodological approach to sugar mill diversification and conversion. *Revista Ingeniería e Investigación*, *32*(2):23-27

Aguilar.R. N., D. A. Rodríguez Lagunes, A. Castillo Morán. (2010) Sugar, co-products and by products on the sugar industry diversification. http://www.revistavirtualpro.com/files/ti02_201011.pdf

Aguilar-Rivera, N., M Galindo, G., M Fortanelli, J., & S Contreras, C. (2009). Why diversify the Mexican sugar agro industry?. *GCG: Revista de Globalización, Competitividad & Gobernabilidad (GCG: Magazine of Globalization, Competitiveness and Governance)*, *3*(1): 62-75

Aguilar Rivera, Noé, Galindo Mendoza, Guadalupe, Fortanelli Martínez, Javier, & Contreras Servín, Carlos. (2011). Competitiveness factors of Mexican sugarcane agroindustry. *Región y sociedad (Region and society)*, *23*(52), 261-297.

Amaya A. 2010. Sugarcane research and technology transfer—strategies for the next decade. Proc. Int. Soc. Sugar Cane Technol., Vol. 27, 2010

Achabou, M. A., & Tozanli, S. (2009). The institutional environment in the strategic decision of the emerging companies: the case of the Algerian sugar industry. *Journal of Multi‐Criteria Decision Analysis*, *16*(5-6), 139-148.

Alonso Pippo, W., Garzone, P., & Cornacchia, G. (2007). Agro-industry sugarcane residues disposal: the trends of their conversion into energy carriers in Cuba. *Waste management*, *27*(7), 869-885.

Almazan, O., Gonzalez, L., Galvez, L., Lalouette, J. A., Bachraz, D. Y., & Sukerdeep, N. (1999). The sugar cane, its by-products and co-products. In *Proceedings of the Third Annual Meeting of Agricultural Scientists, Réduit, Mauritius, 17-18 November 1998.* (pp. xiii-xxv). Food and Agricultural Research Council. p. xiii-xxv

Ahumada M.R. Sugar cane Agroindustrial Diagnosis. Proceedings of XXXII Symposium of Mexican Sugar Technicians Association. Córdoba Ver. 27-28 August 2009

Amores, M. J., Mele, F. D., Jiménez, L., & Castells, F. (2013). Life cycle assessment of fuel ethanol from sugarcane in Argentina. *The International Journal of Life Cycle Assessment*, *18*(7), 1344-1357.

Antony G. (2005). Industry transition and sugarcane farm households in Maryborough. The Annual Conference of the Australian Agricultural and Resource Economics Society, 8-11 February 2005, Coffs Harbour 15 p.

Ashfaq, M., Hassan, S., Naseer, M. Z., Baig, I. A., & Asma, J. (2008). Factors affecting farm diversification in rice-wheat. *Pak. J. Agri. Sci*, *45*(3): 91-94

Avram P. B. Morgen. Benchmarking concept for an integrated sugar, ethanol and co-generation plant. (ISSCT); International Society of Sugar Cane Technologists Congress, Vol.2; 0130-0204; 2005, Guatemala City(GT).

Banerjee S. Determinants of International Competitiveness: A Comparative Study of the Sugar Industry in Australia, Brazil, and the European Union. Thesis for Degree of Masters in Business, School of International Business, Queensland University of Technology, Brisbane, Australia, 2004,136 p.

Bandaranaike S. (2005). Crisis and Change: Engaging Rural Communities in the Sugar Industry. http://www.engagingcommunities2005.org /abstracts/Bandaranaike-Suniti-final.pdf.

Bergquist, D. A., Cavalett, O., & Rydberg, T. (2012). Participatory emergy synthesis of integrated food and biofuel production: a case study from Brazil. *Environment, Development and Sustainability*, *14*(2), 167-182.

Berumen, S. A., & Llamazares Redondo, F. (2007). Usefulness of multiple criteria decision methods (such as AHP) in an environment with growing competitiveness. *Cuadernos de Administración (Books of Administration)*, *20*(34), 65-87.

Bojórquez, L.A. ANP modeling of complex socio-environmental systems: adaptive capacity of smallholder coffee system in Mesoamerica. Proceedings of the International Symposium on the Analytic Hierarchy Process 2009, 1-10 p.

Bosma, U., & Knight, R. (2004). Global factory and local field: Convergence and divergence in the international cane-sugar industry, 1850-1940. *International Review of Social History*, *49*(1), 1-25.

Brumbley, S. M., Purnell, M. P., Petrasovits, L. A., Nielsen, L. K., & Twine, P. H. (2007). Developing the sugarcane biofactory for high-value biomaterials. *International sugar journal*, *109*(1297).

Castellanos Abella, E. C., & Van Westen, C. J. (2007). Generation of a landslide risk index map for Cuba using spatial multi-criteria evaluation. *Landslides*, *4*(4), 311-325.

Chaplin H. (2000). Agricultural diversification: a review of methodological approaches and empirical evidence. Department of Agricultural Economics and Business Management, Wye College, University of London. Working paper 2:70.

Chauhan, M. K., Chaudhary, S., & Kumar, S. (2011). Life cycle assessment of sugar industry: A review. *Renewable and Sustainable Energy Reviews*, *15*(7), 3445-3453.

Cheavegatti-Gianotto, A., de Abreu, H. M. C., Arruda, P., Bespalhok Filho, J. C., Burnquist, W. L., Creste, S., & Ulian, E. C. (2011). Sugarcane (Saccharum X officinarum): a reference study for the regulation of genetically modified cultivars in Brazil. *Tropical plant biology*, *4*(1), 62-89.

Cherubini, F., & Strømman, A. H. (2011). Life cycle assessment of bioenergy systems: state of the art and future challenges. *Bioresource technology*, *102*(2), 437-451.

Contreras, A. M., Rosa, E., Pérez, M., Van Langenhove, H., & Dewulf, J. (2009). Comparative life cycle assessment of four alternatives for using by-products of cane sugar production. *Journal of Cleaner Production*, *17*(8), 772-779.

Corrêa do Lago, A., Bonomi, A., Cavalett, O., Pereira da Cunha, M., & Pinheiro Lima, M. A. (2012). Sugarcane as a carbon source: The Brazilian case. *Biomass and Bioenergy*, 46, 5-12.

Cornland, D. W., Johnson, F. X., Yamba, F., Chidumayo, E. N., Morales, M. M., Kalumiana, O., & Mtonga-Chidumayo, S. B. (2001). Sugarcane resources for sustainable development: a case study in Luena, Zambia. *Stockholm Environment Institute, Stockholm*.

Dale, V. H., Efroymson, R. A., Kline, K. L., Langholtz, M. H., Leiby, P. N., Oladosu, G. A., & Hilliard, M. R. (2013). Indicators for assessing socioeconomic sustainability of bioenergy systems: a short list of practical measures. *Ecological Indicators*, *26*, 87-102.

Dale, V.H., Beyeler, S.C., 2001. Challenges in the development and use of ecological indicators. Ecol. Indic. 1, 3–10

De Oliveira Bordonal, R., Barretto de Figueiredo, E., Aguiar, D. A., Adami, M., Theodor Rudorff, B. F., & La Scala, N. (2013). Greenhouse gas mitigation potential from green harvested sugarcane scenarios in São Paulo State, Brazil. *Biomass and Bioenergy*, *59*, 195-207.

D'Hont D.A., G. Mendes Souza, M. Menossi. Chapter 21 Sugarcane: A Major Source of Sweetness, Alcohol, and Bio-energy Springer, Moore, R. Ming (eds.), *Genomics of Tropical Crop Plants,* 2008, pp. *403-513*

Diogo F. P., T. D. P. Lucas, F. B. de Lima Neto. How to Obtain Fair Managerial Decisions in Sugar Cane Harvest Using NSGA-II, 2007, pp.186-191, 7th International Conference on Hybrid Intelligent Systems.

Domac, J., K. Richards, S. Risovic. (2005). Socio-economic drivers in implementing bioenergy projects. *Biomass & Bioenergy*. 28: 95-266

Doré, T., Clermont-Dauphin, C., Crozat, Y., David, C., Jeuffroy, M. H., Loyce, C., & Valantin-Morison, M. (2009). Methodological progress in on-farm regional agronomic diagnosis: A review. In *Sustainable Agriculture* (pp. 739-752). Springer Netherlands.

Duarte, C. G., Gaudreau, K., Gibson, R. B., & Malheiros, T. F. (2013). Sustainability assessment of sugarcane-ethanol production in Brazil: A case study of a sugarcane mill in São Paulo state. *Ecological Indicators*, *30*, 119-129.

Eakin, H., Bojórquez-Tapia, L. A., Diaz, R. M., Castellanos, E., & Haggar, J. (2011). Adaptive capacity and social-environmental change: theoretical and operational modeling of smallholder coffee systems response in Mesoamerican Pacific Rim. *Environmental management*, *47*(3), 352-367.

Edye, L. A., Doherty, W. O. S., Blinco, J. A., & Bullock, G. E. (2006). The sugarcane biorefinery: Energy crops and processes for the production of liquid fuels and renewable commodity chemicals. *International Sugar Journal*, *108*(1285), 19-27.

Efroymson, R. A., Dale, V. H., Kline, K. L., McBride, A. C., Bielicki, J. M., Smith, R. L., & Shaw, D. M. (2013). Environmental indicators of biofuel sustainability: what about context?. *Environmental management*, *51*(2), 291-306.

Evangelinos, K. I., Allan, S., Jones, K., & Nikolaou, I. E. (2014). Environmental management practices and engineering science: A review and typology for future research. *Integrated environmental assessment and management*. 10(2):153–162

Escobal J. (2004). The determinants of diversification nonfarm income in rural Peru http://ftp.fao.org/docrep/fao/009 /ah500s/ah500s01.pdf

Falconí, F., & Burbano, R. (2004). Economic instruments for environmental management: monocriteriales versus multicriteria decision making. *Revista Iberoamericana de Economía Ecológica (REVIBEC) (Magazine Iberoamericana for Ecological Economics)*, (1), 11-20.

FAOSTAT (2013). http://faostat.fao.org/site/567/default.aspx#ancor

FOLICHTS 2012. World sugar report 1997-2011. 372 p.

Franks, J. R., & Hadingham, B. (2012). Reducing greenhouse gas emissions from agriculture: avoiding trivial solutions to a global problem. *Land Use Policy*, *29*(4), 727-736.

Freitas, A.; Marins, C.; Souza, D. (2006). A multicriteria methodology as a tool for decision-making: a case study. *Revista GEPROS - Gestão da Produção, Operações e Sistemas* 2(3):52-60.

Gálvez T.L. Sugarcane by products handbook. Cuban Research Institute for Derivatives Sugarcane, La Habana Cuba. 3era edición, 2000, 458 p.

Gallardo, L. F., Riestra, D. D., & Aluja, S. A y DJP Martínez. (2002). Factors determining agricultural diversity and production objectives for agroecosystems in the municipality Paso de ovejas, Veracruz, México *Agrociencia*, *36*(004), 495-502.

Gerbens-Leenes, W., & Hoekstra, A. Y. (2012). The water footprint of sweeteners and bio-ethanol. *Environment international*, *40*, 202-211.

Godoy, R., & Bennett, C. (1988). Diversification among coffee smallholders in the Highlands of South Sumatra, Indonesia. *Human Ecology*, *16*(4), 397-420.

Gómez-Merino, F. C., Trejo-Téllez, L. I., & Sentíes-Herrera, H. E. (2014). Sugarcane as a Novel Biofactory: Potentialities and Challenges. In *Biosystems Engineering: Biofactories for Food Production in the Century XXI* (pp. 129-149). Springer International Publishing.

Gómez D. M. Barredo C. J.I., GIS and multicriteria evaluation in regional planning. "2da edición. Edit. Alfa-Omega Ra-Ma. México, 2006, 279 p.

Hahn-Hägerdal, B., Galbe, M., Gorwa-Grauslund, M. F., Lidén, G., & Zacchi, G. (2006). Bio-ethanol–the fuel of tomorrow from the residues of today. *Trends in biotechnology*, *24*(12), 549-556.

Haque T. (2010). Constraints and Potentials of Diversified Agricultural Development in Eastern India. Planning Commission (Government of India) 195 p. http://planningcommission.nic.in/reports/sereport /ser/ser_agridiv1102.pdf

Hilmola O.P., J. Saranen, F. Padilha (2010). Location Criteria and Strategic Factors of Biodiesel Factory Establishment in Finnish Context, Lappeenranta University of Technology Department of Industrial Management Kouvola Research Unit Research Report 219, 66 p.

ICIDCA. (1986). *The sugarcane by products industry*. Editorial Científico-Técnica. 576 pp.

International Sugar Organization. An International Survey of Sugar Crop Yields and Prices Paid for Sugar Cane and Beet. Market evaluation consumption and Mecas (05)05 Statistics Committee 2005, 49 p.

Jamali-Zghal, N., Amponsah, N. Y., Lacarriere, B., Le Corre, O., & Feidt, M. (2013). Carbon footprint and emergy combination for eco-environmental assessment of cleaner heat production. *Journal of Cleaner Production*, *47*, 446-456.

Johnson, F. X., Woods, J., Rosillo-Calle, F., Brown, G., Seebaluck, V., Watson, H., Sobhanbabu, P. R. K. (2007). Cane energy for sustainable development and economic competitiveness in Southern Africa. In *XXVI Congress, International Society of Sugar Cane Technologists, ICC, Durban, South Africa, 29 July-2 August, 2007.* (pp. 1246-1256). International Society Sugar Cane Technologists (ISSCT).

Kamruzzaman M., M. Hasanuzzaman (2007). Factors affecting profitability of sugarcane production as monoculture and as intercrop in selected areas of Bangladesh. *Bangladesh J. Agril. Res.*, 32(3):433-444

Langeveld, J., Dixon, J., & Darnhofer, I. (2010). Development perspectives of the biobased economy: The need for a system approach. In *Building sustainable rural futures: the added value of systems approaches in times of change and uncertainty. 9th European IFSA Symposium, Vienna, Austria, 4-7 July 2010.* (pp. 1530-1537). BOKU-University of Natural Resources and Applied Life Sciences.

Leal, M. R. L., Walter, A. S., & Seabra, J. E. (2013). Sugarcane as an energy source. *Biomass Conversion and Biorefinery*, *3*(1), 17-26.

Lee, A.H.I., Chen, W.C., Chang, C.J., (2008). A fuzzy AHP and BSC approach for evaluating performance of IT department in the manufacturing industry in Taiwan. Expert Systems with Applications 34, pp. 96–107.

Lodi E. O Novo Ciclo da Cana: Study on the sugarcane Agro-industrial system Competitiveness of the and Exploration New Ventures. Brasília: IEL/NC; SEBRAE, 2005, 337 p

Luo, L., Van Der Voet, E., & Huppes, G. (2009). Life cycle assessment and life cycle costing of bioethanol from sugarcane in Brazil. *Renewable and Sustainable Energy Reviews*, *13*(6), 1613-1619.

Junqueira, C. D. Á. R., Molina Junior, V. E., Lossardo, L. F., Felicio, B. D. C., Moreira Junior, O., Foschini, R. C& Lorandi, R. (2009). Identification of potential contamination of aquifers by vinasses in Ribeirão (SP), Brazil. *Revista Brasileira de Geociências*, *39*(3), 507-518.

Lamparelli, R. A., & Dias, H. S. (2009). Methodological development of a quality index for agricultural operations, for corn cultivation using multi-criteria analysis. *Anais da Academia Brasileira de Ciências*, *81*(4), 849-859.

Lapola, D. M., Schaldach, R., Alcamo, J., Bondeau, A., Koch, J., Koelking, C., & Priess, J. A. (2010). Indirect land-use changes can overcome carbon savings from biofuels in Brazil. *Proceedings of the national Academy of Sciences*, *107*(8), 3388-3393.

Marin, F. R., Jones, J. W., Singels, A., Royce, F., Assad, E. D., Pellegrino, G. Q., & Justino, F. (2013). Climate change impacts on sugarcane attainable yield in Southern Brazil. *Climatic Change*, *117*(1-2), 227-239.

Marinoni, O. (2004), Implementation of the analytical hierarchy process with VBA in ArcGIS. *Computers and Geosciences*, 30(6), 637-646.

Mata, T. M., Caetano, N. S., Costa, C. A., Sikdar, S. K., & Martins, A. A. (2013). Sustainability analysis of biofuels through the supply chain using indicators. *Sustainable Energy Technologies and Assessments*, *3*, 53-60.

Menon, V., & Rao, M. (2012). Trends in bioconversion of lignocellulose: biofuels, platform chemicals & biorefinery concept. *Progress in Energy and Combustion Science*, *38*(4), 522-550.

Mishra A. & Hisham, El-Osta (2004). Factors Affecting Farm Enterprise Diversification. *Agricultural Finance Review, Fall* pp.151-166

Moore P. (2009). Sugarcane Biology, Yield, and Potential for Improvement http://www.fapesp.br/pdf/bioen 1803/Bioen_Moore.pdf

Morandini Paoliello J.M. Environmental aspects and energy potential in waste recovery of the sugarcane industry, master's Thesis. Universidade Estadual Paulista. Faculdade de Engenharia, Bauru, 2006,180 p.

Muchow R.C., A. J. Higgins, N.G. Inman-Bamber and P. J. Thorburn. 2001. Towards sustainable sugarcane production using a whole systems analysis approach. Proc. Int. Soc. Suaar Cane Technol., 24: 91-94

Nagesha, N., & Balachandra, P. (2006). Barriers to energy efficiency in small industry clusters: multi-criteria-based prioritization using the analytic hierarchy process. *Energy*, *31*(12), 1969-1983.

Naraine L. (2005). An Examination of Factors Influencing Policy Design and Implementation of Non-Sugar Agricultural Diversification in St. Kitts. Sir Arthur Lewis Institute of Social and Economic Research University of the West Indies 53 p.

Nothard, B. W., Ortmann, G. F., & Meyer, E. (2005). Attributes of small-scale sugarcane contractors that influence their service quality in KwaZulu-Natal. *Agrekon*, *44*(3), 402-422.

Oddershede, A., Arias, A., & Cancino, H. (2007). Rural development decision support using the analytic hierarchy process. *Mathematical and Computer Modelling*, *46*(7), 1107-1114.

Oliveira, D. B. B., Rodrigues, J. P., da Silva, L. F., & Oliveira, P. T. S. (2012). Multi-criteria analysis in the strategic environmental assessment of the sugar and alcohol sector. *Acta Scientiarum: Technology*, *34*(3).

Papalexandrou, M. A., Pilavachi, P. A., & Chatzimouratidis, A. I. (2008). Evaluation of liquid bio-fuels using the Analytic Hierarchy Process. *Process Safety and Environmental Protection*, *86*(5), 360-374.

Paturau, M. J. By-products of the cane sugar industry can introduction to their industrialization. 3ª. Edition. Elsevier Science Publishing Company Inc. New Cork City, N. Y., 1989, 436 p.

Pellegrini, L. F., & de Oliveira Junior, S. (2011). Combined production of sugar, ethanol and electricity: thermoeconomic and environmental analysis and optimization. *Energy*, *36*(6), 3704-3715.

Pérez, Abel (2007). Land tenure and sugar industry. Edit. Porrua, México 214 p.

Pereira, L., & Ortega, E. (2012). A modified footprint method: The case study of Brazil. *Ecological Indicators*, *16*, 113-127.

Pereira, C. L., & Ortega, E. (2010). Sustainability assessment of large-scale ethanol production from sugarcane. *Journal of Cleaner Production*, *18*(1), 77-82.

Pope, R. D., & Prescott, R. (1980). Diversification in relation to farm size and other socioeconomic characteristics. *American Journal of Agricultural Economics*, *62*(3), 554-559.

Ramanujam, V., & Varadarajan, P. (1989). Research on corporate diversification: A synthesis. *Strategic management journal*, *10*(6), 523-551.

Roebeling, P. C., Smith, D. M., & van Grieken, M. (2006). Exploring environmental-economic benefits from agri-industrial diversification in the sugar industry: an integrated land use and value chain approach. In *2006 Annual Meeting, August 12-18, 2006, Queensland, Australia* (No. 25755). International Association of Agricultural Economists, 16 p.

Sabatier, D., Martiné, J. F., Chiroleu, F., Roussel, C., Letourmy, P., Antwerpen, R., & Ney, B. (2014). Optimization of sugarcane farming as a multipurpose crop for energy and food production. *GCB Bioenergy*. doi: 10.1111/gcbb.12133

Sahoo, N. R., Jothimani, P., & Tripathy, G. K. (2007). Multi-criteria analysis in GIS environment for natural resource development-a case study on gold exploration. *Online on: http://www.gisdevelopment.net /application/geology/mineral/geom0002. htm.*

Singh, R. K., Murty, H. R., Gupta, S. K., & Dikshit, A. K. (2012). An overview of sustainability assessment methodologies. *Ecological Indicators*, 15(1), 281-299.

Singels, A., Jones, M., Marin, F., Ruane, A., & Thorburn, P. (2013). Predicting climate change impacts on sugarcane production at sites in Australia, Brazil and South Africa using the Canegro model. *Sugar Tech*, 1-9.

Solomon S. (2000). Sugar production in India by 2000 ad. 1. Constraints and strategies for increasing production and production efficiency. Sugarcane: Research Towards Efficient and Sustainable Production. En. Wilson JR, Hogarth DM, Campbell JA and Garside AL (Eds). CSIRO Division of Tropical Crops and Pastures, Brisbane. pp. 9-11

Summer, D. A., & Wolf, C. A. (2002). Diversification, vertical integration, and the regional pattern of dairy farm size. *Review of Agricultural Economics*, *24*(2), 442-457.

Quintero, J. A., Montoya, M. I., Sánchez, O. J., Giraldo, O. H., & Cardona, C. A. (2008). Fuel ethanol production from sugarcane and corn: comparative analysis for a Colombian case. *Energy*, *33*(3), 385-399.

Qureshi M.E. Application of the Analytic Hierarchy Process to Riparian Revegetation Policy Options 2003, http://espace.library.uq.edu.au/eserv/UQ:8658/n09_Ejaz.pdf

Renouf, M. A., Pagan, R. J., & Wegener, M. K. (2011). Life cycle assessment of Australian sugarcane products with a focus on cane processing. *The International Journal of Life Cycle Assessment*, *16*(2), 125-137.

Renouf, M. A., & Wegener, M. K. (2007). Environmental life cycle assessment (LCA) of sugarcane production and processing in Australia. In *Proceedings of the Australian Society of Sugar Cane Technologists* 29, pp. 385-400).

Rivera, N. A. (2014). Management of Limiting Factors for Diversifying the Sugar Agro-industry. *Revista Venezolana de Gerencia (Venezuelan Magazine of Management)*, 19(65), 66-91.

Saaty, T. L. (1990). How to make a decision: the analytic hierarchy process. *European journal of operational research*, *48*(1), 9-26.

Seebaluck, V., Mohee, R., Sobhanbabu, P. R. K., Rosillo-Calle, F., Leal, M. R. L. V., & Johnson, F. X. (2008). Bioenergy for sustainable development and global competitiveness: the case of sugar cane in Southern Africa. *Thematic Report*, *2*.

Silva JR. A. G. da, Perez, R., Martins, E.L. Analysis of investment in unit production of vegetable oil using the methodology "Analytic Hierarchy Process" with "Expert Choice" software In. 7 Congresso Brasileiro de Agroinformática. Viçosa – MG. 2009, pp. 21-25

Sipahi, S., & Timor, M. (2010). The analytic hierarchy process and analytic network process: an overview of applications. *Management Decision*, *48*(5), 775-808.

Solomon, S. (2011). Sugarcane By-Products Based Industries in India. *Sugar Tech*, *13*(4), 408-416.

Singh,G.B. and Solomon,S. (1995).Sugarcane: Agro-Industrial Alternatives. Oxford & IBH Publishing Co. New Delhi. 556 pp.

Tienwong, K., Dasananda, S., & Navanugraha, C. (2009). Integration of land evaluation and the analytical hierarchical process method for energy crops in Kanchanaburi, Thailand. *Sci. Asia*, *35*, 170-177.

Turcksin L. A Multi-Actor Multi-Criteria analysis to assess the stakeholder support for different biofuel options (2010). http://wwwa.vito.be/bioses/pdf/BIOSES-MAMCA.pdf

Trujillo Murcia E, L. R. García Chávez. (2002). An analysis of efficiency in the Mexican sugar industry in the public and private stages. *Prospectiva Económica (Economic Prospective)*, 1:137-150

Tsiropoulos, I., Faaij, A. P., Seabra, J. E., Lundquist, L., Schenker, U., Briois, J. F., & Patel, M. K. (2014). Life cycle assessment of sugarcane ethanol production in India in comparison to Brazil. *The International Journal of Life Cycle Assessment*, *19*(5), 1049-1067.

Vlosky R. P., F. X. Aguilar, Qinglin Wu. 2005. Demographic profile and spatial analysis of sugarcane growers in Louisiana. Journal American Society Sugar Cane Technologists. 25: 157-172

Waclawovsky AJ, Sato PM, Lembke CG, Moore PH, Souza GM (2010) Sugarcane for bioenergy production: an assessment of yield and regulation of sucrose content. Plant Biotechnol J 8:263–276.

Weiss C. and Wolfgang Briglauer 2002. Determinants and Dynamics of Farm Diversification. Xth EAAE Congress 'Exploring Diversity in the European Agri -Food System', Zaragoza (Spain), 28-31 August 2002, 15 p. http://ageconsearch.umn.edu/bitstream/24929/1/cp02we35.pdf

Wood, A. W. (1985). Soil degradation and management under intensive sugarcane cultivation in North Queensland. *Soil Use and Management*, *1*(4), 120-124.

Wynne A.T., B.J. Milford, E.S. Wallis. (2010) Report on the second ISSCT management workshop, may 2008 Proc. Int. Soc. Sugar Cane Technol., Vol. 27,

Yadav, R. L., & Solomon, S. (2006). Potential of developing sugarcane by-product based industries in India. *Sugar Tech*, *8*(2), 104-111.

Zhang, X., Izaurralde, R. C., Manowitz, D., West, T. O., Post, W. M., Thomson, A. M., & Williams, J. R. (2010). An integrative modeling framework to evaluate the productivity and sustainability of biofuel crop production systems. *GCB Bioenergy*, *2*(5), 258-277

Zhong Min, Tang Yao-Ping. economic benefit evaluation for agricultural products based on analytic hierarchy process, (2009), http://en.cnki.com.cn/article_en/cjfdtotal-hnnk200909041.htm

Zimmermann, B., & Zeddies, J. (2002, July). International competitiveness of sugar production. In *13th International Farm Management Congress, Wageningen, The Netherlands*. http://ageconsearch.umn.edu/bitstream /7000/2/cp02zi02.pdf

In: Sugarcane
Editor: Eleanore Webb

ISBN: 978-1-61122-356-9

Chapter 9

INTEGRATED MANAGEMENT OF SUGARCANE (IMS): USE OF AGRICULTURAL RESIDUE

Adriana C. Lozano*, Maritza Correa and Gloria M. Lopez
Universidad Autonoma de Occidente
Facultad de Ingenieria, Cali, Colombia

ABSTRACT

In this chapter a system of integrated harvest for sugarcane is put forward, with the inclusion of the agricultural residue. This is based on the need of sugar manufacturing companies to obtain new renewable energy alternatives, where the agricultural residue from sugarcane (ARS) can be used as a raw material for this purpose. However, the high costs associated with obtaining it, make refineries resort to burning practices, thereby wasting a potential energy resource.[1] [2].

The system model we developed allows the efficient flow of the operation of the whole crop to be obtained, reducing the management costs of ARS with a major improvement in productivity in sugar manufacturing: linking the activities of collection, transport, and separation of sugarcane and residues in an integrated manner, thereby generating appropriate solution to the local logistic needs in Valle del

* [aclozano; mcorrea; gmlopez]@uao.edu.co

Cauca state, the major sugarcane producer of Colombia, South America. It also allowed us to observe the performance of technological tools to ensure compliance with the demands of the sugarcane industry while at the same time obtain the agricultural residue to generate new products.

INTRODUCTION

At a global scale, for alternative sources of renewable energy has been driven by both negative impacts and effects on ecosystems, due to an overdependence and use of fossil fuels as in typical economical models of developed countries. The most relevant effects have been identified as (i) the volatility of oil prices, (ii) the continued increase of consumption, (iii) the depletion of the resource, and (iv) the generation and release of pollutants into the atmosphere [1].

One of the alternative sources of renewable energy is second generation biofuels coming from waste products from plants of an agricultural origin. This chapter deals with the analysis of agricultural sugarcane residue as a raw material for the production of second-generation bio-fuel materials. This can reduce the impact caused by first generation biofuels, thus contributing to the economic and social development of developing countries.

In Colombia, the Valle del Cauca state has the highest production of sugarcane, (i.e., 80% of the total cultivated area in the country) and is where most refineries that process sugarcane for molasses, sugars and biofuels are concentrated. [2]. According a recent the report [3] of the 200,000 ha cultivated with sugarcane, 85% are harvested in a semi-mechanized way, (i.e., manual cutting is performed with lifting done by mechanical tools). In addition, only 15% is harvested out in a fully mechanized way which results in the production of residue from agricultural sugarcane – ARS - of between 50-150 ton / ha [4]. This represents an opportunity to use ARS as: (a) a raw a material in the framework for the diversification of industrial products, and (b) social recognition in the development of environmental responsibility sugar refineries.

In view of the opportunities offered by the use of ARS, different authors [3] [4] [5] [6] have proposed and evaluated many alternative techniques for its collection. These have included: harvest integration, balers, a grinder, technological adaptations for sugarcane harvesters, and cotton presses. Current research has focused on selecting the best system based on the viability of energy of the residue, leaving aside many important factors like the integration of logistic activities involved into obtaining the maximum yield of ARS product.

In this chapter we present the design of a logistics system that was based on an integrated harvest with a cleaning center for separation of sugarcane from the ARS product. This process was identified as the best alternative with the observed parameters: soil conditions, density of the ARS, percentage of foreign matter and collection capacity per hour [7] [8].

The work was developed in three stages. First, the activities involved in the sugarcane supply-chain were determined under the scheme and principles of the IDEF0 methodology, in order to visualize the elements involved in the generation and disposal of agricultural residues from sugarcane. Second, a logistics system simulation was performed with the software PROMODEL to generate an efficient design of collection, transport, and separation of waste, considering (i) the variability in the time of each activity, (ii) strong interrelation of technical resources (i.e., harvesters, sugarcane road trains, and wagons), (iii) operating costs generated from each unit of technical resource required within a time frame (day), without losing sight of the priority targets in the daily supply of sugarcane. Third, the costs associated with the activities of collection, transport, and separation of the sugarcane and ARS product were established.

1. Description

In order to represent the activities involved in the management of sugarcane, a methodology called Integration Definition for Function Modeling was used, this is commonly known as IDEF0. This allows one to create, analyze, and evaluate various systems, via diagrams represented hierarchically by boxes and arrows that relate and model elements in a clear and accurate manner [9]. To develop the model, we began with the description of the processes involved in the production of sugar, it was based on information obtained during field visits to the pilot refinery and relevant literature [10] [11] [12] [13].

The macro processes of the Integral Management System of Sugar-cane (IMSS), were established in three production stages based on the commonality of their objectives and the elements of the supply-chain of sugarcane:

1.1. Field Management

In this set of subprocesses, the activities that control the conditions for growing and generation of sugarcane are performed. These activities include adaptation, preparation, planting, and lifting the soil.

- Preparation and soil adaptation: Activities related to leveling and preparing the soil before planting, guaranteeing the availability of water, aeration, drainage and nutriments [14].
- Planting: This activity is carried out to renew the crop. Seed that enters the process must have been cut with an age of between 7 and 9 months and come from healthy crops. [13].
- Raising the crop: An activity that supports good germination of the sugarcane, from irrigation to the application of ripeners, amongst others.

In breaking down the field management process we identified input elements, restrictions, and the flow of information and resources necessary to provide sugarcane to the next step in optimal conditions for its harvest.

1.2. Management of the Harvest

This stage the activities comprises necessary to supply sugarcane to refineries for the production of sugar.

- Ripening process and pre-harvest: planning and defining the stages for harvesting[1] based on the conditions of the crop, (i.e., the degree of tipping, ripening and age of the sugarcane) as well as on environmental conditions. In addition, programming the plots to harvest, considering the availability of equipment and human resources.
- Cutting the sugarcane: This depends directly on the crop stage that was programmed in the previous step and on the results of meteorological studies. The products coming out of this stage are chopped -or long- sugarcane depending on the type of cut, and sugarcane residues that are left to germination for subsequent use.

[1] Combination type of cut and state of sugarcane.

When the product of the process is chopped sugarcane, it continues its journey to the shipment stage: this does not happen with long cane.

- Lifting the sugarcane: Once the cane is cut by hand, it is organized by the cutters into stacks, in order to make the collection by the loading machine onto the sugarcane trains more efficient.
- Transport of the sugarcane: Wagons that have been loaded are taken to one side of the lot to continue the operation of the chain, which involves gathering and hooking three to four wagons to a sugarcane train for the subsequent transfer of the sugarcane forecourt in the pilot refinery.

1.3. Production Management

At this stage, activities and operations for the generation of sugar and its derivatives are carried out.

- Weighing: The cane from the field is weighed using electronic weighing scales. Reports resulting from this activity are essential for the transformation process.
- Sampling: Once the load has been weighed, a sample is obtained by a mechanical claw to determine the percentage of impurities in the cane. (i.e., the agricultural residue that was transported with the cane).
- Production: The cane is dropped onto the tables using a wire crane. Once on the table, then the sugarcane goes into the process of production of sugar and its derivatives.

A key element in the production step is the refining rate. (i.e., the number of tons of sugarcane entering the system to comply with programming and the production budget). The results of the impurities report obtained through samplins is crucial to assure the quality of the products derived from sugarcane.

After breaking up the Integrated Sugarcane Management System (ISMS) into its constitute parts, we identified the starting point for the development of a logistic model of the harvest. This begins in the Harvest Management (A2), where the ARS is generated after cutting; the characteristics and quantity vary based on the harvesting system used. Figure 1 shows the breakdown of activities involved [15].

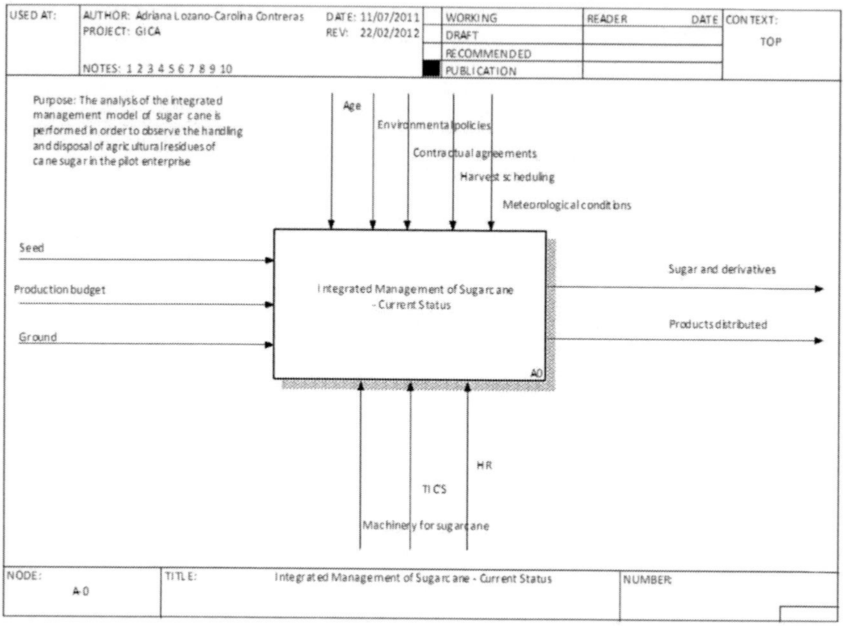

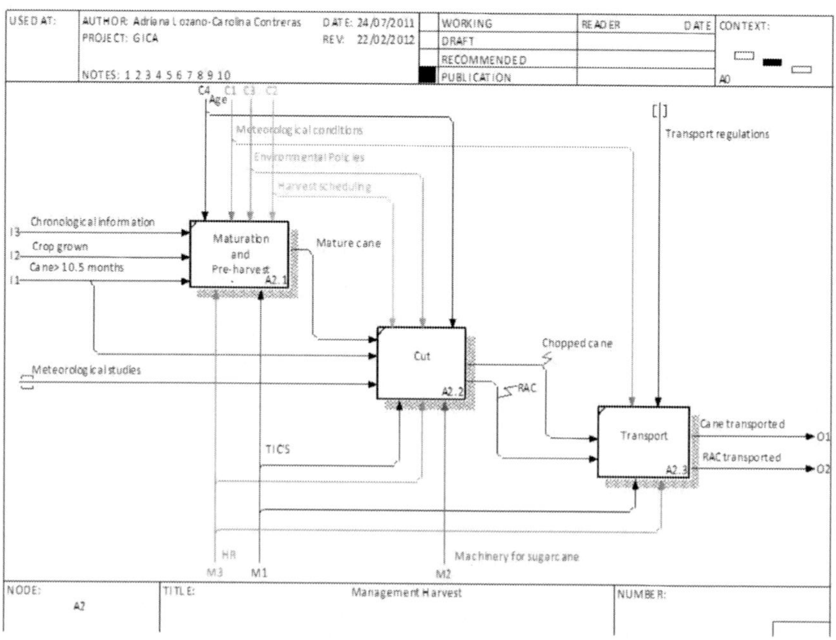

Figure 1. Breakdown of the Integrated Sugarcane Management System (ISMS). (LOZANO & CONTRERAS, 2012).

Using a holistic view, we determinate the starting point for the generation and disposal of ARS as usable raw material within the framework of product diversification and environmental responsibility of the sugar refineries.

Based on the elements characterized in the processes of the ISMS system, and the analysis of alternative harvesting of agricultural residues by the motricity vs dependency matrix, we obtained indicators to evaluate harvesting options, these are [16]:

- Soil conditions: Suitable characteristics for the germination of the stock.
- ARS Density: The quantity of ARS contained in a determined volume, after harvesting.
- Foreign matter: Impurity volume in the collected residuals of agricultural sugarcane.
- Recoverable ARS: The volume of farmed sugarcane residue that can be recovered using harvesting technology measured in tones per hour.

Table 1 shows the results obtained by different authors, using variables with which harvesting alternatives will be evaluated.

Table 1. Summary of elements evaluated

Criteria \ Alternative	Baling	Integrated harvest/cleaning center	Residue Grinder (fodder)/ in bulk	Adapted Harvester	Authors
Condition of the Soil	↓	↑	↓	↓	(MARCHI, S. PIZZINATO, Da ROCHA, & AZEVEDO RAMOS Da SILVA, 2005) (TORRES & VILLEGAS, 2006),
ARS Density (kg/m^3)	231	466 a 333	96	86,4	(AZEVEDO R. DA SIVA , 1998) (RIPOLI & RIPOLI, 2010)
Foreign Matter (%)	7,42	1,39	4,5	-	(RIPOLI & RIPOLI, 2010)

Table 1. (Continued)

Criteria \ Alternative	Baling	Integrated harvest/cleaning center	Residue Grinder (fodder)/ in bulk	Adapted Harvester	Authors
Recoverable ARS (ton/h)	11,5	28	6	-	(MICHELAZZO & BRAUNBECK, 2008)
	13,5	-	5,7	-	(AZEVEDO R. DA SIVA , 1998)
	11,27	-	7,9	-	(RIPOLI & RIPOLI, 2010)

The condition of the soil parameter is referenced to [12]. As mentioned in their study, leaving all the residue in the field causes rot in the roots of the stock and reduced germination. However, Pellegrino & La SCALA cited by [17], concurred that the removal of the majority of the residue can cause problems to the soil, including erosion and land degradation. For this reason, we investigated the appropriate percentage of residue to be left in the field which should be around 50% according to [18].

In this regard, we observed that the soil conditions are favorable for the case of an integrated harvest as they meet the recommendations of the authors cited above, and are unfavorable for the other possible alternatives. For the baler and cutter, this percentage is very low, while with an adapted harvester the percentage approaches zero.

Moreover, in a study of economic, logistical, and energy aspects in the use of ARS, [4] found that the integrated harvesting system is most advantageous. This is due to the fact that a low percentage of foreign matter is obtained from the ARS, compared to other systems presented in the table above. In addition, an integrated harvest system allows effective performance of the harvester and optima charge density also are obtained, which varies from 466-333 kg/m3. Actual lower costs of producing ARS (U.S. $ / ton) which makes it more attractive, (Figure 2).

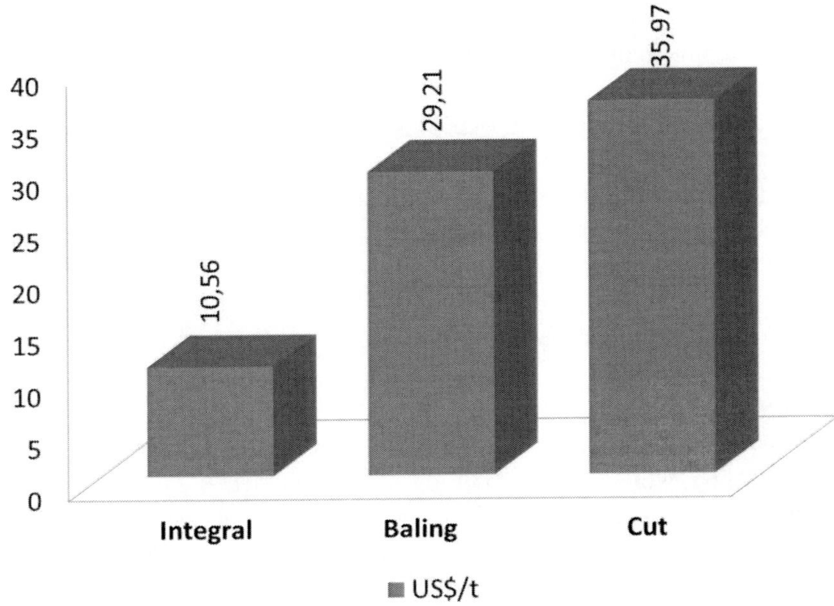

Figure 2. Graph of actual costs at the sugar refinery (RIPOLI & RIPOLI, 2010).

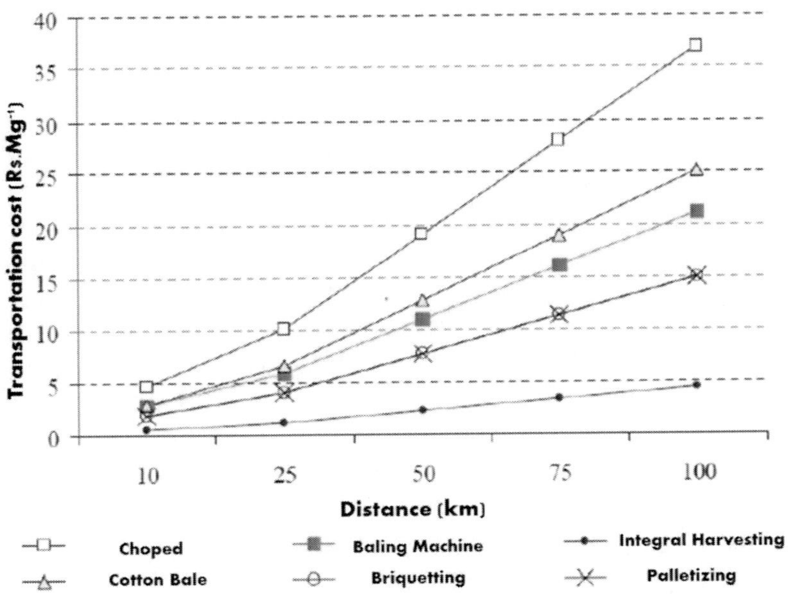

Figure 3. Graphic representing the cost of transport to the factory of ARS depending on distance. (MICHELAZZO & BRAUNBECK, 2008).

Meanwhile, in the comparative analysis of the six systems of residue collection (baling, bulk grinder residue, briquetting, pelleting, cotton baling and integrated harvest) [3], concludes that the integrated harvest system is the best choice because it involves the most efficient method for transporting ARS and also the lowest total cost. Furthermore, it states that the transportation costs to the factory of ARS using integrated harvest have less variation over different distances which do not occur with the other systems tested, (Figure 3).

Given the above, the model of the logistic system for ARS collection will be based on the integral harvest system, using a dry cleaning center.

ARS Collection System: Integrated Harvest - Dry Cleaning Centre

This system has been widely used and researched in countries such as Brazil and Cuba, which consider cutting of the chopped sugarcane, in burned and unburned fields, with the collection of all residual biomass. This is done using a harvester with the primary and secondary extractors turned off allowing at least 50% of the ARS to be left in the field and the rest to be transported to the factory. Once at the factory the separation of the sugarcane residue takes place in a dry cleaning center [19] [20].

According to [19], the many positive and negative effects of this activity in the field, for the harvest and the factory are:

- Effects in the field: The percentage of ARS left in the field provides agronomic benefits: control of weeds, reducing cultivation operations, increasing operational capacity, and reducing internal traffic routes.
- Effects on harvest: Decreased loss of sugarcane of about 1.5 ton / ha, fuel consumption is reduced by 0.12 in the harvest to l / t, increased operational capacity of harvesters going from 23.6 to 14.8 min collection time. Additionally the integrated harvest has the advantage of using of the same sugarcane transport system, so that the existing logistics system of the factory are used [6].
- Effects on the factory: With the implementation of an integrated harvest system, the installation of a dry cleaning system that allows separation of the residue from the sugarcane is required. This can cause delays in delivery times of the sugarcane to the sugar refinery. This system is represented by [18], as a process with four parts: two inputs and two outputs; the result coming from a pneumatic separation process.

After defining the collection technology, programming is done with the necessary resources to carry out the process of lifting the ARS, based on the daily required tons of sugarcane. This then is supplemented into the four harvest systems used in the pilot refinery. However, to implement the new logistics system for collection of ARS, it is necessary to consider that collection technology only applies when the cane is harvested mechanically, in any of the two states, (short and long-cane), therefore, resource scheduling should be a function of this system of harvest.

To meet technical resource requirements in mechanical sugarcane harvesting that meet daily demands of the cleaning center, the following information should be taken into account: Average number of harvest farms, distances between farms, transport and delivery times of the raw material to the factory, capacity and speed of the sugarcane road trains when loaded and empty, efficiency of the harvest collection, available harvestable tons at each farm, setup of the harvest shift front, amongst other elements that make the logistics system a dynamic event model with stochastic and deterministic variables.

2. SIMULATION WITH PROMODEL

Once the problem was recognized, we defined the primary objective, to verifying compliance with daily operating time (discounting supplementary time) for a total of tons of sugarcane-ARS required by the refinery in relation to the mechanical cut. The result gives the minimum number of technical resources for harvesting and transportation, and evaluates the efficient flow of material through the infrastructure system.

2.1. Case Study

As the pilot refinery does not have the required cleaning center needed for utilization ARS, baseline data were taken from sources using technology and methods similar to those used with the selected alternative, which facilitates the simulation model PROMODEL ®, [19] [10] [21]. The input data and assumptions of the simulation model are described below:

Assumptions

- To start the simulation, harvesters and tractors are assigned to farms.
- The harvesters assigned to each of the farms work simultaneously with the same loading time per ton.
- Two tractors for each harvester are considered to ensure the continuous flow of the harvest. These are not part of the simulation.
- The total number of tons of sugarcane-ARS available in assigned farms should be approximately equal to the factory requirement.
- The products separated by the cleaning center are destined for the sugarcane refinery and the processing center (ARS).
- In the harvesting and transport processes a loss of raw sugarcane and residue raw material is not generated.
- The times recorded in the model are stochastic random variables with normal distribution. Deviations are taken based on the experience of the personnel involved in the logistics area of the pilot refinery, because at present no data representing the operations of an integrated harvest exist.

Input Data

The model developed has eight deterministic variables and six stochastic variables which are defined below.

Deterministic Variables

- Because at the present time, the pilot refinery does not have a cleaning center, its location is assumed to be in the same area as the sugar refinery, therefore the average distances (from farm to factory) provided by the refinery, are as those shown in Table 2.
- Simulations are performed under the control of a limit time of 22 effective hours in a day because 10% of time is taken for scheduled stops and shift changes. Unscheduled stops are not considered, so the results are evaluated within ideal scenario conditions.
- Mechanically harvested lines are only taken into account on the refinery's own farms.
- The standard efficiency of the harvester is 35 t / h.
- Sugarcane road train wagons have a capacity of 20 t.
- The processing rate of the cleaning center is 250 t / h.

- The separation efficiency of the cleaning center is 70%.
- The processing time of the cleaning center is 0.3 min / t.

Table 2. Farms cultivated by the pilot refinery

Variables CC 8592 – 8475		
Farm[A]	Average distance per farm (km)	Área (ha)
A	9,2	146
B	7,5	57
C	4,9	335
D	6,7	104
E	4,8	224
F	3,4	74
G	10,8	144
I	4,6	215
J	2,2	75
K	2,2	107
L	11,3	76

[A]Name omitted due to policies of pilot refinery.

Stochastic Variables

- The transport speed has a normal distribution with an average of 20 km/h and a standard deviation of 0.6 when empty. It has an average of 18 km/h and standard deviation of 0.6 when loaded. The delays that occur are due to the poor condition of the roads, climatic factors and technical conditions of the road trains.
- The time the wagons take to travel from the farm to the forecourt area has an average of 4 min, with a deviation of 0.5; this is due to the distances covered by the field tractors with loaded wagons for the operation chain (Operation of hitching wagons).
- Weighing time for the tons of ARS-sugarcane that come to this area has an average of 3 min / t and a standard deviation of 0.4.
- The unloading time per ton of ARS-sugarcane corresponds to 0.533 min / t and has a standard deviation of 0.033.
- The chaining time of full wagons in the forecourt corresponds to a uniform distribution with a minimum value of 240 min and a maximum value of 400, due to the variability of wagons to transport.
- The flow of material to the floor of the cleaning center corresponds to a normal distribution with a mean of 0.3 min / t and a standard deviation of 0.06.

2.2. Construction of the Model in Promodel®

The construction of the model was carried out based on a number of parameters, constraints and assumptions that arise from the study of system entities and their interrelationships. Also it takes advantage of the applications of modeling provided by the simulator, depending on the problem to be solved.

The main idea is to convert the simulation model into a decision support tool at a strategic, tactical and operational level for a system that does not exist yet.

The distribution of the farm, its weighing points, unloading areas and cleaning center, where agricultural activities and production are carried out, are illustrated in Figure 4.

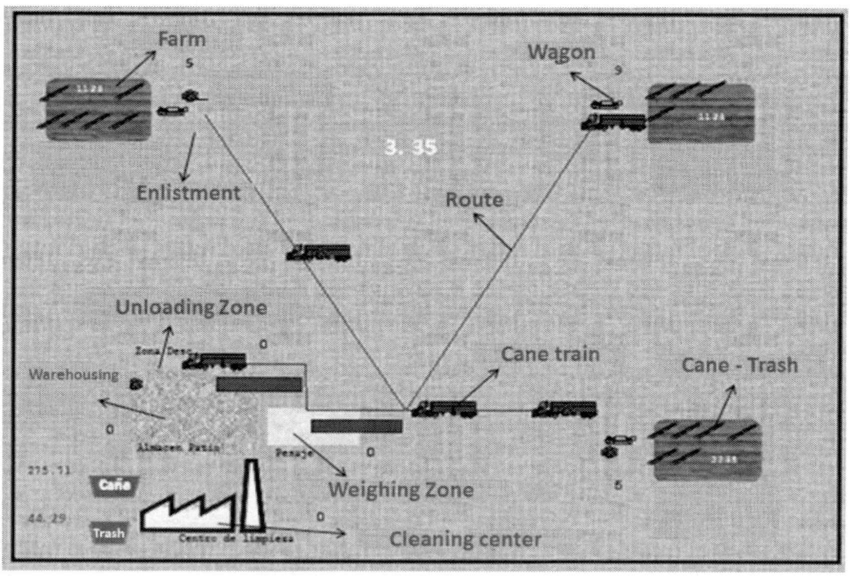

Figure 4. Model lay-out for logistics for collection of farmed sugarcane residue. (Adapted to LOZANO & CONTRERAS, 2012).

2.4. Verification and Validation of the Model

Verification of the model was performed using the same animation, which is monitored and runs using the devised configurations according to the logic established in the analysis of the dynamics of the system. This check is also carried out to ensure coherence across multiple statistics automatically yielded by the simulator, and with additional constructions of variables measuring system performance. Meanwhile, model validation has no external comparable benchmark, since it is adapted to the context of the pilot refinery and is evaluating a proposal for an operational design of a system that does not already exist within the organization. The only close but with poor estimation would be validation against the current system of collection and transportation of sugarcane.

2.5. Implementation of the Model

Against this background, the developed model took into account only one day of operations with a clear goal: to assess the amount of technical resources (equipment) needed and proportionally assigned to meet the requirement of sugarcane supply (in tons) for specific cases on the pilot farm.

The model functions as a terminal system that after ending its run time does not leave work in progress, this in order to assess the amount and efficiency in using high-cost equipment and their respective investment.

In the simulation we assumed a time limit of 22 hours; this did not include the time for supplementary activities in order to meet the quantitative targets of sugarcane supply from the farms to the refinery, and at the same time to provide the proportionate amount of assumed residue for industrial processing.

The implementation of the model involved a sufficient number of replicas -20 - to provide high percent of confidence in the estimate of mean and standard deviations. At this point, it is important to clarify that the sugarcane crop is fully dynamic in time, 24 hours 7 days a week (system without end), where consideration of the concepts of steady state and warm-up time would be necessary if requiring a non-terminal simulation.

In this case, further details from dynamic, valuable data will be necessary to advance a non-terminating system.

In conclusion, we opted for a terminal system to evaluate the feasibility and balance between time and resources in the context of a normal operations day with respective supplementary times.

2.6. Construction of Scenarios

To analyze the possible configurations that occur in the system, three basic scenarios were presented: pessimistic, normal, and optimistic in relation to the distance-volume of three farms. These farms are typical daily supply sources for refineries via mechanized harvesting. Briefly, in each experiment there was the possibility to evaluate the following variables:

- Distances farm-refinery
- Volume to be harvested per farm
- Number of sugarcane road trains to be used
- Number of wagons per road train and farm
- Number of mechanical harvesters in each farm
- Standard efficiency of the harvester (tons of sugarcane-ARS / time)
- Capacity of vehicles (road trains) in the weighing zone (refinery entrance)
- Capacity of vehicles (road trains) in the unloading zone.

The proposed configuration for the scenarios is shown in Table 3.

Table 3. Configuration of scenarios

Scenario	Farm	Distance (km)	Participation percentage (%)	Farm Capacity (t/d)
	A	9,2	20	1128
Pessimistic	G	10,8	20	1128
	L	11,3	60	3385
	B	7,5	33	1862
Normal	C	4,9	33	1862
	D	6,7	34	1918
	F	3,4	20	1128
Optimistic	J	2,2	40	2256
	K	2,2	40	2256

2.7. Optimization

By using SIMRUNNER ®, the following multi-objective function was established: i) minimizing the number of harvesters per farm, in order not to increase the investment costs and / or the harvest operating costs of the refinery; ii) maximizing the utilization rate of the sugarcane road trains, with a reduction in idle time, by controlling the number of road trains used; and iii) complying with the final of delivery of freight to cleaning centers, which must be between 20 and 22 hours, as the effective refining time cannot exceed that which is available per day. As such, the indicated range should meet the prescribed daily refining rate.

Taking into account the above and the established scenarios it is possible to define the multi-objective function, which in SIMRUNNER ® has a structure of weights and weighted resolution, not by hierarchies. Importantly, SIMRUNNER ® not only offers the option to minimize and maximize performance statistics of PROMODEL ® and additional scheduled variables, but also defines target ranges in meeting goals for statistics and variables.

$$MIN\ Z_1 = \sum_{i=1}^{3} Number\ of\ harvester\ in\ the\ farm\ i \qquad (1)$$

$$Target\ Range\ Z_2 = [80 - 100]\ \%\ Utilization\ of\ sugarcane\ road\ trains$$

$$Target\ Range\ Z_3 = [20 - 22]\ hr\ final\ trip\ time$$

Macros created for the simulation of the system are:

- Wagons: 1-4 units
- Harvesters: 1-4 units
- Road Trains 2-10 units
- Unloading Zone: 2-6 stations
- Weighing Zone: 1-3 stations
- Cutting efficiency: 1.68 to 1.74 min / t

After obtaining 20 results in each scenario, three experiments that approximate actual hours of the sugar refinery operation are chosen and also the percentage use of the sugarcane trains.

2.8. Selection of Results

After the optimization step, we choose the results with greatest operating convenience for subsequent economic evaluation. As an intermediate verification step, the chosen structures or configurations in the original model are parameterized to analyze in detail the behavior of the system for each scenario.

3. Cost Analysis

In this step the methodology used by [22] is taken as a reference, because the purpose is similar to the conditions proposed in ARS collection. The steps for analyzing ARS management costs were:

- To determine the current technical capability of the pilot refinery to compare with the resources entered yield by the simulation in order to determine whether it is necessary to make an investment.
- To calculate the cost of sugarcane: ARS ratio per hour, based on the following activities: i) integrated harvest, which includes the cost of loading and transportation of the loaded wagon to the sugarcane road train, ii) transport of loaded wagons to the cleaning center and their return empty to the plot, and iii) separating cane and ARS.

$$\text{Fixed cost per hour (FCH)} = \frac{A\left[1+\frac{(N+1)AT}{200}\right]-VS}{HU} \tag{2}$$

$$\text{Repair and maintenance per hour (RMH)} = \frac{A\times 0{,}8}{HU\times 100} \tag{3}$$

$$\text{Fuel cost per hour (FH)} = 0{,}03 \times PG \times P \tag{4}$$

$$\text{Lubricant cost (LH)} = 0{,}33 + CH \tag{5}$$

$$\text{Salaries and benefits per hour (SBH)} = \frac{1{,}33\times S\times T}{H} \tag{6}$$

$$\text{Variable cost per hour (VCH)} = RMH + FH + LH + SBH \tag{7}$$

$$\text{Total cost per hour (CTH)} = CFH + FCH \tag{8}$$

A: purchase value of machine or implement

N: lifespan in years

AT: Combined rates of interest, insurance, tax, and storage

VS: (10%) of A for tractors y 0 of A for implements

HU: Lifespan of machinery in hours

H: Working hours per year

P: Output en HP

PG: Price of a gallon of fuel

KWH: kilowatt hours consumed

PKWH: Price of kilowatt hours consumed per hour

S: monthly salary of operator

T: working time per number of operators in the machine during the year, in months, multiplied by the number of operators.

- To find the cost ratio represented by harvesting and transporting the ARS to cleaning center, since the costs of these two activities correspond to the entire Sugarcane-ARS product. In the case of separation of the two products, it will be the ARS which assumes 100% of the cost of investment in and operation of the cleaning center, because this process is generated from the necessity of using ARS.
- Finding out the total cost of managing ARS per hour with the following equations:

$$CTMHR = CCR + CTnR + CSR \tag{9}$$

$$CCR = n*CCRCI + m*CCRSI \tag{10}$$

$$CTnR = n*CTnRCI + m*CTnRSI \tag{11}$$

$$CSR = n*CSRCI + m*CSRSI \tag{12}$$

Where:

CTMHR = Total hourly management cost of ARS
CCR= Total cost per hour of ARS at harvest
CTnR= Total cost per hour of ARS in transport
CSR= Total cost per hour of ARS in separation
CCRCI= Total cost per hour with investment in harvesting
CCRSI= Total cost per hour without investment in harvesting
CTnRCI= Total cost per hour with investment in transport
CTnRSI= Total cost per hour without investment in transport
CSRCI= Total cost per hour with investment in separation
CSRSI= Total cost per hour without investment in separation
n= Number of pieces of equipment to buy
m= Number of pieces of equipment that the factory possesses

- The basis of ARS cost per hour is related to costs relative to product handling per day and by ton in dollars.

3.1. Results

In Table 4, we show the results of the simulation model where ARS was collected in conjunction with the sugarcane (integrated harvest), specifying the amount and percentage use of sugarcane road trains, and the optimal number of harvesters and wagons in each scenario.

Table 3. Configurations for scenarios

	% use of Cane trains	# of Harvesters Farm 1	# of Harvesters Farm 2	# of Harvesters Farm 3	# Cane trains	# of wagons in farm 1	# of wagons in farm 2	# of wagons in farm 3
Pessimistic	91,52	4	4	4	6	4	4	4
Normal	88,99	3	3	3	6	4	4	4
Optimistic	94,40	4	4	4	5	4	3	3

From the above configurations, the cost estimation for managing agricultural sugarcane residue by scenario was performed adjusting the methodology used by Astaiza [22] in his study of "Handling residue from sugar cane harvest," where all fixed and variable costs of the operations are considered implicit for the management of the biomass.

Table 5 shows the cost per hour of operation in each of the scenarios, using equations (9), (10), (11) and (12).

Table 5. Total hourly cost of ARS

Activities	Recourse	Pessimistic (U$/hour)	Normal (U$/hour)	Optimistic (U$/hour)
Harvest	Harvester	103,55	67,98	103,55
	Tractor	214,00	160,50	214,00
Transport	Sugarcane road train	53,50	53,50	44,58
	Wagon	0,81	0,81	0,67
Separation	Cleaning center	35,59	35,59	35,59
Total hourly cost of managing ARS		407,44	318,37	398,39

Daily cost was obtained based on the total management costs per hour of ARS-CTMHR per scenario. When divided daily cost by the number of tons of separated residue in the cleaning center, we obtained the total cost of management per ton of ARS (see Table 6).

Table 6. ARS management costs

	Pessimistic	Normal	Optimistic
Total daily management cost(US$)	8963,72	7004,12	8764,60
Total management cost per ton (US$)	11,49	8,98	11,24

From the previous table, it can be seen that in the pessimistic scenario, total management cost per ton of U.S. $ 11.49 for ARS is obtained, with resource configuration of 12 harvesters, 24 tractors, 6 sugarcane road trains, and 24 wagons. In the normal scenario, having a configuration of 9 harvesters, 18 tractors, 6 sugarcane road trains and 24 wagons results in a total management cost of U.S. $ 8.98 per ton. Finally, the optimistic scenario costs U.S. $ 11.24, with a configuration of 12 harvesters, 24 tractors, 5 sugarcane road trains and 20 wagons.

CONCLUSION

The proposed methodology for the analysis of the current system identified critical elements, selecting alternative technologies appropriate to the geographical and environmental conditions of the area of interest, in this case the location of the pilot refinery. In addition, the use of modeling techniques for the IDEF0, revealed, in a hierarchical way, the sets of processes and activities, the resources consumed and the relationship between the composing elements of a integrated management of sugarcane (IMS) in a Colombian-type refinery.

At present, the business model of sugar refineries considers two components: i) agriculture, where sugarcane is seen as the only usable product; and ii) industrial, one where products such as sugars and derivatives originate. That is to say, the use of residue from sugarcane agriculture is not a fundamental function of the business system (pilot refinery). This perception is here reassessed to determine the collection of ARS as a single simultaneous operation of the harvesting of sugarcane, without adding complexity to the system, or burdening the current economic structure. Therefore it is proposed that ARS to be a considered as secondary product of the agricultural stage.

The proposed model contributes to solving the problems associated with the handling costs of the residue from agricultural sugarcane, linking the activities of collection, transport and residue separation in an integrated harvest, thereby generating a suitable logistics proposal to local needs in Valle del Cauca state, Colombia. The cost analysis covers the supply of sugarcane and the ARS to the cleaning center, given that the selected scenery includes collection of both products simultaneously. The cost sensitivity analysis of supplying the residue was made based on the simulation proposed scenarios.

The simulation of the logistic model is a strategic factor that allows observation of the performance of technological resources to ensure compliance with the demand of sugarcane to the factory and at the same time obtains the agricultural residue to generate new products.

REFERENCES

[1] O. Avella. Etanol Celulósico a partir de residuos agrícolas. (Cellulosic ethanol from agricultural residues) http://www.idea.unal.edu.co/eventos/CisdaIV/ponencias/E4_Crisis_Energetica/E4_oscar_r_avella.pdf, accessed 03/ 2012.

[2] O.S. Dias Marina, T. L. Junqueira, et al. Improving second generation ethanol production through optimization of first generation production process from sugarcane, Energy, vol. 43, no.1, July 2012, pp. 246-252, http://www.sciencedirect.com/science/article/pii/S0360544212003325.

[3] Michellazzo, Marcio B, y Oscar A Braunbeck . Análise de seis sistemas de recolhimento do palhiço na colheita mecânica da cana-de-açúcar. (Six waste handling systems analysis of mechanical harvesting of sugarcane.) Revista Brasileira de Engenharia Agrícola e Ambienta. vol 12, nº 5, pp. 546-552. Enero 2008

[4] Ripoli, T.C.C, y M.L.C Ripoli. Aspectos económicos, logísticos y energéticos en el aprovechamiento del RAC en calderas (Financial, logistic and energy aspects in the use of RAC in boilers). Universidade de São Paulo, São Paulo, 2010.

[5] Knudsen González, José, Fernando Marrero Delgado, y Roberto Cespón Castro. Evaluación del nivel de desempeño de la cadena de suministro de los residuos agrícolas cañeros para su uso con fines energéticos en una empresa azucarera. Informe Técnico (Evaluation of performance level supply chain of ARS for use with energy purposes in a sugar factory - Technical Report). Universidad Central Marta Abreu, Las Villas, 2004.

[6] Filho, Luigi Mariani. Utilização da palha de cana-de-açúcar para aumento da capacidade energética de plantas de utilidade (Use straw sugarcane to increase energy capacity plant utility). Thesis, Mechanical Engineering Department, Universidade de São Paul, São Paulo, 2006, pp. 58.

[7] Adriana C. Lozano Riascos, S. Carolina Contreras Gallego, Gloria M. López Orozco. Sistema de recolección de residuos agrícolas cañeros en un ingenio del Valle del Cauca. Modelado y Costeo (Collection system of ARS in a Cauca Valley refinery). Modeling and Costing). Encuentro Nacional de Investigación & Desarrollo ENID 2012. http://www.enid.unal.edu.co/2012/memorias/fscommand/Desarolloor/32.pdf

[8] M., Correa V.; G.M., López O.; A.C., Lozano R. Modeling of Integrated Management System of Sugar Cane: Taking Advantage of Sugarcane Agriculture Residues. En: Annals of Industrial Engineering 2012. Springer. 2014, pp. 155-162.

[9] Im, S.-H., & Jang, K.-J. (2002). Designing performance analysis and IDEF0 for enterprise modelling in BPR. Int. J. Production Economics (76), 121-133.

[10] Amú, L. G. (December 2010). Logística de Cosecha: Evaluación de Tiempos y Movimientos. Indicadores y Control (Logistics Harvest: Evaluation Time and Motion. Control and Indexes) Revista Tecnicaña (26), pp. 25-30.

[11] Nova, J. (2010). Siembra mecanizada de la caña de azúcar en Brasil (Mechanized planting of sugarcane in Brazil). Revista Tecnicaña, pp. 32-36

[12] Torres, J., & Villegas, F. (2006). Sistemas de manejo de residuos y efectos en la producción. San Antonio de los Caballeros (Residue management systems and production effects. San Antonio de los Caballeros). CENICAÑA

[13] Viveros, C. A., & Calderón, H. (1995). El cultivo de la caña en la zona azucarera de Colombia (Farming of sugarcane in colombian sugar zone). CENICAÑA, 131-139.

[14] Rodríguez, C. A., & Daza, O. H. (1995). El cultivo de la caña en la zona azucarera de Colombia (Farming of sugarcane in colombian sugar zone). CENICAÑA, 109-114.

[15] M. Correa, G.M. Lopez, A.C. Lozano (2012). Modeling of integrated management system of sugar cane: taking advantage of sugarcane agriculture residues. 6th International Conference on Industrial Engineering and Industrial Management.

[16] Lozano R. Ac; Contreras G., SC. Diseño de un sistema piloto logístico de recolección de residuos agrícolas cañeros (Design of system for collection and supply of sugarcane green harvesting residues). [Thesis] Cali: Universidad Autónoma de Occidente. 2012

[17] Pacheco, S. (June 2011). Correio Braziliense (Brazilian Mail). Accessed July 30, 2011, http://www.correiobraziliense.com.br

[18] Aguilar, A., Hernández, B., & Oliva, D. (2007). Solución medioambiental sustentable en la cosecha de caña (Sustainable environmental solution in sugarcane harvest). Habana, Cuba: ICINAZ, MINAZ y CUJAE.

[19] Vélez, C. (Diciembre de 2010). Limpieza de caña en seco y aprovechamiento de la materia extraña vegetal como combustible en Brasil (Sugarcane residues dry-cleaning for fuel generation purposes in Brazil). Tecnicaña (26), 10-17.

[20] Verdecia Fonseca, A., Peréz Egusquiza, F., Mata, I., Roque Díaz, P., & Rubio González, Á. (2005). El uso de los residuos agrícolas cañeros RAC, como combustible en cuba. Oportunidades (Using ARS

Agricultural Residues of Sugarcane as fuel in Cuba. Opportunities). Habana, Cuba.

[21] Portilla Yepes, Ana Ximena; Bedoya Martinez, Paola Andrea ―Evaluación del sistema de limpieza en seco de la caña de azúcar y alternativas de aprovechamiento de los residuos generados (Dry cleaning system evaluation of sugarcane and opportunities for to use green residues).‖ Thesis, Engineering department, Universidad del Valle, Santiago de Cali, 2004, 95.

[22] Astaiza, Diego. ―Manipulación de los residuos de cosecha de caña de azúcar y su valor energético (Handling of harvest sugarcane residues and its energy value).‖ Thesis, Facultad de ciencias agropecuarias, Universidad Nacional de Colombia, Palmira, 1998, pp. 109

In: Sugarcane
Editor: Eleanore Webb

ISBN: 978-1-61122-356-9

Chapter 10

DEVELOPMENTS IN MUD FILTRATION TECHNOLOGY IN THE SUGARCANE INDUSTRY

T. J. Rainey*[*], *O. P. Thaval and D. W. Rackemann
Queensland University of Technology, Brisbane, Australia

ABSTRACT

Dirt collected with sugarcane is processed and separated from the juice in the sugar factory by filtration equipment for return to the cane fields. New technologies over the past decade have enabled performance improvements to be obtained for this key unit operation. Filter mud product still contains a reasonable amount of sugar and the transportation of high moisture mud product has considerable cost. Australia's traditional approach has been to use Rotary Vacuum Filters for processing and separating mud and other impurities from juice, but in recent years there has been interest in reducing sugar losses and transportation costs through utilisation of new technologies such as Horizontal Bed Filters, Vacuum Belt Press Filters, Membrane Press Filters and Centrifuges. Increasingly, these alternative equipment are being installed in new factories. This chapter describes the general principles of mud filtration theory and mud conditioning followed by a

[*] Corresponding author: T. J. Rainey. Queensland University of Technology, 2 George St, Brisbane 4000, Australia. E-mail: t.rainey@qut.edu.au.

detailed description and review of the various filtration technologies and analysis of the relative merits associated with the equipment.

INTRODUCTION

The removal of insoluble non-sugar impurities from sugarcane juice is an essential part of raw sugar production and this is commonly achieved through juice clarification. Lime saccharate and polymeric flocculants remove suspended matter which precipitate and leave the clarifier in the underflow and the clarified juice is sent to the evaporators for concentration [1].

The mud underflow from the clarifier typically contains 5–10 % precipitate in suspension and 10–15 % sugar in the juice. Dirt collected with sugarcane is separated from the underflow juice by filtration equipment for return to the cane fields. Historically, Australian factories predominantly use traditional Rotary Vacuum Filters (RVFs) to separate >85 % of the mud solids in juice to produce a filter mud product of 75–80 % moisture [2], however the technology landscape is changing.

A schematic of mud processing typically undertaken in Australian sugar factories with RVFs and their associated equipment is shown in Figure 1.

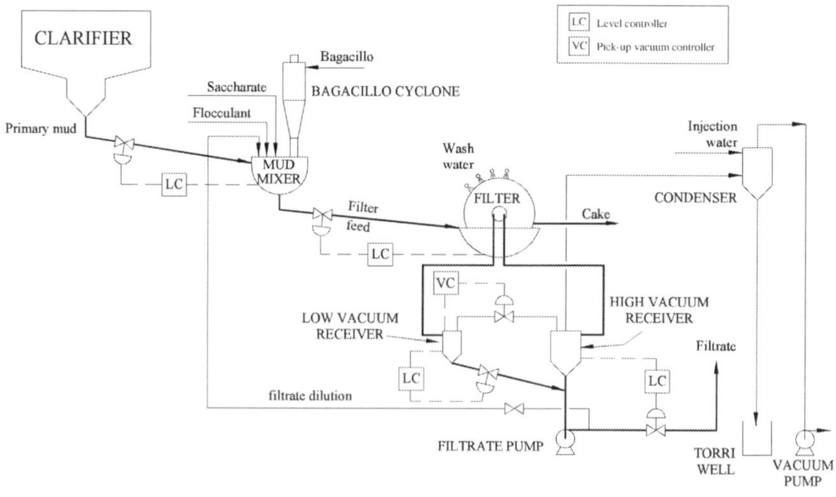

Figure 1. Schematic of sugarcane mud processing with RVFs.

Mud filtration is the last opportunity for the remaining sugar in mill mud to be recovered. The goals of mud filtration are to: (i) recover sugar and return it as part of the filtrate to the process; and (ii) maximise mud solids retention, minimising the amount of mud recycled back into the process. These two goals are achieved by efficient washing with water.

More recently there has been impetus for improving mud distribution back to the cane farms in terms of the transport costs and also the wide spread application of this nutrient-rich product.

As moisture is the major component of filter mud product, processing options and technologies that allow more effective dewatering (while maintaining acceptable levels of sugar recovery) are of most interest to sugar factories.

Although factory mud has nutrient value, the high transport costs to return it to the cane fields impacts on the distances it can be economically transported which also can depend on farm application rates. When taking into account nutrient value, a recent case study found application rates of 100–150 t/ha was more economic than fertiliser application for distances of up to 20 km from the factory [3]. This was based on fertiliser-replacement value although reduced application rates were required to economically extend the treated area. Transportation costs which limit the widespread distribution of factory mud has led to adverse secondary issues such as the accumulation of heavy metals on farms in closer proximity to the factory. This can have a negative environmental impact due to water runoff following high rainfall events.

Increasing the mud solids density in the filter mud product (i.e., reducing moisture content) can extend the economic distribution and application of filter mud over a wider area to reduce heavy metal accumulation (environmental benefit) and fertiliser requirements (economic benefit).

New technologies over the past decade have enabled performance improvements to be obtained for this key unit operation. Such technologies have the potential to improve sugar recovery and/or transport distance whilst improving wash water efficiency. Australia's traditional approach has been to use Rotary Vacuum Filters which are in almost universal use in Australian factories. Internationally, there has been considerable interest in Vacuum Belt Filters (VBFs), Membrane Press Filters (MPFs) and Centrifuges.

Vacuum Belt Filters in particular, are increasingly being installed in new factories, especially in Brazil. For most of these technologies, the mud product can be described as cake, except for Centrifuges which produces granulated material.

FILTRATION THEORY AND ITS APPLICATION TO SUGAR FACTORY MUD

Filter feed conditioning has a significant effect on the filtration properties, such as cake permeability, cake resistance and cake porosity [4]. Commonly, these terms are used interchangeably but incorrectly; they are more closely related for unit operations which are close to steady-state such as in the operation of RVFs. Permeability is related to cake resistance but both are in fact distinct properties from porosity. This distinction becomes important when comparing performance parameters of different technologies.

Permeability is a measure of the ease in which a fluid can flow through a porous medium and is quantified by Darcy's Law. The filtration and washing of mud by RVFs generally obeys Darcy's Law as limited compression of the mud occurs. Darcy's Law relates to steady-state laminar flow through a homogeneous porous media. A common simplified description of Darcy's Law is:

$$\frac{Q}{A} = \frac{K\Delta P}{\mu \Delta L} \tag{1}$$

where Q is the volumetric flow rate (cm^3/s) through a bed of porous material with cross-sectional area A (cm^2), ΔP is the frictional pressure drop (mPa) across the length ΔL (mm), μ is the filtrate viscosity and K is the permeability constant (cm^2). However, the sugar industry generally uses a more functional form of Darcy's Law for quantifying RVF performance. The functional form includes specific cake resistance which is inversely related to permeability [4] as follows:

$$\frac{d\frac{v}{A}}{dt} = \frac{\Delta P}{\mu(\alpha C + R_m)} \tag{2}$$

where v is filtrate volume, t is time, α is the average specific cake resistance, C is the mass of dry solids in the cake and R_m is the hydraulic resistance of the screen.

With alternative separation technologies (Press Filters and Centrifuges), the porous media compresses and so permeability and hence cake resistance changes with porosity according to the Kozeny-Carman relation for capillary flow. The unsteady-state filtration still obeys the principles of capillary theory, namely Kozeny-Carman's equation:

$$K = \frac{1}{kS_v^2}\frac{\varepsilon^3}{(1-\varepsilon)^2} \tag{3}$$

where ε is the porosity (fraction of void volume), S_v is the specific surface area and k is most commonly assumed to be a constant. The underlying theory for the Kozeny Carman relation relates to the assumption of the flow of water through capillaries. The Kozeny-Carman relation can be derived from Poiseulle's Law for flow through long cylindrical pipes [5].

The traditional approach to quantifying cake resistance vis a vis Darcy's Law increases theoretical complexity for new mud filtration technologies that are not operating under steady-state conditions. Dynamic filtration of porous media involves interaction between the compression and permeability of the media; as a medium compresses, the porosity, ε, reduces and the permeability decreases. The variability of permeability (and cake resistance) with porosity is shown in Figure 2 which assumes other factors are constant.

Steady-state compressibility behaviour can be determined by:

$$P_s = M\varphi^N \tag{4}$$

where P_s is the pressure on the solid phase, φ is the solid fraction (i.e., 1-ε), M and N are constants.

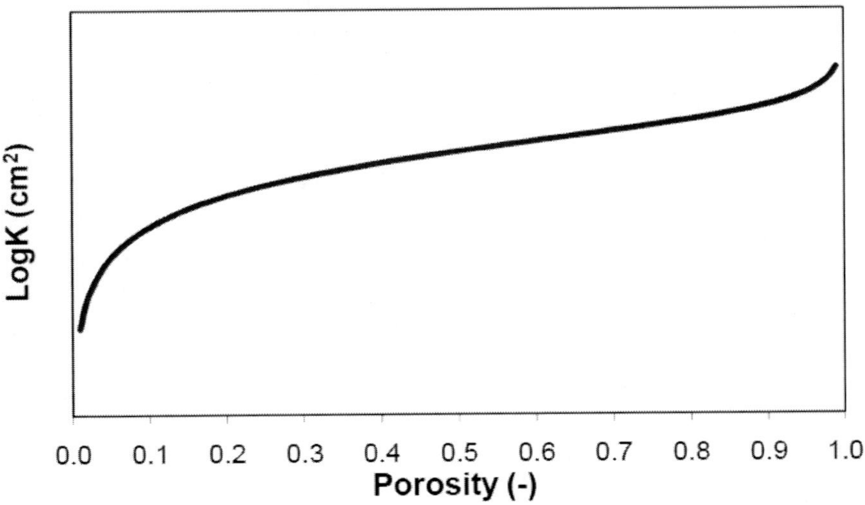

Figure 2. Relation between permeability, K, and porosity, ε, for capillaric porous media with a constant Kozeny factor (k) and specific surface area (S_v).

There is a wide variety of dynamic compressibility models in both soil science literature and bagasse. A general dynamic filtration model proposed by Banks [6] was applied to bagasse by Owen and co-workers [7] and Kent and McKenzie [8]. However, for mud it is possible to quantify and characterise the steady-state permeability and compressibility behaviours separately and combine them into a dynamic model. The theoretical framework for generalised constant rate and constant pressure dynamic filtrations for similar porous media is provided by Landman and co-workers [9, 10].

Mud filter cake contains a significant amount of sugar that is lost from the factory. The primary method of reducing the amount of lost sugar is through the application of sufficient wash water during filtration. However, the effectiveness of the water application depends on the porosity of the cake and the means of application. Provided the wash water is added effectively, increasing the wash water rate will decrease the amount of sugar lost. Most mud filtration technologies rely on displacement washing whereby wash water pushes the juice through capillaries in the mud cake with limited mixing (the basis for the Kozeny Carman relation) rather than diffusing and diluting the juice in the cake. Compressing the cake too much reduces its porosity (resulting in high cake resistance) such that added wash water does not easily penetrate the cake to displace any sugar, while a very porous cake will encourage channelling of the water and a poor washing efficiency will result. Thus, in both cases, sugar recovery would be limited. For RVF operation, if the cake resistance is high, excess water will run off the surface of the cake into the mud boot where less effective dilution washing takes place.

FEED CONDITIONING

Regardless of the technology used, feed conditioning is essential for good separation performance. Mud filter cake resistance is most affected by the quality and quantity of bagacillo. The term bagacillo is given to the very fine fibre particles of bagasse which is the fibrous residue of sugarcane after the juice has been squeezed out. Fine bagacillo improves mud solids retention and its addition improves porosity. In adequate quantities it can increase filtrate rate up to 300 %. Low quality bagacillo can incorporate large pieces of sugarcane rind but these provide no benefit to sugar recovery or porosity (i.e., washing efficiency). Following bagacillo addition, lime saccharate has a major beneficial effect on increasing cake porosity, improving sugar recovery and providing good mud solids retention.

Good RVF performance is achieved when adequate saccharate is added to maintain filtrate at a pH of 8.5. However, adding lime is essentially adding more scale-forming components (i.e., calcium) to the process and can also have the effect of blinding filter screens. This can increase maintenance, and reduce throughput and recovery of the filter station.

Finally, flocculant also has a significant effect on the dewatering capability of filter feed. If used effectively, it reduces cake resistance and improves sugar recovery and mud solids retention. Around 5–10 ppm is required in filter feed and the type of flocculant required varies from factory to factory, depending on the unique soil characteristics of the region. Too much or too little flocculant can be detrimental to the goals of mud solids retention and sugar recovery. To avoid disrupting he flocculated mud particles, the feed to the filter is optimally gravity fed rather than pumped.

Mud density can also impact on filter performance and cake thickness. Lighter mud (i.e., 3–5 % mud solids) which is achieved through dilution with either water or recycling a portion of the filtrate produces a thinner filter feed of lower consistency. This enables a more even and uniform distribution of feed and cake to improve filter performance and reduce feed pipe blockages [11]. The current knowledge on sugarcane mud filtration is based on studies of the operation of RVFs under factory conditions [12-16] or laboratory studies conducted using a batch filter or similar apparatus [4, 17, 18].

There are relatively few recent studies on sugarcane mud filtration reflecting the maturity of the technology. Recent research on sugarcane mud filtration has focused on alternative filtration technologies such as Horizontal Bed Filters [19] and Centrifuges [2]. Current attention has focussed on VBFs, as a dominant technology in the Brazilian sugar industry, although there are surprisingly few publications on this technology for sugarcane mud filtration.

Vacuum Belt Filters have traditionally been used for dewatering in mining, wastewater treatment, food production as well as pulp and paper factory sludge processing due to their large capacity per unit size, high levels of automation and superior filtration performance.

Filtration Technologies

Rotary vacuum filters have changed little over the past 20 years. While they are the predominant filtration technology employed around the world, there is a number of alternative filtration technologies used in the sugarcane industry and these are described in the following.

New technologies may have the potential to decrease the cake moisture while also increasing the sugar recovery and mud solids retention and hence increase profitability for Sugar Mills. This section introduces traditional Rotary Vacuum Filters, and then Vacuum Belt Filters are discussed which are becoming increasingly popular. Other alternative filter technologies follow, namely Membrane Press Filters and Centrifuges.

Rotary Vacuum Filters

Oliver Campbell RVFs are the most widely used filter in sugar factories. The filter is composed of a hollow drum rotating about a horizontal axis and party submerged in the mud feed (i.e., the filter boot) to be filtered (Figure 3). Filter capacity and operation is improved by feeding the mud at multiple points along the length of the filter boot. To avoid settling and stagnation of the mud in the filter boot, an oscillating agitator (2–4 rpm) which breaks the liquid surface is installed, pivoted on the axis of the filter and driven by a separate motor. The filter drum is driven by a small electric motor to provide a drum speed of 4–10 rpm.

The periphery of the drum serves as the filtering surface, divided into 24 independent sections, and extending along the full length of the drum. Each of these sections is connected individually to a vacuum system by a small metal pipe terminating in a distributing valve situated at one end of the drum and applying three different functions:

1 One with connection to atmosphere
2 The second connecting with a chamber where a low vacuum, of the order of 20 40 kPa (abs) is maintained
3 The third connecting to a chamber where higher vacuum is maintained, of the order of >60 kPa (abs)

The filter screens are copper, brass or stainless steel and perforated with ~100–120 holes/cm^2 which are 0.5 mm in diameter [20]. With due care, the screens suffer very little wear and may be used for several seasons without being replaced. They are, however, fragile and delicate.

Operation

As the filter rotates, the section which first enters the mud boot is immediately connected with low vacuum.

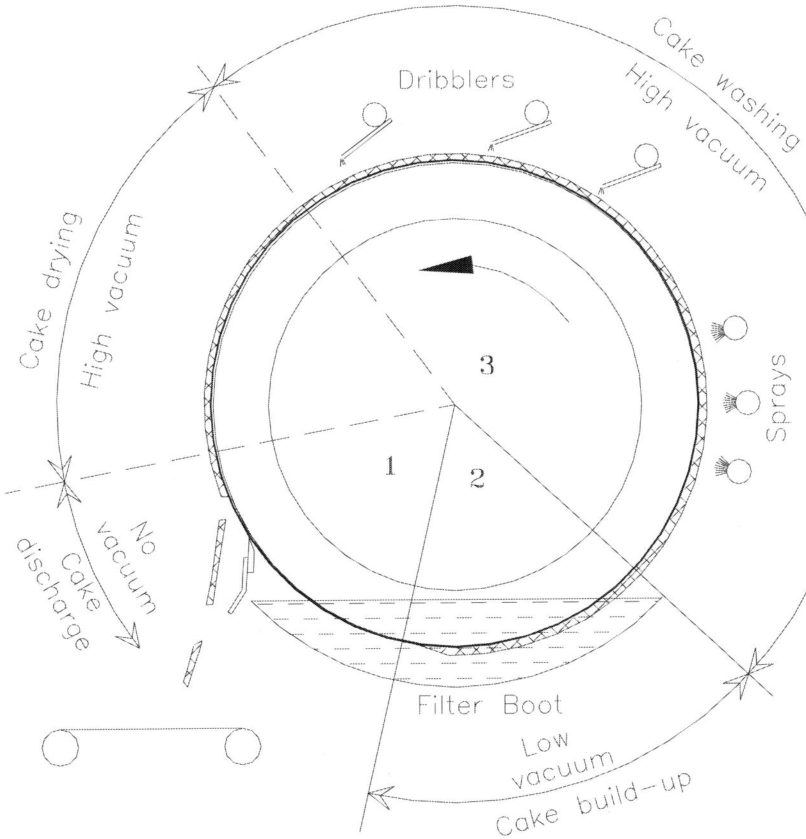

Figure 3. Diagrammatic operation of continuous rotary vacuum filter.

The liquid (filtrate) passes through the perforations which become coated with fine bagacillo and suspended matter. The first filtrate is sent to the low vacuum receiver (Figure 1). The cake continues to build, until the filter section emerges from the mud in the filter boot.

After the screen rotates and emerges from the filter boot, it encounters high vacuum. The juice passing through the cake is more effectively filtered by its own impurities and by the fine bagacillo which it contains, forming the necessary filtering surface which builds the cake on the surface of the drum. This filtrate is sent to the high vacuum receiver.

The filter section is then sprayed with hot wash water. High temperature wash water is essential since the viscosity of juice is inversely proportional to temperature and lower viscosity favours filtration [21].

The vacuum draws the water through slowly, and is designed in such a way that the water has just sufficient time, to pass through the cake and to displace the juice.

Drying commences in the final stage of rotation as the water is drawn towards the inside of the cake without further addition of water applied.

Finally the filter screens contact a scraper to remove the cake from the drum. When the filter section is about to reach the scraper, the distributor valve breaks the vacuum which has held the cake against the filter surface. The slightest contact of the scraper detaches the cake dropping into a screw or belt conveyor transporting the mud product to a storage bin prior to distribution back onto cane fields.

The thickness of the cake is variable, generally 5–20 mm and depends on the rotational speed of the drum and mud feed conditioning. The final filter cake typically contains 0.5–3 % sugar, corresponding to a pol loss of 0.2–1.2 %.

The mud feed for filtration and wash water should be >80 °C to avoid the risk of waxes blocking the filter screens and to prevent microbial growth.

Filter Sizing and Performance

The required filter area should be related to the mud solids entering the factory. Generally for sugar factories with milling trains, two-thirds of the mud solids entering the factory remain in the juice with the remainder leaving the factory with bagasse. Historical RVF filter area ratings of 0.5–0.8 m^2 per t/h cane were standard for milling factories with lower ratings for diffuser factories [11]. However, improved filter station capacity ratings provide 85–100 m^2 of filter area per 1 t/h of mud solids loading Pol loss in cake at this value is expected to be around 0.3 % for wash water % cake of 200. The filter mud solids loading (MSL, t/h/100 m^2) is expressed in terms of (dry) mud solids output per unit of filter area [16]:

$$MSL = \frac{100 \times Mud\ solids\ rate\left(\frac{t}{h}\right)}{Filter\ area\ (m^2)} \tag{5}$$

A similar term is used to rate the wash water application to the filters expressed as wash water loading (WWL, t/h/100 m^2), which is analogous to mud solids loading and is defined as:

$$WWL = \frac{100 \times Wash\ water\ rate\left(\frac{t}{h}\right)}{Filtr\ area\ (m^2)} \tag{6}$$

The sugar recovery achieved on the filter stations is linked heavily with filter station capacity. Wright and co-workers [16] proposed a correlation for pol loss in cake versus mud solids loading using data from a pilot filter.

The work was conducted using a constant wash water % mud solids value of around 1200. The regression expression for these data was found to be:

$$Pol\ loss\ in\ cake = 0.15 + 0.1334 \times MSL^{1.75} \tag{7}$$

where

MSL is mud solids loading, t/h/m^2; and

$$Pol\ loss\ in\ cake = \frac{Pol\ \%\ MS \times MS\ \%\ cane}{Pol\ \%\ cane} \tag{8}$$

where

$Pol\ \%\ MS$ is the pol % mud solids in cake.
$MS\ \%\ cane$ is mud solids in cake % cane

The relationship between pol % MS and the wash water % mud solids in cake ($WW\ \%\ MS$) is given by the expression:

$$Pol\ \%\ MS = 176.9 \times WW\ \%\ MS^{-0.21} - 34.04 \tag{9}$$

Using the empirical relationship based on pilot data predicts a pol % MS value of 5–6 for typical industry wash water rates ($WW\ \%\ MS$ of 1200). Both expressions can be combined to provide an economic model to estimate the value of pol % MS. The combined expression is given below:

$$Pol\ \%\ MS = (0.15 + 0.1334 \times MSL^{1.75}) \times \left(\frac{pol\ \%\ cane}{MS\ \%\ cane}\right) \times 100 \times (32.17 \times (WW\ \%\ MS)^{-0.21} - 6.19) \tag{10}$$

Wright and co-workers [16] further developed their performance relationship to fit factory data for estimating pol loss based on operating parameters. As the $WW\ \%\ MS$ is dependent on MSL, the regression was simplified to determine the pol % MS as a function of mud solids loading and wash water loading on the filters as follows:

$$Pol\,\%\,MS = 1.265 \times MSL^{2.5} + 30.78 \times WWL^{-0.25} - 12.43 \quad (11)$$

For data collected from a typical sugar factory, the estimated pol % MS was plotted using the above equation for various mud solids and wash water loadings to yield the trends shown in Figure 4.

With increasing wash water application rates there is a diminishing return on improvements in sugar recovery. Increased amounts of wash water also increases the amount of filtrate which when recycled to the process dilutes the juice and increases evaporation loads.

Feed Conditioning

Steindl [4] undertook trials to measure the cake resistance for various levels of feed conditioning. Filtration tests were performed using an apparatus consisting of a piece of filter screen of approximately 100 mm diameter connected through a condenser to a measuring cylinder. During a test, filtrate passes through the screen and is cooled in the small, water cooled condenser before collecting in the measuring cylinder. The whole apparatus is connected to the wash filtrate receiver as the vacuum source.

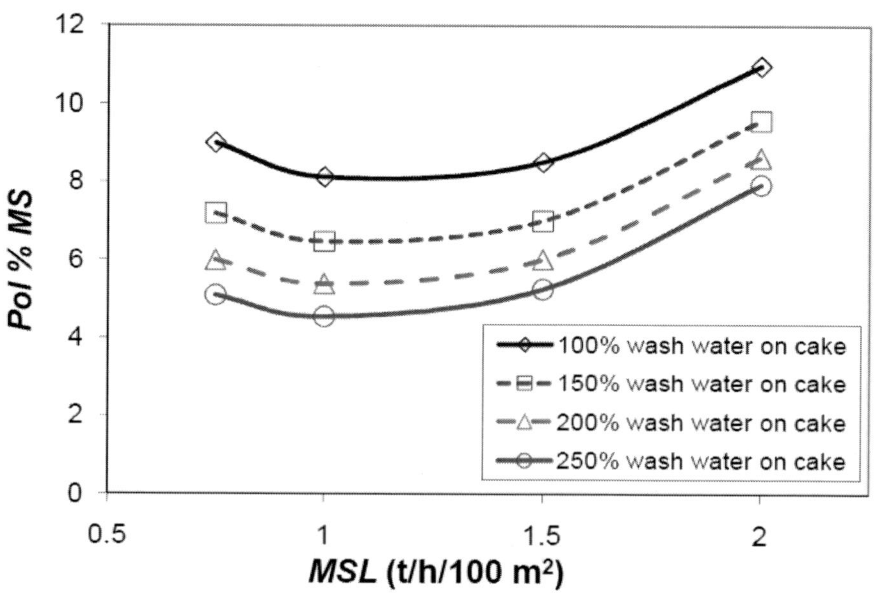

Figure 4. Prediction of RVF performance for various application rates and filter station performance.

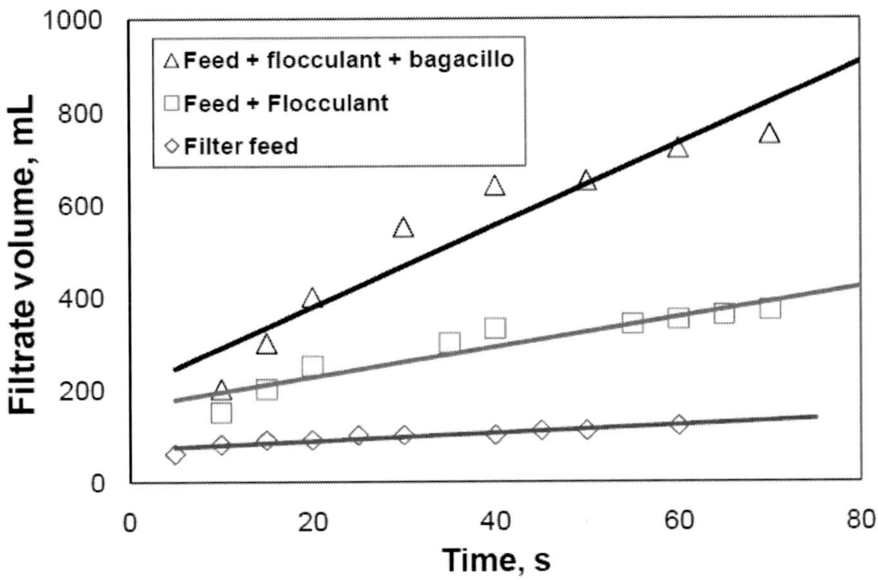

Figure 5. Filtration rate data showing the effects of added flocculant followed by the addition of extra bagacillo [4].

The apparatus allows filtrate collection rates to be measured over short time intervals to provide a gauge of cake resistance and filter performance. The effect of bagacillo and flocculant addition on filtration performance is shown in Figure 5.

Vacuum Belt Filters

Vacuum Belt Filters have been used in other industries to produce very dry filter cakes and have been used in Australian sugar mills for ash dewatering. In the 1980's VBFs were assessed by Crees and Willersdorf [19] and Kruger [22] for mud filtration and showed that reasonable filtration performance could be achieved with sufficient feed conditioning. The addition of flocculant was essential in reducing pol loss.

Crees and Willersdorf [19] found the pol loss was minimised with flocculants at dosage rates of 300–350 ppm on mud solids.

Crees and Willersdorf [19] were able to produce cakes of 6–7 pol % mud solids with wash application rates of 150 % on cake (~1000 % on mud solids) when operating with vacuum levels of ~ 40 kPa (g).

Kruger [22] found the optimum operation of the VBFs was operation with thin cakes of ~6 mm that allowed more effective cake washing at lower vacuum levels (20–25 kPa (g)) to achieve cakes of 0.5 % pol and cake moistures of 70–75 %. These performance figures are slightly better than typical RVF operation. Further work on VBFs was discontinued in the mid 1980s due to higher capital and maintenance costs relative to RVFs at the time.

In the 1990's, Technopulp Industrial (Brazil) developed a Vacuum Belt Press Filters (VBPF) which incorporates an additional press section to the VBF. Technopulp currently have over 600 units installed in South America. The Technopulp VBPF is shown in Figure 6 and Figure 7. Following feed distribution, a filter unit is comprised of three distinct sections:

- Gravity drainage section. Filtrate is allowed to drain from the fresh cake. This filtrate is the cleanest of the three separate filtrates from the filter. About 70 % of the filtrate is recovered from this section. Wash water is added to this section.
- Vacuum drainage section. A low vacuum of up to 20 kPa (g) is applied to draw off water prior to the press section. About 15–25 % of the filtrate is recovered from this section. The last of the wash water is added at the start of the vacuum section. It is more a case of needing air flow through the cake for drying rather than vacuum.
- Cake press section. The filtrate is squeezed between two belts as it passes over a series of rollers. The filtrate from this section is about 2–3 brix and is of the lowest quality.

Operational and performance data provided by Technopulp include:

- Values of pol loss % pol in cane of 0.35–0.5 (cake pols of 1.25–1.5 %).
- Moisture of the cake is typically 60–65 %.
- Wash water application rates of 150 % on cake (~1200 % on mud solids).
- Recommended cake thickness of 8–12 mm.
- Retention is typically about 96 %.
- Flocculant dosage at around 3 to 6 ppm on cane (400–450 ppm on mud solids).
- Addition of lime required to maintain filtrate pH of 7.8–8.0.
- No bagacillo required if the level of bagacillo in juice is above 0.2 %.

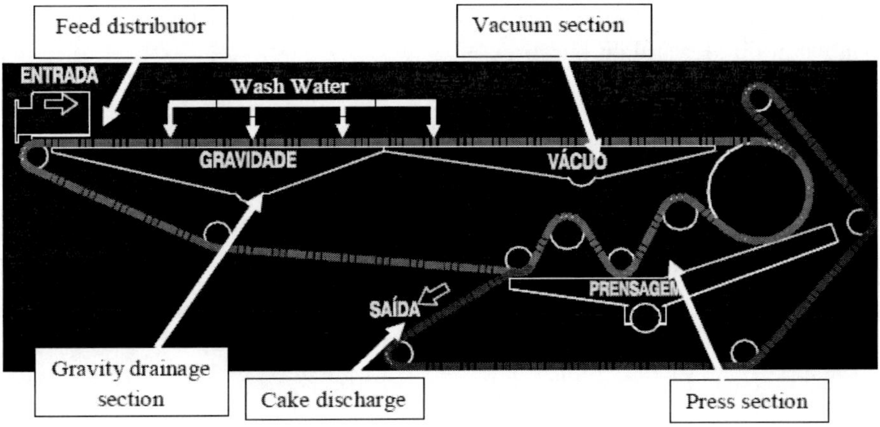

www.technopulp.com.br.

Figure 6. Schematic of the Technopulp VBPF showing the different processing sections.

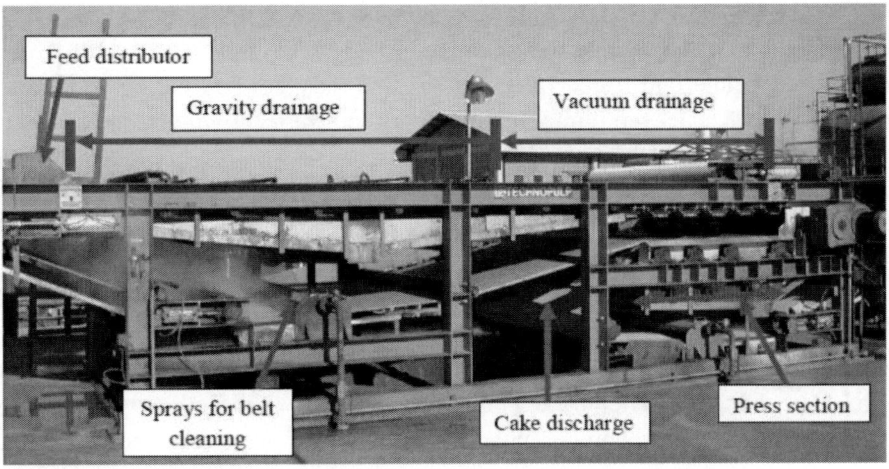

www.technopulp.com.br.

Figure 7. Photograph of an installed Technopulp VBPF installed in Brazil.

VBPFs can process higher solids loadings per unit of filter area than RVF (approximately 3 x higher). The largest Technopulp VBPF unit has a processing capacity of 1.2–1.5 t/h of mud solids (37 m^2 of filter area). Current designs provide for multistage counter current washing for reduced water usage and higher pol recovery. The VBPFs require slightly similar smaller footprint area (and unit weight) compared to equivalent RVF processing capacity and would require ancillary equipment such as a larger filter drive, wash water and flocculant pumps.

Membrane Chamber Press Filters

Historically, plate and frame presses were the first technology applied for mud filtration but lost favour in the 1960's due to high footprint area required, high operating costs (labour and maintenance) and unsatisfactory washing efficiency [22]. However, the technology has been revisited in recent years.

A modification of simple plate and frame filter press is the Membrane chamber Press Filter (MPF) system.

The chamber press consists of a frame and plates that are held in place during operation by a hydraulic ram. The plates are recessed, forming chambers that fill with mud (or wash water) during operation with one inner membrane (diaphragm) forming the end of each chamber. The inner membrane is able to expand to squeeze the mud cake at higher pressures using compressed air. The membrane squeezing step differentiates the MPF from conventional plate and frame filter presses that only employ the filtration and washing steps. The membrane squeezing step reduces the required feed pressure (from 15 bar down to 6 bar), can produce cake of lower moisture and allows shorter cycle times. The chamber plates including the membrane are covered with porous cloths that are used to filter the mud. The area behind the cloth collects the filtrate and channels it out of the plate for removal.

Allen and Wimmler [23] reported on a Netzsch MPF that was able to produce cakes with around 10 pol % mud solids which is slightly higher sugar loss than typical "good practice" RVF operation. The high pol losses resulted from poor washing efficiency. The high levels of compaction of the mud within the membrane chamber would help to trap pol in pockets and small pores by increasing the specific cake resistance and hence limit displacement washing mechanisms. Early designs also incorporated a feed directed into one quadrant of the chamber rather than uniformly across all quadrants with similar addition for wash water (but a different quadrant to the mud feed).

This sub-optimal design leads to non-uniform cake production over the entire chamber and poorly distributed contact between cake and water and coupled with the non-uniform cake thickness, non-uniform washing results.

Modifications to equipment design have helped to improve washing efficiency. While the MPF operates in batch cycles, the process can be automated and multiple units allow a measure of continuous operation to avoid build up of mud stocks within the factory. One of the advantages of MPFs other than their ability to produce low moisture cakes are their cleaner operation with less vapour and dust produced compared to RVFs that are open to the environment.

For equivalent processing capacity to RVFs, the MPF would require almost double the filter area and weight but this is accommodated in a much smaller and compact footprint area.

Additional ancillary equipment required includes a small hydraulic device, membrane inflation compressor, large feed pump and cloth washing plant.

The MPF operates under the following steps which are illustrated in Figure 8:

1 Filling. The plate pack of the press is closed and sealed by a hydraulic cylinder. The membrane chambers are then filled with filter feed (between the two cloths) by pumping the feed to the unit.
2 Filtration. The filter feed is pumped into the membrane chambers under gradually increasing pressure (to a maximum of 5–6 bar). Filtrate is forced through the end and inner membrane cloths and is collected. Over time, the formed cake builds up and the feed pressure rises to overcome the increasing resistance of the thicker cake. The cake grows from the filter cloth inwards until it fills the entire chamber.
3 Washing. Water is added under pressure to the inner membrane. The water then passes through the mud cake displacing sugar as it is directed through only the end cloths where the filtrate is collected.
4 Squeezing. Compressed air (15–20 bar) is added to the inner membrane which expands to squeeze the mud cakes between the cloths. Filtrate is forced through the end cloths and the inner membrane cloths where it is collected.
5 Drying/flushing. Compressed air is added under pressure into the membrane chambers to flush out any remaining fluid inside the mud cake and through the end cloths where it is collected. It also removes any remaining filtrate. A drying/flushing step may also be performed both before and after the washing step to flush out the water and feed from the inlet and filtrate pipes.
6 Cake discharge and cleaning. The pressure is released and the chambers are opened. This allows the cake to be discharged by gravity or mechanical means. The membrane cloths can then be

washed if necessary. The cake discharge step is fully automated in commercial installations.

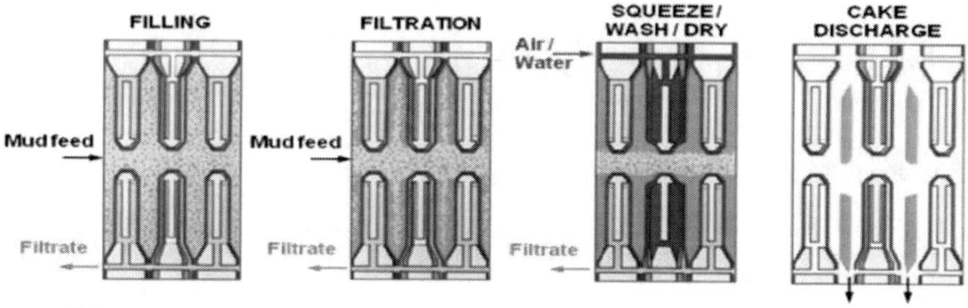

www.andritz.com.

Figure 8. Diagram of the MPF operation.

The washing, squeezing and drying steps are optional but help produce low moisture cakes in shorter cycles.

Full scale MPF units have been installed and operated in China and Asia. Reported operational conditions and performance data include:

- Cycle times of two hours with filtration to a maximum of 4 bar pressure and cake washing at pressures of 4–6 bar. Membrane squeezing was performed at 7–8 bar. Lower pressures reduce the wear and tear on the filter and membrane cloths.
- Cake pols were ~4 % and cake moistures of ~ 50 % were produced.
- Wash water rates were similar to RVFs.

A typical industrial MPF is shown in Figure 9.

www.andritz.com.

Figure 9. Typical industrial MPF manufactured by Andritz.

Centrifuge

Centrifuge technology has not received the same widespread acceptance as RVFs and VBFs but this technology has merit as it can produce a product with less than 50 % moisture and can efficiently recover sugar.

There have been numerous investigations by centrifuges in the past 40 years by two research organisations SRI/QUT and BSES [2, 24-27].

The first series of trials were conducted in 1973-1976 and the centrifuge was installed at a factory over a number of seasons.

The technology was later reviewed in 2003-2005. The centrifuges used in the 1970s investigations had limited capacity and the final economic assessment favoured RVFs. Sugar recovery was aided by adding dilution water to the feed or by adding wash water to the bowl, although adding wash water had a negative impact on solids retention.

Previous Studies

Early investigations identified that although good sugar recovery was possible, mud solids retention and erosion were key issues.

- Mud solids: Two separate early investigations found that increasing the fibre ratio to 0.4 and using flocculant improved mud solids retention to 65 %; some trials operating at a pH of 10 resulted in mud solids retention of 95 % [24]. The quality of the bagacillo also plays a major role with finer bagacillo greatly improving mud solids retention.
- Erosion: The erosion was controlled by removing the coarsest components (i.e., sand) using hydro-cyclones prior to the centrifuge [25-27].

Although technically competitive at the time, centrifuges were noted to be less economical than RVFs and Stewart and co-workers [25-27] foresaw that the technology should be reviewed as their capacity increased. To this end, the technology was reviewed in 2003-2005.

Operational Principle

The decanter centrifuge consists of a rotating inner screw contained within a rotating casing. A cross-sectional view of a centrifuge during operation is shown in Figure 10. The mud is fed through the inner screw and is discharged into a horizontal zone called the centrate pool. The insoluble mud solids are separated from the liquid (centrate) by centrifugal forces.

The mud solids are transported through to the discharge by a screw conveyor which rotates at a slight speed differential to the bowl. The mud solids concentrates through an incline to a region known as the 'beach' drying region and the centrate overflows adjustable weirs at the opposite end.

The speed differential between the bowl and the scroll plays a vital role in adjusting the final product moisture. The weir depth changes the amount of solids in the centrate and also affects final product moisture.

Other parameters affecting performance include the feed rate, the level of flocculate addition and how the feed mud has been conditioned.

Internal baffles can also be installed to press the dry product and further reduce moisture content.

Performance Characteristics

Two Alfa Laval centrifuges with a capacity of 10–12 t/h were investigated. These were a P3400 and a G2 40 unit (Figure 11).

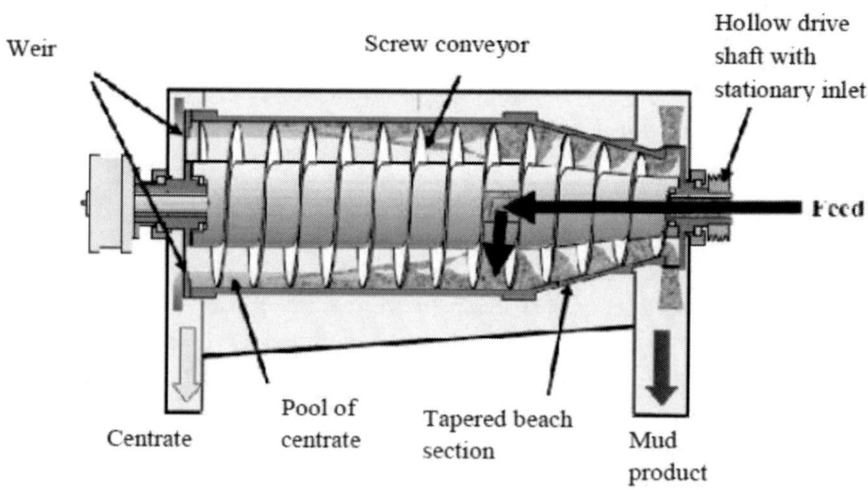

Courtesy of Alfa Laval.

Figure 10. Cross-sectional view of a decanter centrifuge.

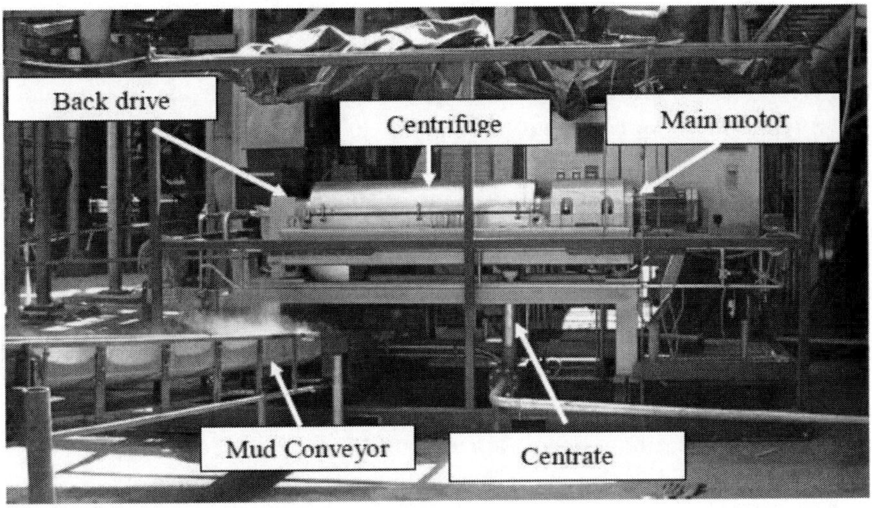

Figure 11. Pilot G2 40 decanter centrifuge [2].

Trials with the G2 40 unit were more successful mainly because of the higher level of instrumentation, particularly torque monitoring instrumentation which was useful in providing information about the degree of separation between solids and liquid within a centrifuge. Final cake moistures of 50 % were possible. The investigation ultimately found that the centrifuge was slightly more expensive than an equivalently sized RVF and the technology was not implemented in Australia. Since the 2003-2005 study, Alfa Laval have developed a G3 unit which it claims is capable of 10 % higher throughput (or dryer cake) and 40 % lower energy costs [28].

Trials showed that with no flocculant addition, increasing the fibre ratio increases mud solids retention (Figure 12) but reduces the final moisture (Figure 13). However using relatively high flocculant levels (~1000 ppm on mud solids), the centrifuge was able to achieve virtually 100 % mud solids retention and product of 53–58 % moisture for a feed fibre ratio of 0.3–0.4. At high flocculant rates, the feed rate to the centrifuge had little impact on mud solids retention.

Distribution of Mud Product

One of the interesting features of the VBFs and Centrifuge is the spreadability of the final mud product.

The cake produced from VBFs was easy to spread with existing trucks as the cake granulated and crumbled readily based on its low moisture. Spreading the drier mud product from a Centrifuge was investigated using two methods: (i) a truck which is conventionally used to spread gypsum (these trials generated a lot of dust, see Figure 14, which could be potentially reduced by using more moist mud); and (ii) a standard mud truck which discharges by tipping the mud from the back.

Figure 15 shows the auger distributor at the back and base of a truck bed that helps distribute the mud product during spreading. The mud is evenly distributed onto cane fields, see Figure 16. Little dust was generated during unloading from standard mud trucks although as the mud dried a small amount of dust was created. Other mud transport trucks such as fertiliser trucks with rotating discs or slat conveyors rather than the auger arrangement shown in Figure 15 are utilised in Brazil for distributing drier mud product.

Centrifuges produce a product which has relatively low bulk density. Compared to RVFs, trucks carrying centrifuged mud would need to be slightly larger (~10 %) to take the same number amount of mud solids whilst still carrying the same weight. On the other hand, the mud cake produced by VBFs was of similar density to RVF cake and due to lower moisture would require less number of trucks to transport the same amount of mud solids.

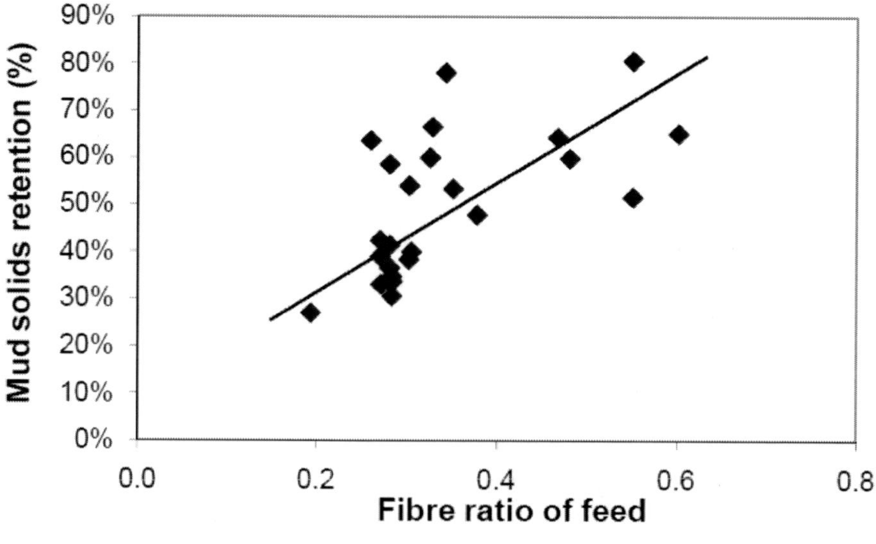

Figure 12. Effect of fibre ratio on mud solids retention (no flocculant).

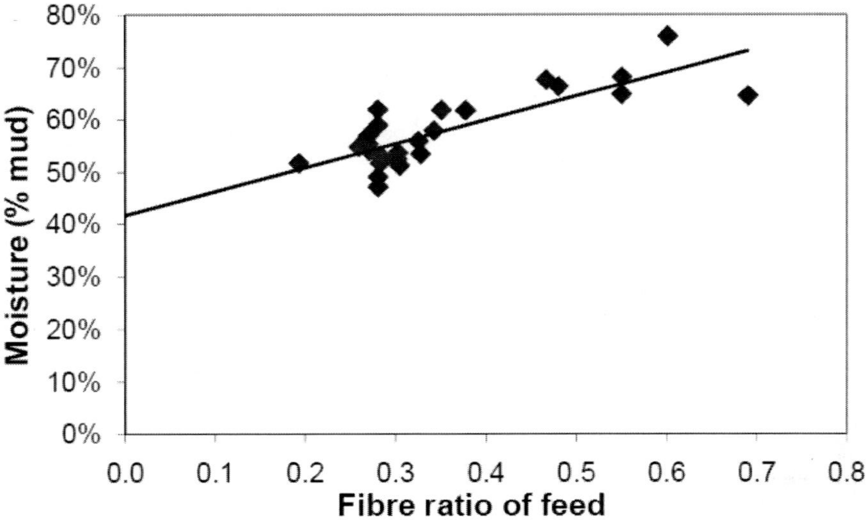

Figure 13. Effect of fibre ratio on the final moisture of mud product (no flocculant).

Figure 14. Centrifuge mud spread by gypsum truck during trials [2].

The distribution of mud product onto cane fields is not the only transport issue. The conveyors, storage and hopper systems are one area that currently restricts the transport of drier mud products, at least in Australia due to current equipment configurations.

Some mills add additional water to the conveyors and hopper systems to improve mud transportability and flowability into the mud transport trucks.

Figure 15. Auger distributor at the back base in the bed of the mud truck.

Figure 16. Centrifuge mud spread by a conventional mud truck.

Further Developments Impacting on Mud Separation

Mud Solids Recycling

The juice clarification system employed in the sugar factory is limited by its capacity to process mud solids. Mud filtration aims to retain as much of the mud solids in the cake but often the filtrate is recycled back to the process adding 15–20 % of mud solids to the clarification system.

In some overseas factories a separate clarifier for filtrate is utilised to produce filtrate that is forward processed within the factory to improve juice clarification capacity and performance [29]. This processing strategy is largely based on the need to improve overall factory sugar quality and recovery that results from the detrimental effects of recycling filtrate backwards to the process. Filtrate can also be added to the mud to condition the filter feed for improved filtration and reduced sugar losses [11].

Diluted filter feed can produce cakes of high porosity enabling more effective washing but typically 10–20 % filtrate is recycled in this manner.

Automation

The filter station is one of the last remaining sections of the factory that relies heavily on operators to maintain optimum performance. While the monitoring of mud conditioning characteristics such as mud solids loading and fibre ratio prior to the filter would provide pre-emptive advantages for optimising control of the filter stations, such instruments are not currently commercially available. An alternative is the monitoring of operating parameters after the clarifier to provide an early indication of filter processing problems. Key parameters requiring continual operator monitoring and manual adjustment include the mud-juice interface in the clarifiers and the conditioning of the feed to the rotary vacuum filters. The mud level can change dramatically in minutes depending on the quality of the cane supply.

Smoothing out the flow of mud solids within the factory with increased levels of automation is a good strategy for improving the performance of the clarifier and filter station. The performance of the clarifier is directly linked to the performance of the filters through the recycle of soluble and insoluble impurities in the filtrate and the operation of the clarifier is a critical factor in achieving good quality sugar. In recent years ultrasonic sonar, fibre optics, guided wire radar and penetrating pulse devices have been considered for the measurement of the mud-juice interface level in clarifiers.

However these instruments need to overcome the harsh and specific operating conditions with a requirement for the mud level transducer to withstand temperatures in excess of 100 °C while fully submerged in cane juice, the system must be capable of identifying a mud-juice interface with a density differential of no more than 10 kg/m^3 and the system must providing a signal that is suitable for control, without interference from suspended mud particles, fibre and moving or stationary clarifier components.

Similarly microwave technology has been considered for indirectly measuring density and concentration as a gauge of mud conditioning [30].

These two examples show how if these technologies can be utilised, control procedures could be developed to provide optimal performance of the clarifier and filter stations.

Comparison of Mud Filtration Technologies

Table 1 presents a brief comparison between the technologies. The list is presented in the order of the most prominent technology in the industry (RVFs) to the least prominent (i.e., Centrifuges).

The filtration performance of RVFs is very well established and so are the issues, such as large footprint, high mud moisture content, high initial investment costs and known maintenance issues. RVFs will continue to be utilised in new factories based on risk minimisation although the main interest in RVFs is in improving performance of existing units.

Vacuum Belt Filters are becoming increasingly popular due to their lower initial investment cost, lower final moisture content and excellent mud solids retention but they have higher maintenance costs.

The authors expect the proliferation of VBFs to continue in their various embodiments (with and without vacuum and additional press sections) and that performance will continue to improve. Membrane Press Filters produce cake with excellent final moisture content and they perform with good mud solids retention and a slight cost advantage over RVFs.

However, this is evaluated against higher sugar losses and maintenance costs. It is a little unclear as to what their future uptake will be, but recent trends suggest further reduction in capital costs and improved designs to reduce sugar losses will increase the installation of MPF technology in the sugar industry.

Finally Centrifuges also produce excellent final moisture content, produce an easily distributed mud and have a small footprint, although electricity costs are high and maintenance requirements are uncertain.

Centrifuges are becoming more cost-competitive with RVFs but may not become mainstream due to the recent proliferation of VBFs as the main alternate technology. They may find a niche in factories where footprint is a primary concern or the company has a certain strategy for distributing mud.

Table 1. Comparison of mud separation technologies

Technology	Sugar recovery	Final moisture content	Mud solids retention	Cost	Ancillaries	Footprint	Maintenance	Distribution of mud
Rotary Vacuum Filters	Reasonable (0.3–0.4 pol loss % pol in cane)	75–80 %	>90 %	Can be expensive	Medium electricity consumption -vacuum pump -wash sprays	Largest	Long life expectancy. Issues with screen maintenance which is dependent on feed conditioning.	Limiting
Vacuum Belt Filters	Reasonable if high flocculant used (0.3–0.35 pol loss % pol in cane)	60–70 %	>95 %	Relatively cheap especially if manufactured in low cost country	Higher electricity consumption -vacuum pump - floc pump -wash facilities	Slightly less than RVFs	Slightly more maintenance than RVFs. Need to replace belts every 2-4 years	Reasonable
Membrane Press Filters	Fair (0.6–0.7 pol loss % pol in cane)	55–60 %	>95 %	Slightly cheaper than RVFs	Medium electricity consumption -feed pump - compressor -wash facilities	Medium but heavy weight (structural requirements may be needed)	Higher costs than RVFs.	Good
Centrifuges	Comparable to RVF. For good recovery, require internal basket wash	50–60 %	>85 %	More expensive than RVFs but continually improving	Very high electricity consumption (approx double RVF usage)	Smallest	Reasonable with sufficient feed conditioning. Long term costs not known due to limited application	Excellent although dust potentially an issue

When comparing mud separation technologies, a factor for consideration is the availability of expertise for the selected technology within the sugar company and at cooperating nearby sugar factories.

Acknowledgments

The authors acknowledge the work of Mr. Rod Steindl and Dr. Floren Plaza for their contributions on the various filtration technologies.

The authors are not associated with distributors of any filtration technology and there is no conflict of interest.

References

[1] Doherty, W. O. S., et al., Flocculation and sedimentation of cane sugar juice particles with cationic homo- and co-polymers. *Journal of Applied Polymer Science,* 2003. 90(1): p. 316-325.

[2] Steindl, R. J., T. J. Rainey and F. Plaza, Low moisture mud from solid bowl decanters, In: *Proceedings of the International Society of Sugarcane Technologists*, 2010: Veracruz, Mexico.

[3] Qureshi, M. E., M. K. Wegener and G. M. Mason, *Mill mud case study in Mackay.* 2000, Cooperative Research Centre for Sustainable Sugar Production: Townsville.

[4] Steindl, R. J., Reducing the specific cake resistance to reduce the pol loss in filter cake, In: *Proceedings of the International Society of Sugarcane Technologists,* 2001. p. 392-394.

[5] Rainey, T. J., A study into the permeability and compressibility properties of Australian bagasse pulp, In: Faculty of Built Environment and Engineering. 2009, *PhD Thesis*, Queensland University of Technology: Brisbane.

[6] Banks, P. J., *Theory of constant-rate expression and subsequent relaxation, in Chemeca*, 1984. p. 813-822.

[7] Owen, D. R. J., S. Y. Zhao and J. G. Loughran. An overview of crushing theory investigations at Swansea. In: *Proceedings of the Australian Society of Sugarcane Technologists*, 1994. p. 264-270.

[8] Kent, G. A. and N. J. McKenzie, A simple test for determining some important material properties for bagasse. *Proceedings of the Australian Society of Sugarcane Technologists*, 2003. cdrom.

[9] Landman, K. A., C. Sirakoff and L. R. White, Dewatering of flocculated suspensions by pressure filtration. *Physics of fluids A*, 1991. 3(6): p. 1495-1509.

[10] Landman, K. A., L. R. White and M. Eberl, Pressure Filtration of Flocculated Suspensions. *AIChE Journal*, 1995. 41(7): p. 1687-1700.

[11] Rein, P., *Cane sugar engineering*. 2007, Berlin: Bartens.

[12] McGrath, G. J., Filtration and washing aspects of filter mud, In: *Proceedings of the Queensland Society of Sugarcane Technologists*, 1971. p. 261-267.

[13] Noble, A. G., The use of polyelectrolytes in mud filtration, In: *Proceedings of the Queensland Society of Sugarcane Technologists*, 1971. p. 251-259.

[14] Noble, A. G., The effect of flocculants on filter performance. *Proceedings of the Queensland Society of Sugarcane Technologists*, 1972. p. 303-314.

[15] Wright, P. G., Modelling the effects of soil and other factors on mud filtration, In: *Proceedings of the Australian Society of Sugarcane Technologists*, 2006. p. 507-517.

[16] Wright, P. G., C. C. Steggles and R. J. Steindl, The balance between capacity and performance of rotary mud filters. *Proceedings of the Australian Society of Sugarcane Technologists,* 1997. p. 434-440.

[17] Hale, D. J., E. Whayman and A. L. Willersdorf, Mud filtration-preliminary studies, In: *Proceedings of the Queensland Society of Sugarcane Technologists*, 1971. p. 269-279.

[18] Lionnet, G. R. E., Mud conditioning for good filter operation, In: *Proceedings of the South African Sugar Technologists Association*, 1984. p. 39-41.

[19] Crees, O. L. and A. L. Willersdorf, An evaluation of a horizontal vacuum filter, In: *Proceedings of the Australian Society of Sugarcane Technologists,* 1983. p. 215-218.

[20] Hugot, E. and G. H. Jenkins, *Handbook of cane sugar engineering*. Sugar Series, ed. G. H. Jenkins. Vol. 7. 1986: Elsevier.

[21] Kulkarni, D. P., *Cane sugar manufacturing in India*. Vol. 1. 2009: The Sugar Technologist Association of India (STAI).

[22] Kruger, G. P. N., The application of a horizontal vacuum belt filter to smuts dewatering and cane mud filtration, In: *Proceedings of the South African Sugar Technologists Association*, 1984. p. 34-38.

[23] Allen, C. J. and W. Wimmler, The membrane chamber filter press- an alternative to rotary vacuum filters, In: *Proceedings of the Australian Society of Sugar Cane Technologists*, 1994. p. 340-343.

[24] Hale, D. J., et al., Mud Centrifugation- Preliminary Trials, In: *Proceedings of the Queensland Society of Sugarcane Technologists*, 1974. p. 243-248.

[25] Stewart, P., et al., Evaluation of a solid bowl centrifuge, In: *Proceedings of the Queensland Society of Sugar Cane Technologists*, 1974. p. 249-259.

[26] Stewart, P., A. G. Noble and G. A. Brotherton, Centrifuge perofmance in the treatment of cane mud, In: *Proceedings of the Queensland Society of Sugar Cane Technologists*, 1975. p. 319-329.

[27] Stewart, P., A. G. Noble and G. A. Brotherton, Improved mud centrifugation at Mossman, In: *Proceedings of the Queensland Society of Sugar Cane Technologists*, 1975. p. 235-240.

[28] Alfa Laval. Aldec G3. 2014 [cited 2014 10/06/2014]; Available from: http://www.alfalaval.com/products-and-solutions/pages/products-and-solutions.aspx.

[29] Singhal, S., et al., Treatment of filtrate from rotary vacuum filters to avoid recycling, In: *Proceedings Of the International Society of Sugar Cane Technologists,* 2005. p. 170-171.

[30] Theisen, K. H. and I. Geyer, Microwave technology: An ubiquitous in line measurement instrument for density, total solids, concenrtation, especially brix in today's sugar factory. *International Sugar Journal*, 2007. 1301(109): p. 310-319.

In: Sugarcane
Editor: Eleanore Webb
ISBN: 978-1-61122-356-9

Chapter 11

ENVIRONMENTAL IMPACTS OF SUGARCANE PRODUCTION, PROCESSING AND MANAGEMENT: A CHEMIST'S PERSPECTIVE

Solomon Omwoma[1*], Moses Arowo[2], Joseph O. Lalah[3] and Karl-Werner Schramm[4]

[1]Department of Chemistry, Maseno University, Maseno, Kenya
[2]Department of Chemical & Process Engineering, Moi University, Eldoret-Kenya
[3]Department of Chemical Science and Technology, Technical University of Kenya, Nairobi, Kenya
[4]Molecular Exposomics, Helmholtz Zentrum Muenchen, The German National Research Centre for Environmental Health, Neuherberg, Munich, Germany

ABSTRACT

Sustainable sugarcane production and processing requires intensification of benefits and minimization of both short term and long term loses. Identification of long term loses/benefits from sugarcane production and processing is a difficult venture that entails critical scientific analysis based on collected scientific data, historical events and laboratory experiments. Moreover, most companies do not invest in research activities geared towards identifying critical long term loses or

* Corresponding author: E-mail: solomwoma@yahoo.com.

benefits. The long term effects of agronomic activities in sugarcane farming are herein discussed. Disposal of processed and unprocessed wastes from sugarcane production and processing activities are also discussed with various possible technical solutions and scientific techniques of effectively generating profits from such wastes summarized. It is hoped that the diverse green technologies of sugarcane production and processing explored herein can be of significant contribution to the management of this vital sector of the economy.

1. INTRODUCTION

Sugarcane (*Saccharum* spp.) remains to be the world's largest cash crop with estimates of 23.8 million hectares in more than 90 countries, and with a worldwide harvest of approximately 1.69 billion tons by the year 2010 [1]. There are approximately 10 species of sugarcane with *saccharum officinarum*, *saccharum robustum* and *saccharum spontaneum* being dominant [2]. Environmental conditions necessary for sugarcane cultivation include tropical or temperate climatic conditions with plentiful supply of water: approximately for more than six months annually - either from rainfall or irrigation; with frost not favoring its growth and up to altitudes of 1,600 m [3]. Although this type of climate exists between 22 °N and 22 °S and some up to 33 °N and 33 °S, other regions outside this range such as the Natal region of South Africa still grow sugarcane due to anomalous climatic conditions such as warm ocean currents that sweep down the coast [3].

Sugarcane is one of the most efficient photosynthesizers in the plant kingdom categorized as a C4 plant due to its ability to convert up to one percent of solar energy into biomass [4]. In fact, Rolph [3] simply refers to it as '*Sugarcane is nothing more nor less than a concentrated sunshine'*. As such, sugarcane is referred to as a carbon crop since sugar and biomass are harvested rather than its protein-rich grains. Sugarcane is grown on different types of soils including the highly fertile well drained mollisols, heavy cracking vertisols, infertile acid oxisols, peaty histosols, rocky andisols, etc. with plentiful sunshine and water supplies increasing its production [3]. For this reason, arid countries with good irrigation schemes such as Egypt have emerged as excellent sugarcane producers [5].

Although there are documented guidelines for effective and sustainable agronomic practices in sugarcane cultivation [6], there are certain long term effects that can be determined through effective record keeping of events and analytical data of every step involved in cultivation and processing. This chapter attempts to critically analyse some of the data and scientific reports collected over a long period of time with great emphasis on long term effects of sugarcane production and processing to the environment.

2. Impact on Atmospheric Carbon Dioxide Concentration

Global warming, the unequivocal and continuing rise in the average temperature of earth's climate system, is due to the heat retaining phenomena of greenhouse gases such as CO_2, CH_4, H_2O, O_3, N_2O, etc. These gases generate an increase in the earth's temperature by about 33 °C, thus in their absence, the average earth temperature would be -19 °C (currently it is 14 °C) [7]. Initially, the gases were naturally generated and maintained through processes such as the water and carbon cycles. However, anthropogenic activities such as land-fills, burning fossil fuels, clearing of forest cover, industrial processes, power stations, etc. are currently increasing their levels at an alarming rate [8]. Consequently, there has been a remarkable increase in the average earth's temperature by 0.8°C, and if this continues unabated, then there is high risk of extreme severe consequences of global warming such as rising sea level, decreased snow cover in the northern hemisphere, species extinction, shutdown of thermohaline circulation, etc. [9,10].

Emission of CO_2 to the environment is considered as a primary factor in causing global warming [11]. In 1999, Schoen reported an increase of 80 ppm in atmospheric CO_2 within a time frame of 200 years, with most increment occurring in the past 50 years, as compared to the previous 80 ppm recorded over 10 000 years [12]. For this particular reason, efforts have been put in place to control the release of CO_2 into the environment [11-15] as well as its removal from the environment [16]. However, it should be noted that the best CO_2 sequestarators remains to be plants with sugarcane being the most favorable due to its economic importance, early maturity and higher photosynthesizer abilities of category C4 in the plant kingdom [4,17].

Studies have been done through incorporation of radioactivity ($^{14}CO_2$) into sugarcane leaves as a function of time in order to determine a steady state of photosynthesis under physiological conditions of concentration of carbon dioxide and light intensity [18,19]. The results showed the presence of large proportions of ^{14}C in 3-phosphoglycerate, hexose monophosphates and sucrose. The ^{14}C appeared first in C-4 of the dicarboxylic acids and C-1 of 3-

phosphoglycerate. The labelling pattern in hexoses were consistent with their formation from 3-phosphoglycerate. The reaction giving rise to C4 dicarboxylic acid appeared to be the only quantitatively significant carboxylation reaction in sugarcane leaves. This research findings based on successful incorporation of ^{14}C into the C4 dicarboxylic acid pool and its subsequent transfer to sugars via 3-phosphoglycerate (Figure 1) prove that sugarcane is a C4 photosynthesizer [18].

Technically, C4 photosynthesizers fix atmospheric carbon at the β-position of phosphoenolpyruvate (PEP) by the action of phosphoenolpyruvate carboxylase (PEPC) in the cytoplasm of mesophyll cells. The oxaloacetate so formed is then reduced to malate in the chloroplasts by NADP-malic dehydrogenase (NADP-MDH) or transformed to aspartate by transamination. These acids are then exported to the bundle sheath cells, where decarboxylation occurs (via malic enzyme or PEP carboxykinase) to yield CO_2 that is re-fixed by the reductive pentose phosphate (RPP) pathway operative in these cells. The other three carbon atoms are recycled to the mesophyll cells in the form of pyruvate or alanine, where PEP is generated by the chloroplast enzyme pyruvate, Pi dikinase (PPDK) [20].

However, it should be noted that sugarcane farming activities including use of bio solids as fertilizers increase soil carbon stock hence increasing the release of the same carbon dioxide gas being fixed by plants [21-23], although this is considered insignificant as compared to the fixing rate.

3. Impact on Soil pH

Continued use of agronomic inputs such as nitrogenous fertilizers in sugarcane farming eventually lowers the soil pH [22,24]. In Papua New Guinea, the pH of top soils under sugarcane cultivation decreased from 6.5 to 5.8 between 1979 and 1996 [24]. In Fiji, a decline in soil pH from 5.5 to 4.6 was recorded over the first 6 years of cane cultivation while in Philippines, a reduction from 5.0 to 4.7 was recorded over 19 years [24]. This change in soil pH is mainly due to the use of acidifying nitrogenous fertilizers such as urea and ammonium phosphates, coupled with nitrate leaching that occurs under the high rainfall conditions that often prevail in cane cultivation areas.

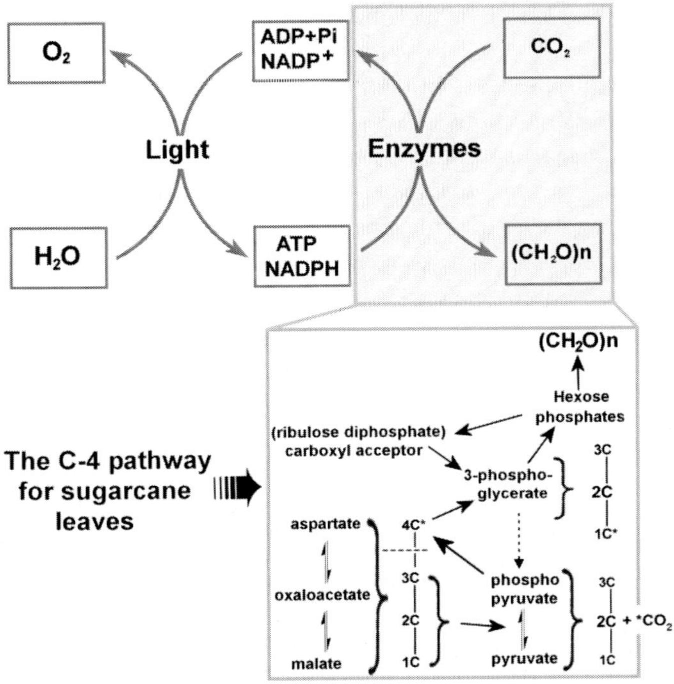

Figure 1. The C4 photosynthetic pathway for fixation of atmospheric carbon dioxide in sugarcane leaves. The dotted arrow indicates a minor pathway [18].

Changes in soil pH are usually accompanied by other chemical changes in both the soil and the receiving waters. For instance, a reduction in soil pH makes heavy metals more soluble, bioavailable and mobile [25]. Hence, continued use of nitrogenous fertilizers affects the acidity of soils and makes heavy metals readily available for transportation from soils to the aquatic environment; especially during long rainy season accompanied with large surface runoffs, leaching and erosion of the soils [26,27]. Furthermore, the use of biosolids as fertilizers in sugarcane farming increases the total organic carbon of the soil hence influencing heavy metal transport in soil [28].

In addition, it has been scientifically proven that nitrogenous fertilizers are contaminated with heavy metals that accumulate in soil with repeated application [22,29]. Lawrence and Brian from Oregon State University point out that contamination of nitrogenous fertilizers with heavy metals has drawn the attention of farmers, environmental organizations, consumers, and public policymakers [30]. In his book, Alloway clarifies that inorganic fertilizers contain traces of heavy metals that accumulate in soils with repeated applications [25].

Most importantly, the use of these fertilizers affects the soil pH, which is one of the chemical conditions of soil and a significant secondary determinant of heavy metal transport and fate at the application site [31]. Firstly, ionization of metals increases at low pH thereby increasing their water solubility and mobility. Secondly, hydroxonium ions (H_3O^+) displace most other cations on negative surface charges. These mechanisms have been clearly shown to reduce metal adsorption by cation exchange and organic complexation [32].

Once heavy metals find their way into aquatic environments, a large amount of them get deposited into the sediments due to other factors like dilution factor, sedimentation and precipitation [33]. Heavy metal analysis of sediments and water samples collected from contaminated sites show the concentration in sediments to be of several orders of magnitude greater than in water [33]. Sediment associated heavy metals pose a direct risk to detrital and deposit feeding benthic organisms, and may also represent a long-term source of contamination to higher trophic organisms. Bioaccumulation and bio concentration of toxic heavy metal residues in aquatic environments can result in their transfer into food chains putting terrestrial consumers including humans and birds at risk [34-37]. Contaminated food webs can also cause health and economic disadvantages to people as contaminated commercial foods like fish become restricted or banned due to high metal burdens [37].

Heavy metal solubility can be affected by several factors including temperature and pH changes [38]. However, the presence of heavy metals in aquatic environment will affect its electrical conductivity, chemical oxygen demand and dissolved oxygen [39]. Adverse effects including death of animals due to lack of oxygen may arise if the above physicochemical parameters exceed the allowable limits [40,41].

However, another major concern is the fact that not all the applied fertilizers are utilized in the soils. Most of the inorganic nitrogen and phosphates applied find their way into aquatic environments due to surface runoffs and leaching into ground water. Previous studies have shown that addition of these nutrients into water systems results in large proliferations of algae and other aquatic weeds such as water hyacinth, which have detrimental effects on the water quality [42,43]. Algal blooms and water hyacinth deplete oxygen supply in the water system and are also harmful to other aquatic species. Additionally, nutrients cause taste and odor problems that result in reduced recreational use, and increased water treatment costs [42].

3.1. The Process of Soil Acidification

Soil acidification (declining soil pH due to net proton (H^+) accumulation) is a natural process that occurs during pedogenesis and is often associated with high rates of leaching. However, the rate of acidification can be accelerated through farming activities with the result that the soil resource may become significantly degraded. Under intensive agricultural production, continual acidification of these soils is likely to occur through the use of high inputs of ammonium-based nitrogen fertilizer, the high level of base removal as a result of crop uptake and subsequent removal and the generally high rainfall environment of the region which facilitates losses of basic cations through leaching [44]. Acidification of soils is usually accompanied with fertility loss and declining productivity [45], hence placing the products of farming systems operating in acidifying environments subject to scrutiny under the recent ISO 14000 treaty [44].

Nitrification of ammonium based fertilizers such as urea and organic N in crop residues (Organic Matter) is an acidifying reaction which occurs through the microbial conversion of NH_4^+ to NO_3 with the consequent production of protons (H^+). The extent of acid generation by fertilizers is a function of the fertilizer type, environmental and edaphic factors [46]. Estimates of potential net acidity generated by frequently used fertilizer sources in the sugar industry are presented in Table 1, with diammonium phosphate being the most acidifying and urea the least on the basis of N per kg fertilizer applied. In view of the potential acidity generated by nitrogenous fertilizers, the equivalent amount of $CaCO_3$ required to neutralize this acidity is also shown (Table 1). The theoretical amount of $CaCO_3$ required to neutralize the acidity generated by application of 180 kg N ha $^{-1}$y^{-1}, typical for the NO_3^- from ammonium-based fertilizer is a significant source of acidity generation in these production systems. In contrast, the addition of basic nitrate fertilizers such as $Ca(NO_3)_2$ causes little change in pH due to the absence of nitrification and may in some cases result in an increase in soil pH [46,47].

Although the production of NO_3^- through nitrification process for nitrogenous fertilizers is a net proton accumulating reaction, the subsequent leaching of nitrate can lead to a significant decline in exchangeable bases because Ca^{2+} and Mg^{2+} will move downwards as counter ions for the very mobile NO_3^-, resulting in an accumulation of protons at the point of nitrification [48]. Consequently, there is spatial disjunction between the production of NO_3^- and its subsequent uptake by the plant. The result is an accumulation of H^+ at the point of production and net alkalization due to the uptake of NO_3^- by the plant at some other point in the profile.

The uncoupling of these two processes results in net proton accumulation at one point in the profile and net alkalization at some other point. As long as nitrate is taken up at the point of production, the outcome will always be neutral [47].

3.2. Scientific Evidence of Accelerated Acidification under Sugarcane Production

Soil acidification rates can be measured in terms of absolute changes or relative to some control soil. In the former case, acid addition rates can be estimated from analyses of soils before and after a given period of acidification- long-term study [47]. However, relative rates of acidification can also be derived from survey data (e.g. fence line contrasts of developed and undeveloped sites). This approach has been used in a number of studies in the sugar industry [44,49]. The use of fence line comparisons essentially results in a conservative estimation of the net acidification rate.

Table 1. Estimated potential acidity produced as a result of nitrogenous fertilizers in sugarcane plantations [47]

Source	Nitrification reaction	Net potential acidity generated (kmols H^+.kgN)	Potential acidity generated from 180 kg N ha^{-1} (kmoles H^+ ha^{-1})	[a]Amount of $CaCO_3$ required to neutralize acidity generated by an application of 180 kg ha^{-1} (kg ha^{-1})
Urea	$(NH_4)_2CO + 4O_2 = 2H^+ + CO_2 + H_2O$	0.072	13.0	650
Diammonium Phosphate	$(NH_4)_2HPO_4 + O_2 = 3H^+ + 2NO_3^- + H_2PO_4 + H_2O$	0.0107	19.3	965

a: assuming that 1 kmol requires 50 kg $CaCO_3$ to neutralize.

As part of a wider study on the possible role of changes in soil properties over time on sugar yield decline [50], differences in soil chemical properties between new land and land which had been under sugarcane monoculture for more than 20 years were examined in detail by Bramley et al. [49]. They found that there was no consistent effect of time under sugarcane monoculture on soil chemical properties across sites located in the Burdekin, Herbert and Tully Districts of North Queensland, Australia; either when the distribution of properties through the soil profile, or property values at specific depths were considered. However, marked effects were observed in some sites with respect to some soil properties and these were generally consistent with soil acidification.

The changes included reduced topsoil pH in old land (0.5-1.5 pH units); increased exchange acidity and aluminum saturation in soils under old land, especially in Herbert where percent Al saturation increased from approximately 5 to 20 %. However, although little change was noted in the Tully soils where much higher Al % (45 %) in both old and new land soils were measured, decreased cation exchange capacity (CEC) and increased anion exchange capacity (AEC) were observed in old land soils in some sites of each region. Wood obtained similar results in a paired sites study in the Herbert River District, North Queensland, Australia; where he found that many of the differences in soil chemical properties between new cane land and land that had been under sugarcane for several years could be associated with soil acidification [46]. Sugarcane soils were found to have a lower pH, lower cation exchange capacity and lower levels of exchangeable base cations (calcium, magnesium and potassium).

In addition, the analysis of soil samples from 1064 sites in the Herbert sugarcane area, taken as part of the CSR (Central Sugar Refinery, in Malaysia) Herbert River soil survey has shown that mean topsoil (0-20cm) pHw (pH in 1:5 soil:water) is 4.97 and mean subsoil (40-60cm) pHw is 5.28 [51]. Schroeder et al. [52] have also reported marked acidification in a range of sugarcane soils in South Africa. The effects of soil acidification on sugarcane growth have received little attention from Australian researchers, presumably because Hetherington [53] concluded that cane was tolerant of low pH-induced aluminum toxicity. However, Schroeder et al. [52] reported marked differences between South African varieties in terms of their response to lime application suggesting that not only was cane affected by the effects of low soil pH, but that these effects may be variety specific. In general, most crops perform better when the soil pH is approximately 5.6 to 6.0 [54].

From the previous discussion it can clearly be seen that soil acidification is a continuing problem confronting the Australian sugar industry and that the largest potential acidifying component is the contribution derived from the use of nitrogen fertilizers [47]. The contribution to the total proton pool arising from the export of millable cane is relatively smaller compared to that derived from the nitrification of ammonium and urea based fertilizer sources, assuming that significant leaching occurs [47]. Most researches on acidification of sugarcane farms have been on sugarcane production with little being done on environmental effect. Alloway [25], points out that reduction in acidity of soil leads to naturally occurring heavy metals in the soils becoming more soluble, bioavailable and enhances their mobility.

Another long term impact of using fertilizers in sugarcane production is the production of N_2O that is a greenhouse gas causing global warming. Specifically, production of N_2O gas from soils in sugarcane plantations occurs after application of vinasse fertilizer followed by urea [55,56].

4. Impact of Surface Runoffs from Sugarcane Fields on Aquatic Ecosystem Health

The health of aquatic systems within and or near sugarcane production and processing zones is increasingly threatened due to exposure to pollutants from both agronomic inputs and processing wastes [27,57,58]. The Australia's Great Barrier Reef and Lake Victoria are examples of such threatened aquatic systems within sugarcane plantation catchments being affected by a range of pollutants such as nutrients and pesticides [26,57,59]. Among other aquatic effects, eutrophication that is associated with uncontrolled aquatic plant growth; algal blooms; biodiversity loss and/or death of aquatic animals; destruction of water transport; and poor quality of domestic water are the major challenges facing sugarcane production and processing. Several management strategies have been suggested to reduce pollutant loading into aquatic systems. Drewry et al. proposed reducing N fertilizer applications, reducing tillage and changing management of fallows prior to planting between sugarcane crops [60]. Omwoma et al. have demonstrated an efficient way of trapping these nutrients from canals draining the sugarcane plantations before they are transferred into aquatic systems [26]. Fulcher et al. have also shown that application of pesticides according to label recommendations greatly reduces pesticides wash-offs [61]. The use of recommended levels of both pesticides and fertilizers [62] is very important and should be accompanied with precise timings so as to avoid rainfall wash offs.

Although the suggested methods do not take care of ground water pollution due to leaching [63] and base flow discharge [64], a great pollutant reduction to other aquatic systems such as rivers and lakes can be avoided as runoffs are the major transport agents of these pollutants. However, factories should invest in weather forecast equipment and personnel, critical soil analytical methods to determine region specific fertilizer suitable requirements and strict management to achieve high crop production within a clean environment [65].

In the Everglades Agricultural Area of Florida, soil subsidence and phosphorus runoff from sugarcane fields to aquatic systems are serious problems being addressed in order to have sustainable sugarcane farming. The management uses rice as a rotational crop in sugarcane fields in order to remove PO_4^{2-} from soils. The rice is grown on sugarcane fields every 4 to 5 years under flooded conditions. Specifically, rice crop absorbs excess phosphorus, and flooding the fields halts subsidence, controls pests, and provides a wetland habitat for native animal species [66].

The weighted global average indicates that only 50 % of N fertilizer applied in crop fields are consumed by crops [67]. The reasons provided for the low fertilizer use uptake include high soil nitrification rates (Table 1) and extreme weather conditions that promote N leaching and denitrification processes [67]. It should be noted that sugarcane plants have higher preference for NH_4^+-N uptake than NO_3^- -N [67,68]. Therefore, there are research activities directed towards breeding sugarcane cultivars with enhanced capacity to use nitrate as well as undertaking agronomic measures that will reduce nitrification in soil [67]. Low nitrate content in shoots of sugarcane cultivars accompanied with low translocation rates of ^{15}N-nitrate to shoots has been reported and it indicates inefficiencies in the uptake and root-to-shoot transfer of nitrate [67]. Even though nitrate or assimilatory products of ammonium and amino acids act as regulatory signals in plants, there is negative feedback arising from endogenous nitrate on transport systems and can be inferred from the whole plant or organ studies which show negative correlation between nitrate concentration and uptake rates [69,70]. For instance, nitrate concentration levels in cytosol of barley roots were found to be 4 mM whereas vacuolar nitrate concentrations increased from 4 to 75 mM when plants were supplied with 0.01 to 10 mM nitrate [71].

Experimental results have indicated that nitrate uptake is inhibited in N-replete sugarcane and this is correlated with the increasing nitrate content in roots [67]. The advantage of nitrate as a N source is the uncoupling of N supply and demand, while ammonium causes toxicity in cells hence the rapid assimilation and is limited by the carbon supply to roots [69]. It is therefore assumed that many nitrophile species will exhibit efficient use of nitrate through rapid transport and storage of nitrate which is considered an evolutionary advanced character in angiosperms [72]. Thus, N uptake in excess of demand and the resulting storage of nitrate occurs when excess ammonium and nitrate are supplied to sorghum and maize but not sugarcane and other related species. These findings suggest a broader spectrum of N use among the studied crops and wild species than previously recognized and question the assumed efficient use of nitrate in sugarcane crop systems.

The problems associated with plant utilization of synthetic N fertilizers use might be solved by the use of microbes that are capable of biological N_2 fixation (bio fertilizers) [73]. Diazotrophic bacteria of the genera Gluconacetobacter [74], Azospirillum [75], Burkholderia [76] and Herbaspirillum [77] have been isolated from intercellular spaces, roots and rhizosphere of sugarcane [78-80]. Nevertheless, there arises some inconsistent responses of crop cultivars and growth locations that might limit the success of 'bio fertilizers' based on diazotrophic and plant growth-promoting rhizobacteria [81]. However combining both technologies might limit losses of synthetic N to the environment leading to eutrophication of aquatic systems [82].

5. Green Utilization of Sugarcane Bagasse

5.1. Bioethanol Production

Large amount of bagasse is generated as a result of industrial processing of sugarcane. Bagasse is the residue obtained after juice extraction through milling of cane. It corresponds to about 25 % of the total fresh weight of the sugarcane plant and it contains between 60 – 80 % of carbohydrates [83]. The abundant chemical composition of sugarcane bagasse include cellulose (32 – 44 %), hemicellulose (27 – 32 %), lignin (19 – 24 %), and small amounts of extractives and mineral salts [84,85]. The most economical and ecofriendly way of disposing this waste should be through enzymatic conversion of the lignocellulosic biomass to bioethanol, though most factories prefer discarding it as agricultural waste or burning it for energy supply [86,87].

The main obstacle in the generation of bioethanol from sugarcane bagasse is supposedly the close association and complexity of the carbohydrate–lignin complex in sugarcane bagasse. Many efforts have been made to overcome this problem and make the process economically feasible. For instance, the development of an efficient pre-treatment step and optimization of enzymatic cocktails for cell wall deconstruction have been investigated. It is more environment-friendly to use enzymes in the bioconversion processes than using chemical processes. The use of enzymes ensures product specificity and minimizes wastes thus making the process more eco-friendly [88]. However, enzymatic hydrolysis requires biomass in small particle sizes and the removal of lignin with phenolic compounds which show more inhibition than non-phenolic compounds [89,90]; hence the need for bagasse-pretreatment prior to enzymatic hydrolysis reactions.

It is difficult to hydrolyze the cellulose present in pretreated lignocellulosic materials due to both enzyme- and substrate-related factors. Changes in cellulose porosity and surface area, accumulation of lignin, and changes on its crystallinity and degree of polymerization can be summarized as substrate-related factors [91,92]. End-product inhibition due to glucose and cellobiose accumulation [93,94], thermal denaturation of enzymes after long periods of mechanical agitation [95], and irreversible adsorption of enzymes onto lignin and/or lignin-carbohydrate complexes [96,97] are classified as enzyme related factors. Enzymatic hydrolysis of cellulose can be achieved through a mixture of microbial hydrolases composed of three different classes of enzymes: 1) endo-β-1,4-glucanases, 2) exo-β-1,4-glucanases or cellobiohydrolases, 3) β-1,4-glucosidases; otherwise called ‘‘the cellulose complex’’. In summary, the enzymatic hydrolysis reaction (by fungal strains like Trichoderma reesei) is achieved through all the three classes of hydrolases that act synergistically for the complete enzymatic hydrolysis of cellulosic substrates. 1) Cellulose reducing and non-reducing chain ends are formed by the action of endo-β-1,4-glucanases, 2) cellobiohydrolases act on these chain ends releasing mostly cellobiose, with cellobiohydrolases I working progressively from the reducing ends while cellobiohydrolases II works from the non-reducing ends, 3) β-1,4-glucosidases complete this process by converting cellobiose to glucose [98].

Therefore, for effective hydrolysis of sugarcane bagasse by enzymes, the following pretreatments of lignocellulosic materials are used: Steam explosion [99-101], Dilute acid hydrolysis [102], Alkaline pretreatment [102,103], Wet oxidation [104], Ammonia fiber expansion [105], Enzymatic pretreatment [106], Organosolvents [107] etc.

5.1.1. Alkaline pretreatment

The alkaline pretreatment methods show less sugar degradation and furan derivatives formation [108] than thermal and acid pretreatment methods [109]. Alkaline pretreatment process may be improved through the application of ultrasound [110], as the ultrasonic treatment of aqueous media produces cavitation, which generates high temperature, pressure and extreme shear forces [111]. As such, the decomposition of water molecules into free radicals by cavitation will aid in breaking the linkages in lignin and xylan networks [112,113]. Ultrasonic energy combined with proper solvents allows destruction of the recalcitrant lignocellulosic structure, fractionation of biomass components, and then assists many thermochemical and biochemical reactions, with increased equilibrium yields of sugars, bio-ethanol and gas products by 10-300 % [111]. Sonication promotes hydrolysis, esterification and transesterification in biodiesel synthesis, reduces reaction time by 50-80 %, lowers reaction temperature and reduces the amounts of solvent and catalyst [111]. Pretreatment of bagasse with alkali/alkaline peroxide and ultrasound for the extraction of hemicellulose from sugarcane bagasse has been reported to yield 90 % hemicellulose and lignin removal [114]. More recently, the use of ultrasound assisted ammonia pretreatment method [115] and the use of ultrasound-assisted alkaline pretreatment of sugarcane bagasse for fermentable sugar production [116,117] have been reported. Under these technologies, the influence of sugarcane bagasse particle size, liquid ammonia concentration, sonication time, temperature and liquid to solid ratio on cellulose recovery and delignification have been evaluated with maximum cellulose recovery and delignification observed at the optimum conditions of particle size 0.274 mm, sonication time 45 min, ammonia concentration 10 %, liquid to solid ratio of 10 mL/g and temperature 80 °C.

Alkaline ethanolysis and sequential enzymatic hydrolysis for production of glucose and subsequent lactic acid has also been demonstrated using physico-chemical treatments, that is, ultrasonic bath and ultrasonic probe and compared to mechanical stirring [118]. The experimental results indicated the highest glucose yield with least contamination of xylose to be obtainable from acid ethanolysis fractionation of 5 N H_2SO_4 + 50 %, v/v ethanol when stirred at 90 °C for 4 h. The alkaline ethanolysis was accomplished with the release of high amount of both glucose and xylose, although it was not a favorable substrate for homofermentative lactic acid bacteria. An interesting one-step treatment of sugarcane bagasse with 80 % acetic acid and 70 % nitric acid mixture under sonication has also been reported to yield 43.0 - 43.6 % of pure cellulose fraction which contain minor amounts of bound hemicelluloses (3.2 - 4.3 %) and are relatively free of associated lignin (0.2 - 0.6 %) [119].

5.1.2. Steam explosion

The process of bagasse pretreatment with steam explosion involves pretreatment of bagasse with saturated steam at 160–240 °C for about 20 to 30 min in the absence (autohydrolysis) or presence of an acid catalyst [94,120-122]. The high pressure steaming results in partial hydrolysis of hemicelluloses and lignin to water-soluble monomers and oligomers while cellulose is modified partially from its crystallinity and degree of polymerization. This process improves cellulose susceptibility to solvation and enzymatic hydrolysis. However, this process sometimes generates biological inhibitors such as furan compounds and organic acids. Nonetheless, most of these inhibitors can be eliminated from the fibrous material by water washing [123,124] although this detoxification step increases production cost due to energy consumption [125]. Another detoxification process can be through extraction of the steam treated material with boiling solvents in order to remove other inhibitory compounds such as phenolic acids [101,126]. Successful steam explosion pretreatment reproduction and alkaline delignification reactions for ethanol production from different varieties of natural sugarcane bagasse, pretreated bagasse and delignified pretreated bagasse have been documented [87].

5.1.3. Ammonia fiber expansion

The use of ammonia fiber expansion process as a pretreatment procedure for bagasse hydrolysis to bioethanol has been described. In this process, liquid ammonia is added to the biomass under moderate pressure (100 to 400 psi) and temperature (70 to 200 °C) before rapidly releasing the pressure [105]. In particular, this process decrystallizes the cellulose, hydrolyses hemicellulose, removes and depolymerises lignin and increases the size and number of micropores in the cell wall, hence significantly increasing the rate of enzymatic hydrolysis [123]. It has been experimentally shown that the process can improve accessibility of cellulose and hemicelluloses during enzymatic hydrolysis by breaking down the ester linkages and other lignin carbohydrate complex bonds in bagasse [127]. Furthermore, maximum glucan conversion of the ammonia fiber expansion process pretreated bagasse and cane leaf residue by cellulases is approximately 85%, and the supplementation with hemicellulases during enzymatic hydrolysis improves the xylan conversion to 95-98 %.

5.1.4. Enzymatic Pretreatment

Pretreatment of sugarcane bagasse can also be achieved using enzymes prior to the hydrolysis reactions in a similar manner as the bleaching process of wood pulp with ligninolytic enzymes [106]. This process is advantageous due to: i) the mild reaction conditions, ii) higher product yields and fewer side reactions, iii) less energy demand, iv) minimal corrosion and pressure build up in the reactors [128]. Naturally, lignin decomposition is primarily attributed to metabolism by organisms especially microorganisms such as white-rot [129]. White-rot produces several ligninolytic enzymes including laccases, manganese peroxidases and lignin peroxidases that catalyze one-electron oxidation of lignin units, producing aromatic radicals [130]. The microbial lignin degradation can be mainly attributed to secondary metabolism or to restricted availability of nitrogen, carbon, or sulphur and it is not degraded as sole carbon and energy sources, requiring additional co-substrates such as cellulose, hemicellulose or glucose [131]. Generally, most white-rot fungi preferentially attack lignin more readily than hemicellulose and cellulose [132,133]. Among these groups of fungi include the *Ceriporiopsis subvermispora*, *Phellimus pini*, *Phlebia* spp. and *Pleurotus* spp. However other White-rot fungi exhibit a pattern of simultaneous decay characterized by degradation of all cell wall components and in this group we have *Trametes versicolor*, *Heterobasidium annosum* and *Irpex lacteus* as examples [134]. This technique has recently been patented by Kumar et al. [135] where they claim a process for one step production of L-Lactic Acid from lignocellulosic biomass using thermophilic bacteria *Paenibacillus macerans* IIPSP3 (MTCC5569), which is not only capable of hydrolyzing cellulose to glucose but also further ferments it to L-Lactic Acid under aerobic conditions without any growth inhibition in the presence of lignin. The invention further provides a process which has less chances of contamination as the fermentation is carried out at higher temperatures and is economically attractive, since preferably no external enzyme loadings are required.

5.1.5. Wet Oxidation Pretreatment Method

Wet oxidation pretreatment method involves hydrothermal treatment which is the process of treating bagasse with water and air or oxygen at temperatures above 120 ºC [136]. The two types of reactions that occur during wet oxidation include 1) a low-temperature hydrolytic reaction, 2) a high-temperature oxidative reaction [137]. It has been demonstrated experimentally that alkaline wet oxidation at 195 ºC for 15 min yields a solid material with nearly 70% of cellulose, with a solubilization of approximately 93 % of hemicelluloses and 50 % of lignin, and an enzymatic cellulose convertibility of about 75 % [137]. However, it should be noted that a significant part of the polysaccharides is lost, and the enzymatic convertibility of the pretreated material is poor.

5.1.6. Organosolvents

This pretreatment technology involves the use of an organic liquid and water, with or without the addition of a catalyst which can either be an acid or alkali. The mechanism involved in removal of lignin from lignocellulosic materials is the partial hydrolysis of lignin bonds to give a pulp rich in cellulose. Addition of a catalyst enhances the selectivity of the solvent with respect to lignin with most of the hemicellulose sugars being reported to be solubilized by this process [138,139]. The advantages associated with this technique over the aqueous based processes include the recovery of lignin and polyoses from the liquor which is easily achieved through distillation [107]. Specifically, lignin is separated as a solid material while polyose fraction is obtained in aqueous solution. Nevertheless, pretreatment of sugarcane bagasse with organosolvents has some limitations such as i) the pretreated solids need to be washed with organic solvent to avoid the re-precipitation of the dissolved lignin and ii) the reactions occur at higher pressures hence not economically viable.

Mesa et al. experimentally showed that the combination of a dilute-acid pretreatment followed by organosolvent pretreatment (with NaOH) under optimized conditions of 60 min, at 195 °C and 30 % v/v ethanol was efficient for the fractionation of sugarcane bagasse with the subsequent enzymatic hydrolysis yielding a residual solid material containing 67.3 % (w/w) glucose [139]. Novo et al. reports a process of using glycerol- water mixtures to obtain a pulp with a residual lignin amount lower than 8 % with delignification being close to 80 % and residual cellulose content higher than 80 % [107]. The use of glycerol here presents some advantages such as: 1) low solvent cost as the crude glycerol produced in the transesterification process for biodiesel production can be applied instead of pure glycerol, 2) the pretreatment can be performed under atmospheric pressure, decreasing the energy consumption, 3) due to its highly polar structure, glycerol can easily penetrate the bagasse tissue, providing an effective reaction medium for delignification. However, the high energy consumption for solvent recovery may decrease the attractiveness of using glycerol [140,141].

5.2. Lignin Production

Lignin extracts from sugarcane bagasse have been utilized as natural and potent substances for coating and preserving fresh fruits [142]. This property of lignin arises out of its inability to be degraded by most forms of biological means and the fact that lignin has antimicrobial and antifungal activities [143]. The best way of extracting lignin without interfering with its chemical composition is by soaking it in dilute (0.5 % v/v) phosphoric acid for 4 h followed by steam explosion at 180 °C for 10 min prior to ethanol extraction as shown in **Figure 2** [144].

5.3. Biohydrogen Production

Sugarcane bagasse has been utilized as a feedstock in biohydrogen production. Biohydrogen is produced biologically via biophotolysis, dark-fermentation and photo-fermentation of sugars such as glucose, fructose, galactose, arabinose, lactose and sucrose [145-148]. Rai et al. have integrated sugarcane bagasse in the dark-fermentation process by *Enterobacter aerogenes* and photo-fermentation by *Rhodopseudomonas* [149]. Sugarcane bagasse was hydrolyzed using sulphuric acid and the hydrolysate detoxified by passing it through adsorbent resin column to remove the inhibitory furfural before being subjected to dark-fermentation process. The cellulosic residue from acid hydrolysis was hydrolyzed by the new isolate *Cellulomonas fimi* to release sugars for H_2 production by *E. aerogenes*, through simultaneous saccharification, filtration and fermentation. Optimum concentration for acid hydrolysis by H_2SO_4 was found to be 2 % (v/v) for 60 min and cumulative H_2 production during dark-fermentation by *E. aerogenes* and simultaneous saccharification, filtration and fermentation was 1000 ml/L and 613 ml/L, respectively. Alternatively, bagasse can be substituted with molasses [150].

5.4. Hemicelluloses Production

There is emerging interest in the industrial use of hemicelluloses from sugarcane bagasse as water-soluble polymers that could see synthetic polymers being replaced [151]. Films and coatings made from hemicelluloses enjoy numerous applications in the food and medicinal industries such as active food packaging, wound dressings and drug capsules [152]. Banerjee et al. have demonstrated successful extraction of xylan-rich hemicellulose components from sugarcane bagasse by the use of pressurized hot-water extraction and alkaline peroxide method (Figure 3) [153]. The extracted hemicelluloses contained mainly arabinoxylans with varying substitutions and a classical structure with a backbone of β-(1→4)-linked xylosyl residues substituted with arabinose at C-2 and C-3 of the main chain. The main difference occurs in the distribution of branches along the xylan backbone.

5.5. Production of Adsorbent Materials

The use of biosorbent as efficient pollutant removal from industrial waste water and ground water has attracted much attention. Bagasse has been demonstrated to be an efficient biosorbent for heavy metals [154,155], manganese [156], hexavalent chromium [157], methylene blue and gentian violet [158], etherdiamine [159], etc. The versatility of these benign environmental cleaning technique lies in the ability of bagasse to adsorb large amounts of the pollutant from aqueous media before its further processing techniques such as burning it in boilers to generate steam energy [160,161] and the resultant ash separated into individual industrial products. More so, the bagasse ash has recently found application in clay bricks formation by replacing natural clay for up to 20 wt.% [162] and in production of glass-ceramics with silicates as the major crystalline phases [163]. Furthermore, sugarcane bagasse can be modeled into nanomaterials in the form of long, straight, tubular structures with smooth walls and axially-uniform diameters, which is the characteristic of carbon nanotubes. These materials have typical lengths in the order of 50 nm and diameters in the range of 20 to 50 nm [164,165] and can be an alternative source of carbon nanotubes that have been reported as the best adsorbent materials [166].

5.6. Animal Feed

Besides its use for sugar production, sugarcane is a fodder resource increasingly used as a reserve for feeding ruminants during the dry season [167]. However, there arises a few environmental impacts such as release of CH_4 gas by animals fed on sugarcane forage even though it may be a more preferred way of disposing off the forage waste than burning it (in terms of CH_4 release to the environment). Ruminant CH_4 emission depends on the amount of organic matter fermented in the rumen and on the carbohydrate fermentation pattern [168]. Cell wall carbohydrate causes larger amounts of CH_4 emission than starch [169], although the effect of dietary soluble sugars on methane emission is not well known. According to a meta-analysis, fermentation of C_4 tropical grasses (grasses grown with a C_4 metabolism will result in higher CH_4 emission (10–17 %) than that of C_3 temperate grasses [170]. Archimède et al. [168] used whole sugarcane plant forage with higher levels of total sugar, (300–500 g/kg DM, [171]) as compared to conventional C4 grass (26–122 g/kg DM, [172]) which is the most used tropical grass with lower total sugar, to study the amount of CH_4 gas released: DM = dry matter. They reported a higher CH_4 emission of Black belly rams consuming whole sugarcane forage compared with *Dichanthium* sp. hay.

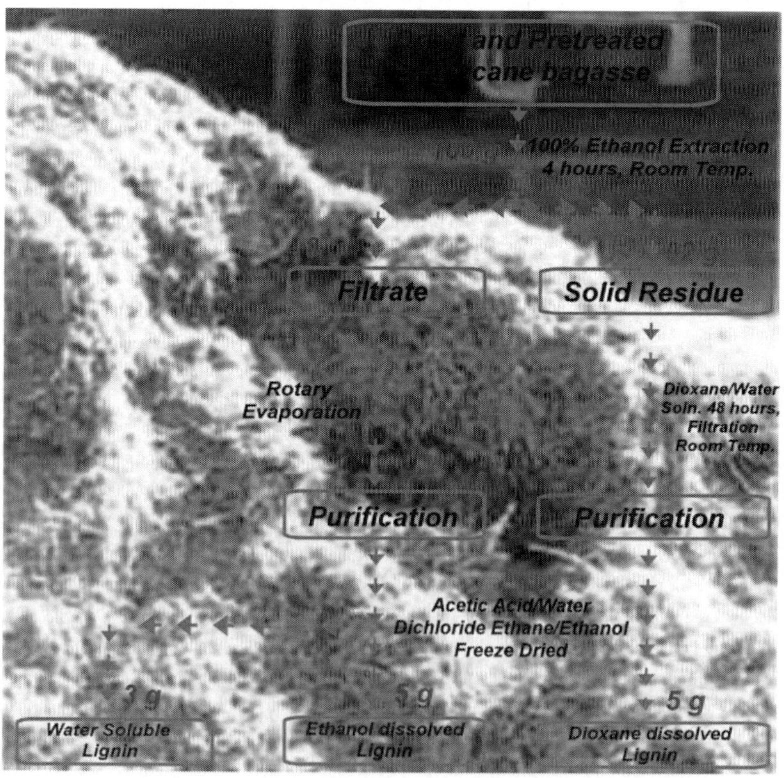

Figure 2. Lignin extraction from sugarcane bagasse [144].

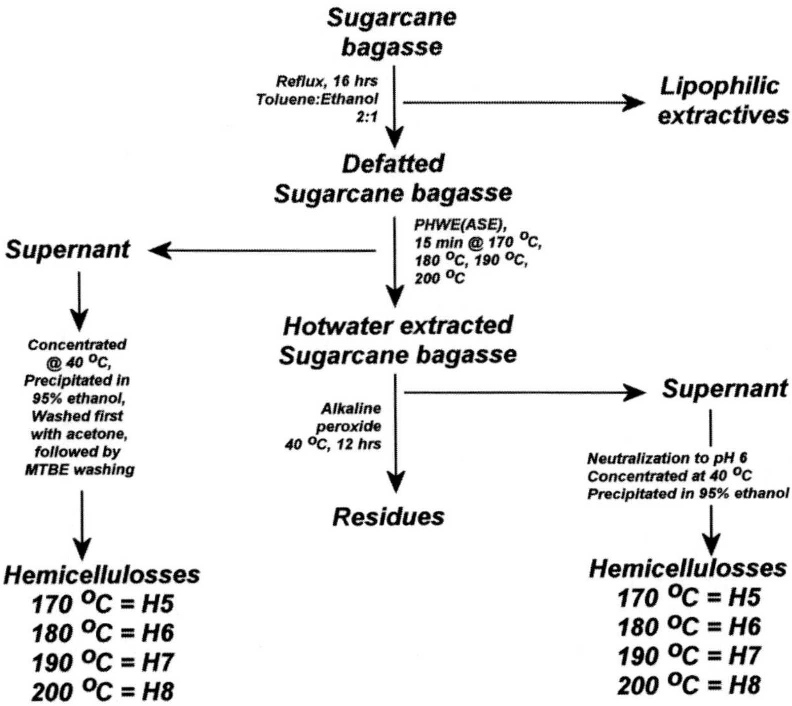

Figure 3. Scheme for the isolation of hemicelluloses by sequential extraction of sugarcane bagasse with pressurized hot water (PHWE) and alkaline peroxide [153].

6. Green Utilization of Sugarcane Molasses

6.1. Antioxidants Dietary Source

Sugarcane by-products like molasses are long term dietary source of antioxidants and phytochemicals such as phenolics, flavonoids, triterpenoids, phytosterols, etc. [173-175]. Phenolics and flavonoids extracted from fresh sugarcane and molasses show antioxidant, anti-inflammatory, antimutation and tyrosinase inhibitory capabilities in laboratory experiments [174,176]. Among other extraction technologies, supercritical carbon dioxide fluid extraction with piecewise distillation separation seems to offer the best results in obtaining antioxidants from sugarcane molasses [177]. The operating conditions for this method include: i) extraction pressure of 33.3 MPa, ii) temperature of 43.3 °C, iii) time of 86.7 min, iv) 90 % ethanol content of sugarcane molasses, v) flow rate of CO_2 of 20 L/h. The advantages associated with the above method include its inability to extract high polar harmful compounds such as sylvite and sodium salt from the molasses [178]. Additionally, antioxidants from sugarcane molasses can be achieved through solvent extraction of steam exploded lignocellulosic biomass method [126]. Under this method, boiling solvent extraction shows higher solid and phenolic yields than room temperature extraction and solubilities of phenolics and sugars are higher in anhydrous ethanol and deionized water than in ethyl acetate under each individual extraction condition. Antioxidants can also be extracted from sugarcane bagasse. Mandelli et al. evaluated the enzymatic production of xylooligosaccharides and antioxidant compounds from sugarcane bagasse using XynZ from Clostridium thermocellum, a naturally chimeric enzyme comprising activities of xylanase and feruloyl esterase along with a carbohydrate binding module [179].

6.2. Production of Alternative Energy

Sugarcane molasses have provided efficient raw materials for the production of agrochemicals such as butanol, lipids, acetate, butarate, ethanol, hydrogen gas, etc. [150]. Of more environmental concern, is the use of sugarcane molasses as feed stocks for production of alternative energy sources to fossil fuels. Fossil fuels are non-renewable and may not be available for our children's children to use. For this reason, coupled with the increase in prices of petroleum based fuels, future depletion of worldwide petroleum reserves and environmental policies to reduce CO_2 emissions, have stimulated research towards the development of biotechnology to produce clean energy from renewable resources that are environmental-friendly [180-187].

The production of biochemicals from sugarcane molasses and bagasse are promising alternative energy sources and will continue to impact positively on the agro chemistry industry. Sugarcane molasses is an important organic waste due to its high sugar content (55 %) and high volume of production. It is even more viscous and has higher total sugar content than beet molasses. The availability and cost of sugarcane molasses make it an attractive feedstock for use in many countries. The main contents of sugarcane molasses are shown in Table 2.

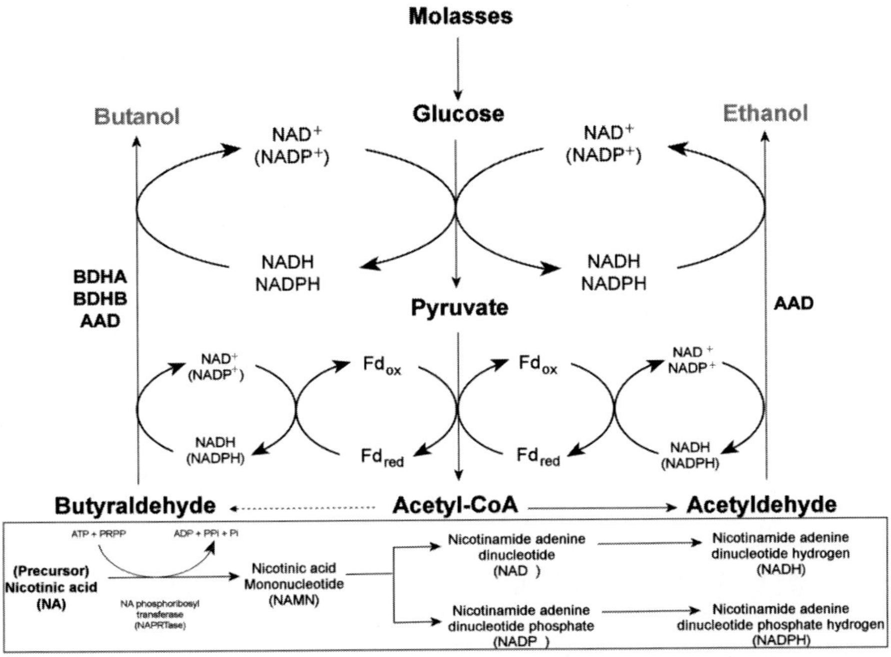

Figure 4. A summary of chemical processes in a butanol plant; showing relationships between alcohol synthesis and NADH/NADPH regenerations. The reducing cofactors (NADH/NADPH)-dependent enzymes are abbreviated as: AAD - alcohol/aldehyde dehydrogenase, BDHA - butanol dehydrogenaseI and BDHB - butanol dehydrogenaseII [188].

Table 2. Selected major constituents of Sugarcane Molasses [150]

Constituent	Mg/g
Total Sugars	388
Total Proteins	29
Total nitrogen	4.6
Sodium	0.85
Potassium	24.34
Calcium	5.3
Magnesium	1.9
Phosphorus	0.78
pH	5.5

The production of agrochemicals such as butanol and ethanol in large scale from sugarcane molasses have been summarized in Figure 4. In these mechanisms, availability of reducing factors (e.g., NADH and NADPH) play important role in improving the efficacy of products conversion in cofactor-dependent production systems. Recently, Li et al. used nicotinic acid (NA), the precursor of NADH and NADPH, to supplement the growth medium of a wild-type *Clostridium* sp. strain BOH3 and achieved an increase in the levels of NADH and NADPH [188].

The use of lipids from single-cell oil microorganisms to produce biodiesel has been noted to increase its production and is of low ecosystem impact [189]. Microorganisms such as oleaginous yeasts and fungi have also been considered as potential oil sources for biodiesel production because they accumulate large amounts of lipids. Among these microorganisms, *Epicoccum purpurascens* [190], *Mortierella isabelina* [191], etc. [192] have attracted special attention.

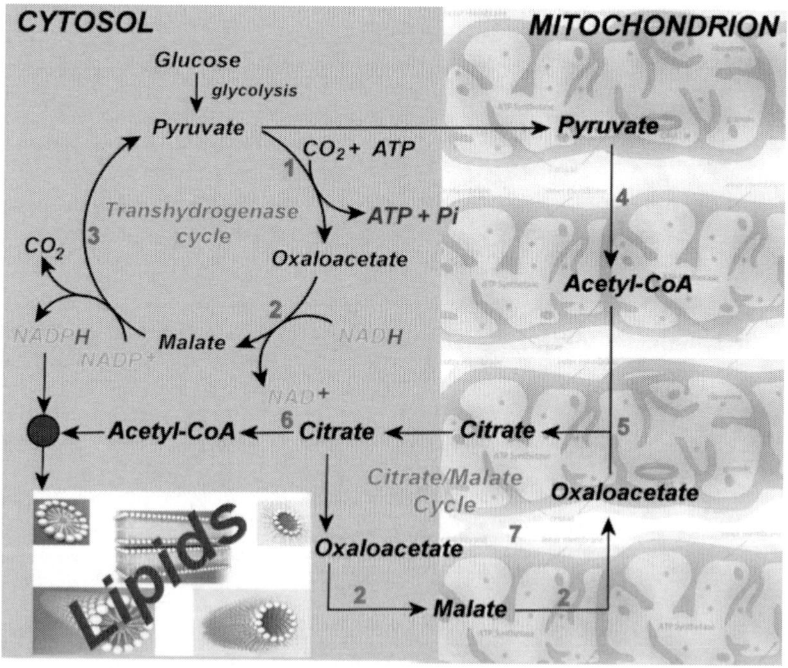

Figure 5. An overview of metabolic pathways involved in lipid biosynthesis by oleaginous fungi. The enzymes involved include: 1, pyruvate decarboxylase; 2, malate dehydrogenase; 3, malic enzyme; 4, pyruvate dehydrogenase; 5, citrate synthase; 6, ATP: citrate lyase; 7, citrate/malate transport [150].

The generation of lipid by fermentative oleaginous fungi is accompanied by the formation of organic acids as metabolic products [193] that accumulate leading to a sharp drop in culture pH and subsequent inhibition of fungal growth. However, Baggy et al. utilizes this drawback by filtering the spent media and using it for H_2 gas production in the second step [150]. The first stage involves the isolation of oleaginous fungi: *Alternaria alternata*, *Cladosporium cladosporioides*, *Epicoccum nigrum*, *Fusarium oxysporum*, *Aspergillus parasiticus* and *Emericella nidulans var. lata* from the culture media after biosynthesis of lipids as shown in Figure 5. The isolated dry fungal biomass is then esterified to produce biodiesel. In the second stage, the spent medium of fungal culture is used as the fermentation medium for hydrogen production by *Clostridium acetobutylicum* (ATCC 824) as shown in Figure 6.

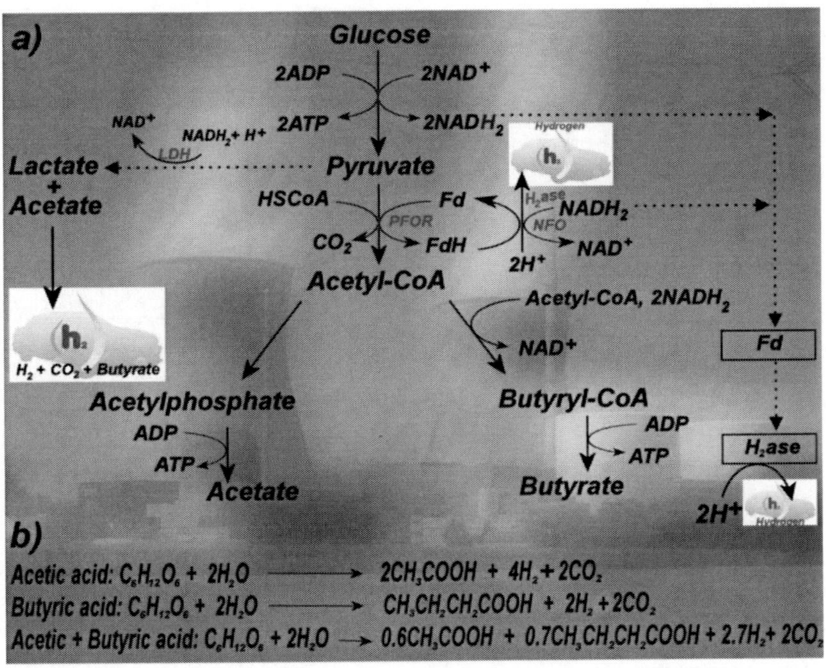

Figure 6. a). An overview of the metabolic pathways of glucose fermentation for biohydrogen production by *Clostridium acetobutylicum* and **b)**. Stoichiometric relations between glucose and the products formed during carbohydrate fermentation. Dashed lines indicate hypothetical pathways. Enzymes: PFOR, Pyruvate ferredoxin oxidoreductase; LDH, Lactate dehydrogenase; NFO, NADH: ferredoxin oxidoreductase; H_2ase, Hydrogenase [150].

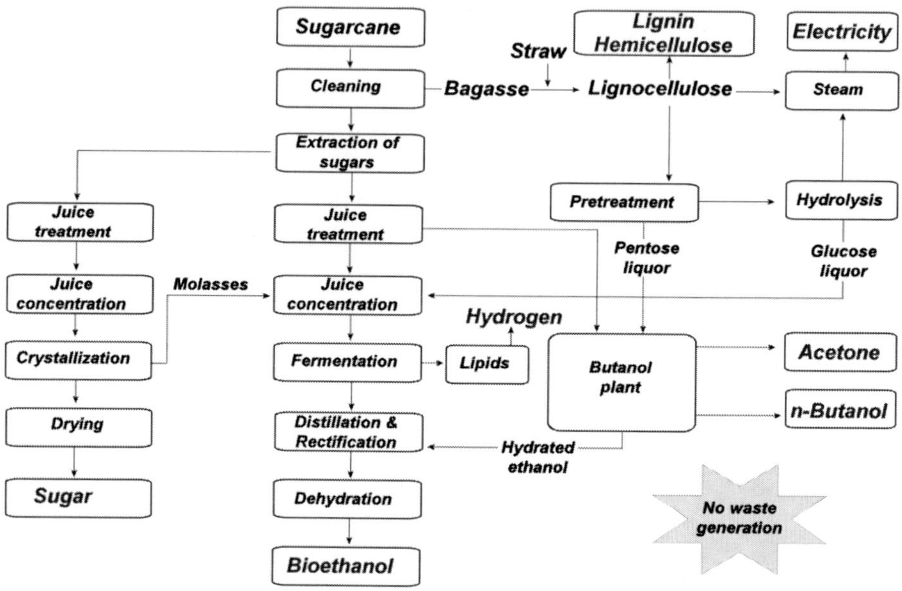

Figure 7. Summary of mechanisms involved in ensuring negligible waste generation from sugarcane processing.

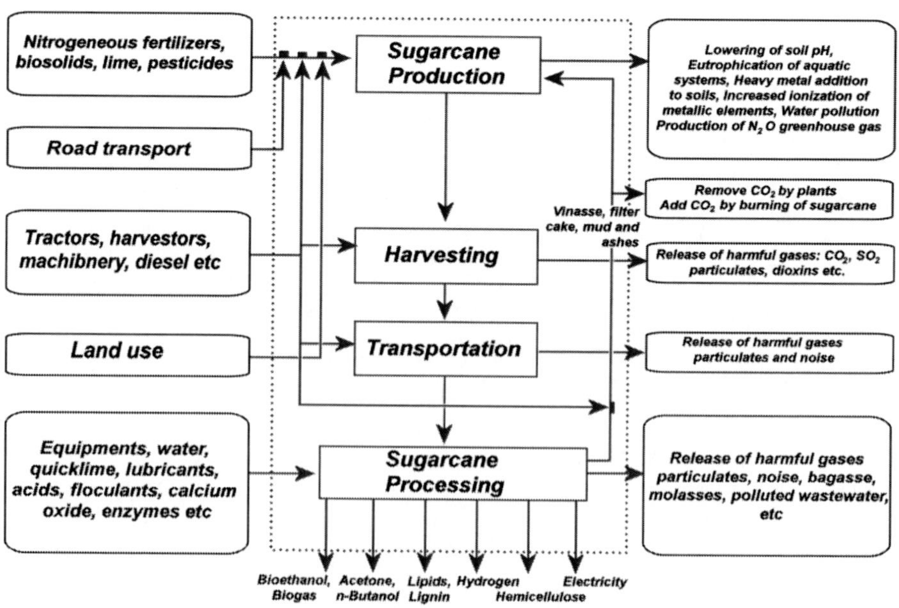

Figure 8. Summary of environmental impacts of sugarcane production and processing.

The maximum total H_2 yield is obtained with the spent medium of *E. nigrum and A. alternata*. These results demonstrated the possibility of interlinking the biodiesel production technology by fungi with hydrogen production by *C. acetobutylicum* ATCC 824 in order to exploit the residual sugars in the spent media and therefore increase the economic feasibility of biofuel production from molasses.

CONCLUSION

In summary, a flow chart (Figure 7) is used to show the no waste generation policy from sugarcane production and processing, and another flow chart (Figure 8) to show the environmental impacts arising out of sugarcane production and processing. To insure sustainable sugarcane production and processing, there is need to include the following strategies in management:

1. It is important to realize that sugarcane is an important plant that helps in sequestration of carbon dioxide; a greenhouse gas.
2. In order to manage soil acidification that arises out of nitrogenous fertilizer use in sugarcane production, the use of 650 kg/ha or 965 kg/ha of $CaCO_3$ to neutralize acid generated by 180 kg/ha of urea or diammonium phosphate respectively is recommended. Alternatively, use of basic nitrate fertilizers such as $Ca(NO_3)_2$ is advised.
3. In order to avoid contamination of aquatic systems within sugarcane plantation zones, use of pesticides according to label recommendations is paramount. In addition, correct weather forecast accompanied with precise timings in pesticide and fertilizer application will greatly reduce aquatic contamination due to surface runoffs and wash offs. Furthermore, region/site specific fertilizer requirement is significantly important both economically and for aquatic health protection.
4. Sugarcane bagasse is economically important in industrial production of bioethanol, biohydrogen, lignin, hemicelluloses and activated carbon and as effective pollutant adsorbents.
5. Sugarcane forage can be used as an effective fodder resource for feeding ruminants during dry seasons.
6. Sugarcane molasses are effective in production of antioxidants, butanol, lipids, acetate, butyrate, ethanol, hydrogen, etc.

ACKNOWLEDGMENT

This project was partly funded by the Alexander von Humboldt Foundation of Germany (Group Linkage Project, Ref. No. 3.4-Fokop-DEU/1064266). Financial support from the IAEA CRP Project 13695/RO and the Peregrine Fund (USA) are highly appreciated.

REFERENCES

[1] FAO. *(Food and Agriculture Organization of the United Nations) "Crop production"*; http://faostat.fao.org/site/567/DesktopDefault.aspx?PageID = 567#ancor Retrieved 2014/3/15.

[2] Van Antwerpen, R., Berry, S., Van Antwerpen, T., Smithers, J., Joshi, S. & Van Der Laan, M. (2013). *Sugarcane as an Energy Crop: Its Role in Biomass Economy*, In: Biofuel Crop Sustainability, Singh B. P. (Ed.); Wiley Balackwell.

[3] Rolph, G. M. (1917). Something about sugar; its history, growth, manufacture and distribution John J. Newbegin Publishers: San Francisco.

[4] Whitmarsh, J. *The Photosynthetic Process*; http://www.life.illinois.edu/govindjee/paper/gov.html#58: Retrieved 2014/3/15.

[5] Hassan, S. F. & Nasr, M. I. (2008). *Sugar Tech.*, *10*, 204-209.

[6] TAS. *(Thai Agricultural Standard), Good Agricultural Practices for Sugarcane* http://www.acfs.go.th/standard/download/eng/GAP_sugarcane. pdf: Retrieved 2014/3/15.

[7] Schwarzenbach, R., Muller, L., Rentsch, C. & Lanz, K. (2012). *For Climate's Sake!: A Visual Reader of Climate Change*; Lars Muller Publishers Zurich, Switzerland.

[8] IPCC Climate Change *2013. The Physical Science Basis - Summary for Policymakers, Observed Changes in the Climate System. Date of Access: 2014/1/25*; http://www.climatechange2013.org/images/uploads/WGI_AR5_SPM_brochure.pdf.

[9] Vaughan, D. G. & Spouge, J. R. (2002). *Clim. Chang.*, *52*, 65-91.

[10] National Research Council (2011). *America's Climate Choices*; The National Academies Press: Washington, DC.

[11] Arakawa, H., Aresta, M., Armor, J. N., Barteau, M. A., Beckman, E. J., Bell, A. T., Bercaw, J. E., Creutz, C., Dinjus, E., Dixon, D. A., Domen,

K., DuBois, D. L., Eckert, J., Fujita, E., Gibson, D. H., Goddard, W. A., Goodman, D. W., Keller, J., Kubas, G. J., Kung, H. H., Lyons, J. E., Manzer, L. E., Marks, T. J., Morokuma, K., Nicholas, K. M., Periana, R., Que, L., Rostrup-Nielson, J., Sachtler, W. M. H., Schmidt, L. D., Sen, A., Somorjai, G. A., Stair, P. C., Stults, B. R. & Tumas, W. (2001). *Chem. Rev.*, *101*, 953-996.

[12] Schoen, D. (1999). *Environ. Sci. Technol.*, *33*, 160A-163A.

[13] Zhao, Y., Shen, Y., Ma, G. & Hao, R. (2014). *Environ. Sci. Technol.*

[14] Xie, Y., Wang, T. T., Liu, X. H., Zou, K. & Deng, W. Q. (1960). *Nat. Commun.* 2013, *4*.

[15] Stewart, C. & Hessami, M.-A. (2005). *Energ. Convers. Manage.*, *46*, 403-420.

[16] Fast, A. G. & Papoutsakis, E. T. (2012). *Curr. Opin. Chem. Eng.*, *1*, 380-395.

[17] Ludwig, M. (2013). *Photosynth. Res.*, *117*, 147-161.

[18] Hatch, M. D. & Slack, C. R. (1966). *Biochem. J.*, *101*, 103-111.

[19] Hatch, M. D. & Slack, C. R. (1967). *Biochem J.*, *102*, 417-422.

[20] Iglesias, A. A., Gonzalez, D. H. & Andreo, C. S. (1986). *Biochem. Educ.*, *14*.

[21] Teixeira, D. D. B., Bicalho, E. d. S., Panosso, A. R., Cerri, C. E. P., Pereira, G. T. & Scala Júnior, N. L. (2013). *Sci. Agric.*, *70*, 195-203.

[22] Omwoma, S., Lalah, J. O., Ongeri, D. M. & Wanyonyi, M. B. (2010). *Bull. Environ. Contam. Toxicol.*, *85*, 602-608.

[23] Silva-Olaya, A. M., Cerri, C. E. P., La Scala Jr, N., Dias, C. T. S. & Cerri, C. C. (2013). *Environ. Res. Lett.*, *8*, Doi: 10.1088/1748-9326/1088/1081/015014.

[24] Oliver, D. C. (2004). *Environmental impacts of sugar production*; CABI Publishers: Wallingford Oxfordshire United Kingdom.

[25] Alloway, B. J. (1995). Heavy metals in soils, Second Edition (B.J. Alloway, Ed.); Blackie: New York.

[26] Omwoma, S., Nyaigoti Omwoyo, W., Alwala, J. O., Ongeri, D. M. K., Sylus, L. C. & Lalah, J. O. (2012). *The Environmentalist*, *32*, 494-502.

[27] Omwoma, S., Lalah, J. O., Ongeri, D. M. K. & Schramm, K.-W. (2013). *Environ. Earth Sci.*, Doi: 10.1007/s12665-013-2824-y.

[28] Wang, X. S. (2007). *Environ. Geol.*, *54*, 269-273.

[29] Mortvedt, J. J. (1995). Heavy metal contaminants in inorganic and organic fertilizers: nutrient cycling in agroecosystems; Springer: Netherlands.

[30] Lawrence, R. C. & Brian, W. S. (2002). *Heavy Metal in Fertilizers: Considering for Setting Regulations in Oregon*; Department of Environment and Molecular Toxicology Oregon State University Corvallis Oregon.

[31] Alloway, B. J. (1995). Soil processes and the behavior of heavy metals in Soils, Second Edition (B.J. Alloway, Ed.); Blackie: New York.

[32] Ford, R. G., Scheinost, a. C. & Sparks, D. L. (2001). *Adv. Agron.*, *74*, 41-62.

[33] Odiete, W. O. (1999). *Environmental Physiology of animals and pollution.,* Diversified resources Ltd: Lagos.

[34] Wang, W. (1987). *Environ. Int.*, *13*, 437-457.

[35] Gough, L. P. & Herring, J. R. (1993). *Agric. Ecosyst. Environ.*, *46*, 55-68.

[36] Driscoll, C. T., Otton, J. K. & Iverfield, C. K. (1994). "Trace Metal speciation and cycling." In Molden, B. and Eemy, J. (Eds), Biogeochemistry of smallcatchments: A Tool for environmental research John Wiley and sons: New York.

[37] Ongley, E. D. (1996). *Control of water pollution from agriculture: FAO irrigation and drainage paper 55*; Food and Agriculture Organization of the United Nations: Rome.

[38] Martínez, C. E., Jacobson, A. & McBride, M. B. (2001). *Environ. Sci. Technol.*, *35*, 908-916.

[39] Duinker, J. C., Nolting, R. F. & Michel, D. (1992). *Thalassia Jugosl.*, *18*, 191-201.

[40] Soltanpour, P. N. & Raley, W. L. (1989). *Fact sheet no. 4.908, Evaluation of drinking water quality for livestock.*; Colorado State University Cooperative: Colorado.

[41] Richard, S. (1991). Phosphorus and Nitrogen Removal from Municipal Wastewater: Principles and Practice. 2nd edition; Lewis publishers: USA.

[42] Fried, S., Brendan, M. & Erin, N. (2003). *Tillers*, *4*, 21-24

[43] Kayombo, S. *Experience and Lessons Learned Brief*; http://projects.inweh.unu.edu/inweh/inweh/content/2405/IW%20LEARN/lakevictoria_2005s.html: Retrieved 2014/3/15, 2005.

[44] Moody, P. W. & Aitken, R. L. (1997). *Aust. J. Soil Res.*, *35*, 163-173.

[45] Ahern, C. R., Isbell, R. F. & Weinand, M. M. G. (1995). Acid surface soil distribution and extent in Queensland, Australia, 147-153. In R. A. Date, N. J. Grundon, G. E. Rayment and M. E. Probert (Eds.) Plant Soil

Interactions at Low pH. ; Kluwer Academic Publishers: The Netherlands.
[46] Wood, A. W. (1985). *Soil Use Manage.*, *1*, 120-124.
[47] Wood, A. W., Noble A. D. & B, R. G. V. (2003). Sugar Research and Development Corperation. Final Report project No. CSR024.
[48] Haynes, R. J. (1983). *Grass Forage Sci.*, *38*, 1-11.
[49] Bramley, R. G. V., Ellis, N., Nable, R. O. & Garside, A. L. (1996). *Aust. J. Soil Res.*, *34*, 967.
[50] Garside, A. L., Bramley, R. G. V., Bristow, K. L., Holt, J. A., Magarey, R. C., Nable, R. O., Pankhurst, C. E. & Skjemstad, J. O. (1997). *Proc Australas Soc Sug. Rechnol.*, *19*.
[51] Wood, A. W. & Bramley, R. G. V. (1996). Soil survey – a tool for better fertilizer management in the Australian Sugar Industry. In Sugarcane: Research towards efficient and sustainable production, (Eds. J.R. Wilson, D.M. Hogarth, *J.A. Campbell and A.L. Garside)*; CSIRO Division of Tropical Crops and Pastures Brisbane.
[52] Schroeder, B. L., Turner, P. E. T., Meyer, J. H. & Robinson, J. B. (1996). Advances in quantifying soil acidity and acidification rates in the South African sugar industry. In Sugarcane: Research towards efficient and sustainable production, (Eds. J.R. Wilson, D.M. Hogarth, J.A. Campbell and A.L. Garside); CSIRO Division of Tropical Crops and Pastures: Brisbane.
[53] Hetherington, S. J. (1986). Effects of aluminium toxicity on the growth of sugarcane. M.Sc. thesis, Department of Agriculture; University of Queensland Brisbane, Australia.
[54] Tisdale, S. L., Nelson, W. L. & Beaton, J. D. (1985). *Soil fertility and fertilisers, 4th Edition.*; Macmillan New York.
[55] Silva Paredes, D., Lessa, A. C. d. R., Sant'Anna, S. A. C., Boddey, R. M., Urquiaga, S. & Alves, B. J. R. (2013). *Nutr. Cycl Agroecosys.*, *98*, 41-55.
[56] Signor, D., Cerri, C. E. P. & Conant, R. (2013). *Environ. Res. Lett.*, *8*, Doi: 10.1088/1748-9326/1088/1081/015013.
[57] Biggs, J. S., Thorburn, P. J., Crimp, S., Masters, B. & Attard, S. J. (2013). *Agric. Ecosyst. Environ.*, *180*, 79-89.
[58] Smith, S. V., Swaney, D. P., Talaue-Mcmanus, L., Bartley, J. D., Sandhei, P. T., McLaughlin, C. J., Dupra, V. C., Crossland, C. J., Buddemeier, R. W., Maxwell, B. A. & Wulff, F. (2003). *Bioscience*, *53*, 235-245.
[59] Mitchell, C., Brodie, J. & White, I. (2005). *Mar. Pollut. Bull.*, *51*, 23-36.

[60] Drewry, J., Higham, W. & Mitchell, C. (2008). *Water Quality Improvement Plan: Final Report for Mackay Whitsunday Region.* ; Reef Catchments Mackay Whitsunday Inc.: Mackay.
[61] Fulcher, J. M., Wayment, D. G., White, P. M., Jr. & Webber, C. L., 3rd. (2014). *J. Agric. Food Chem.*, *62*, 2141-2146.
[62] Lofton, J. & Tubaña, B. (2013). *J. Plant Nutr.*, Doi: 10.1080/01904167.01902013.01828752.
[63] Armour, J. D., Nelson, P. N., Daniells, J. W., Rasiah, V. & Inman-Bamber, N. G. (2013). *Agric. Ecosyst. Environ.*, *180*, 68-78.
[64] Rasiah, V., Armour, J. D. & Nelson, P. N. (2013). *Agric. Ecosyst. Environ.*, *180*, 103-110.
[65] Thorburn, P. J., Biggs, J. S., Attard, S. J. & Kemei, J. (2011). *Agric. Ecosyst. Environ.*, *144*, 1-12.
[66] Heather, K. (1999). *Sustainable Sugercane Farming in Florida*; http://www.agroecology.org/ Case%20Studies/sugarcane.html.
[67] Robinson, N., Brackin, R., Vinall, K., Soper, F., Holst, J., Gamage, H., Paungfoo-Lonhienne, C., Rennenberg, H., Lakshmanan, P. & Schmidt, S. (2011). *PLoS One*, *6*, e19045.
[68] Hajari, E., Snyman, S. J. & Watt, M. P. (2014). *Plant Cell Tiss. Org.*
[69] Subbarao, G. V., Wang, H. Y., Ito, O., Nakahara, K. & Berry, W. L. (2006). *Plant Soil*, *290*, 245-257.
[70] Cardenas-Navarro, R., Adamowicz, S. & Robin, P. (1999). *J. Exp. Bot.*, *50*, 613-624.
[71] Miller, A. J. & Smith, S. J. (2008). *Ann. Bot.*, *101*, 485-489.
[72] Stewart, G. R. & Schmidt, S. (1999). Evolution and ecology of plant mineral nutrition. In Physiological Plant Ecology (ed MC Press, JD Scholes, MG Barker) Blackwell Science.
[73] Carvalho, T. L. G., Ferreira, P. C. G. & Hemerly, A. S. (2011). *Trop.Plant Biol.*, *4*, 31-41.
[74] Cavalcante, V. & Dobereiner, J. (1988). *Plant Soil*, *108*, 23-31.
[75] Baldani, J., Caruso, L., Baldani, V. L. D., Goi, S. R. & Döbereiner, J. (1997). *Soil Biol. Biochem.*, *29*, 911-922.
[76] Govindarajan, M., Balandreau, J., Muthukumarasamy, R., Revathi, G. & Lakshminarasimhan, C. (2006). *Plant Soil*, *280*, 239-252.
[77] Baldani, J. I., Pot, B., Kirchhof, G., Falsen, E., Baldani, V. L. D. & Olivares, F. L. e. a. (1996). *Int. J. Syst. Bacteriol.*, *46*, 802-810.
[78] James, E. K. & Olivares, F. L. (1998). *CRC Crit. Rev. Plant Sci.*, *17*, 77-119.
[79] James, E. K. (2000). *Field Crops Res.*, *65*, 197-209.

[80] Fischer, D., Pfitzner, B., Schmid, M., Simoes-Araujo, J. L., Reis, V. M. & Pereira, W. e. a. (2012). *Plant Soil*, *356*, 83-99.
[81] Figueiredo, M., Seldin, L., Fernando Araujo, F. & Mariano, R. (2011). Plant growth promoting rhizobacteria: fundamentals and applications. In Plant Growth and Health Promoting Bacteria, Maheshwari, D.K. (ed.). Springer: Berlin, Germany.
[82] Paungfoo-Lonhienne, C., Lonhienne, T. G. A., Yeoh, Y. K., Webb, R. I., Lakshmanan, P., Chan, C. X., Lim, P.-E., Ragan, M. A., Schmidt, S. & Hugenholtz, P. (2014). *Microbial Biotechnology*, *7*, 142-154.
[83] Vargas Betancur, G. J. & Pereira Jr, N. (2010). *Electron. J. Biotechnol.*, *13*.
[84] Rezende, C. A., de Lima, M. A., Maziero, P., Deazevedo, E. R., Garcia, W. & Polikarpov, I. (2011). *Biotechnol. Biofuels*, *4*, 54.
[85] Karp, S. G., Woiciechowski, A. L., Soccol, V. T. & Soccol, C. R. (2013). *Braz. Arch. Biol. Technol.*, *56*, 679-689.
[86] Pandey, A., Soccol, C. R., Nigam, P., Brand, D., Mohan, R. & Roussos, S. (2000). *Biochem. Eng. J.*, *6*, 153-162.
[87] Rocha, G. J. M., Gonçalves, A. R., Oliveira, B. R., Olivares, E. G. & Rossell, C. E. V. (2012). *Ind. Crops Prod.*, *35*, 274-279.
[88] Rozzell, J. D. (1999). *Bioorg. Med. Chem.*, *7*, 2253-2261.
[89] Dasari, R. K. & Eric Berson, R. (2007). *Appl. Biochem. Biotechnol.*, *137-140*, 289-299.
[90] Pan, X. (2008). *J. Biobased Mater. Bioenergy*, *2*, 25-32.
[91] Gupta, R. & Lee, Y. Y. (2009). *Biotechnol. Bioeng.*, *102*, 1570-1581.
[92] Nazhad, M. M., Ramos, L. P., Paszner, L. & Saddler, J. N. (1995). *Enzyme Microb. Technol.*, *17*, 68-74.
[93] Laser, M., Schulman, D., Allen, S. G., Lichwa, J., Antal, M. J. & Lynd, L. R. (2002). *Bioresour. Technol.*, *81*, 33-44.
[94] Ramos, L. P. & Saddler, J. N. (1994). *Appl. Biochem. Biotechnol. 45-6*, 193-207.
[95] Azevedo, H. & Bishop, D. Cavaco-Paulo, A. (2002). *Appl. Biochem. Biotechnol.*, *101*, 61-75.
[96] Emmel, A., Mathias, A. L., Wypych, F. & Ramos, L. P. (2003). *Bioresour. Technol.*, *86*, 105-115.
[97] Palonen, H., Tjerneld, F., Zacchi, G. & Tenkanen, M. (2004). *J. Biotechnol.*, *107*, 65-72.
[98] Teeri, T. T. (1997). *Trends Biotechnol.*, *15*, 160-167.
[99] Hernandez-Salas, J. M., Villa-Ramirez, M. S., Veloz-Rendon, J. S., Rivera-Hernandez, K. N., Gonzalez-Cesar, R. A., Plascencia-Espinosa,

M. A. & Trejo-Estrada, S. R. (2009). *Bioresour. Technol.*, *100*, 1238-1245.
[100] Prado, J. M., Follegatti-Romero, L. A., Forster-Carneiro, T., Rostagno, M. A., Maugeri Filho, F. & Meireles, M. A. A. (2014). *J. Supercrit. Fluids.*, *86*, 15-22.
[101] Aguiar, R. S., Silveira, M. H., Pitarelo, A. P., Corazza, M. L. & Ramos, L. P. (2013). *Bioresour. Technol.*, *147*, 416-423.
[102] Hendriks, A. T. W. M. & Zeeman, G. (2009). *Bioresour. Technol.*, *100*, 10-18.
[103] Zhang, Q. & Cai, W. *Biomass Bioenergy*, 2008, *32*, 1130-1135.
[104] Rosgaard, L., Pedersen, S. & Meyer, A. S. *Appl. Biochem. Biotechnol.* 2007, *143*, 284-296.
[105] Bals, B., Rogers, C., Jin, M., Balan, V. & Dale, B. (2010). *Biotechnol. Biofuels.*, *3*, 1.
[106] Kuhad, R. C., Singh, A. & Eriksson, K. E. L. (1997). Microorganisms and enzymes involved in the degradation of plant fiber cell wall. In: Eriksson, KEL, editor. Biotechnology in the Pulp and paper industry. Advances in biochemical engineering biotechnology., Springer Verlag: Berlin.
[107] Novo, L. P., Gurgel, L. V., Marabezi, K. & Curvelo, A. A. (2011). (1997). *Bioresour. Technol.*, *102*, 10040-10046.
[108] Gonzalez, G., Lopes-Santin, J., Caminal, G. & Sola, C. (1986). *Biotechnol. Bioeng.*, *28*, 288-293.
[109] Fengel, D. & Wegener, G. (1984). *Wood - Chemistry, Ultrastructure, Reactions*; Walter de Gruyter: Berlin, New York.
[110] Filson, P. B. & Dawson-Andoh, B. E. (2009). *Bioresour. Technol.*, *100*, 2259-2264.
[111] Luo, J., Fang, Z. & Smith, R. L. (2014). *Prog. Energy Combust. Sci.*, *41*, 56-93.
[112] Chuanyun, D., Bochu, W., Huan, Z., Conglin, H., Chuanren, D., Wangqian, L., Toyama, Y. & Sakanishi, A. (2004). *Colloids Surf. B*, *34*, 7-11.
[113] Yaldagard, M., Mortazavi, S. A. & Tabatabaie, F. (2008). *J. Chem. Eng.*, *25*, 517-523.
[114] Sun, J. X., Sun, R., Sun, X. F. & Su, Y. (2004). *Carbohydr. Res.*, *339*, 291-300.
[115] Ramadoss, G. & Muthukumar, K.(2014). *Biochem. Eng. J.*, *83*, 33-41.
[116] Velmurugan, R. & Muthukumar, K. (2012). *Bioresour. Technol.*, *112*, 293-299.

[117] Velmurugan, R. & Muthukumar, K. (2011). *Bioresour. Technol., 102*, 7119-7123.
[118] Sakdaronnarong, C., Srimarut, N., Lucknakhul, N., Na-songkla, N. & Jonglertjunya, W. (2014). *Biochem. Eng. J., 85*, 49-62.
[119] Sun, J., Suna, X. F., Zhaoa, H. & Sunb, R. C. (2004). *Polym. Degrad. Stabil., 84*, 331-339.
[120] Brownell, H. H. & Saddler, J. N. (1987). *Biotechnol. Bioeng., 29*, 228-235.
[121] Brownell, H. H., Yu, E. K. C. & Saddler, J. N. (1986). *Biotechnol. Bioeng., 28*, 792-801.
[122] Oliveira, F. M., Pinheiro, I. O., Souto-Maior, A. M., Martin, C. & Goncalves, A. R., Rocha, G. J. (2013). *Bioresour. Technol., 130*, 168-173.
[123] Mosier, N., Wyman, C., Dale, B., Elander, R., Lee, Y. Y., Holtzapple, M. & Ladisch, M. (2005). *Bioresour. Technol., 96*, 673-686.
[124] Ramos, L. P. (2013). *Quim. Nova*, 2003, *26*, 863-871.
[125] Moncada, J., El-Halwagi, M. M. & Cardona, C. A. (2013). *Bioresour. Technol., 135*, 533-543.
[126] Li, J., Lin, J., Xiao, W., Gong, Y., Wang, M., Zhou, P. & Liu, Z. (2013). *Bioresour. Technol., 130*, 8-15.
[127] Krishnan, C., Sousa, L. C., Jin, M., Chang, L., Dale, B. E. & Balan, V. (2010). *Biotechnol. Bioeng., 107*, 441-450.
[128] Lee, J. (1997). *J. Biotechnol., 56*, 1-24.
[129] Falcón, M. A., Rodríguez, A., Carnicero, A., Regalado, V., Perestelo, F. & Milstein, O., De la Fuente, G. (1995). *Soil Biol. Biochem., 27*, 121-126.
[130] Giardina, P., Cannio, R., Martirani, L., Marzullo, L., Palmieri, G. & Sannia, G. (1995). *Ap. Enviro. Microbiol., 61*, 2408-2413.
[131] Silva, I. S., Menezes, C. R. d., Franciscon, E., Santos, E. d. C. d. & Durrant, L. R. (2010). *Braz. arch. biol. technol., 53*, 693-699.
[132] Kamei, I., Hirota, Y., Mori, T., Hirai, H., Meguro, S. & Kondo, R. (2012). *Bioresour. Technol., 112*, 137-142.
[133] Khuong, L. D., Kondo, R., De Leon, R., Kim Anh, T., Shimizu, K. & Kamei, I. (2014). *Int. Biodeterior. Biodegrad., 88*, 62-68.
[134] Wong, D. W. S. (2009). *Appl. Biochem. Biotechnol. 157*, 174-209.
[135] Dilip Kumar, A., Jayati, T. & Deepti, A. (2014). *Espacenet Bibliographic data: WO2014013509 (A1).*
[136] McGinnis, G. D., Wilson, W. W. & Mullen, C. E. (1983). *Ind. Eng. Chem. Prod. Res. Dev., 22*, 352-357.

[137] Martín, C., Klinke, H. B. & Thomsen, A. B. (2007). *Enzyme Microb. Technol.*, *40*, 426-432.
[138] Sun, Y. & Cheng, J. (2002). *Bioresour. Technol.*, *83*, 1-11.
[139] Mesa, L., González, E., Cara, C., González, M., Castro, E. & Mussatto, S. I. (2011). *Chem Eng J.*, *168*, 1157-1162.
[140] Saberikhah, E., Mohammadi, R. J. & Rezayati-Charani, P. (2011). *Cellulose Chem. Technol.*, *45*, 67-75.
[141] Zhao, X., Cheng, K. & Liu, D. (2009). *Appl. Microbiol. Biotechnol.*, *82*, 815-827.
[142] Jonglertjunya, W., Juntong, T., Pakkang, N., Srimarut, N. & Sakdaronnarong, C. (2014). *LWT - Food Sci. Technol.*, *57*, 116-125.
[143] Doherty, W. O. S., Mousavioun, P. & Fellows, C. M. (2011). *Ind. Crop Prod.*, *33*, 259-276.
[144] Zeng, J., Tong, Z., Wang, L., Zhu, J. Y. & Ingram, L. (, Tong). *Bioresour. Technol.*, *154*, 274-281.
[145] Kumar, N. & Das, D. (2000). *Process Biochem.*, *35*, 589-593.
[146] Lin, C. Y. & Jo, C. H. (2003). *J. Chem. Technol. Biotechnol.*, *78*, 678-684.
[147] Rai, P. K., Singh, S. P. & Asthana, R. K. (2012). *Appl. Biochem. Biotechnol.*, *167*, 1540-1549.
[148] Rajagopalan, G., He, J. & Yang, K. L. (2014). *Bioresour. Technol.*, *154*, 38-43.
[149] Rai, P. K., Singh, S. P., Asthana, R. K. & Singh, S. (2014). *Bioresour. Technol.*, *152*, 140-146.
[150] Bagy, M. M. K., Abd-Alla, M. H., Morsy, F. M. & Hassan, E. A. (2014). *Int. J. Hydrogen Energy*, *39*, 3185-3197.
[151] Gatenholm, P. & Tenkanen, M. (2004). *ACS Symp. Ser.*, *864*, 1-2.
[152] Hansen, N. M. L. & Plackett, D. (2008). *Biomacromolecules*, *9*, 1493-1505.
[153] Banerjee, P. N., Pranovich, A., Dax, D. & Willfor, S. (2014). *Bioresour. Technol.*, Doi: 10.1016/j.biortech.2014.01.020
[154] Chao, H.-P., Chang, C.-C. & Nieva, A. (2013). *J. Ind. Eng. Chem.*, *Doi:* 10.1016/j.jiec.2013.12.027.
[155] Gurgel, L. V. A., Freitas, R. P. d. & Gil, L. F. (2008). *Carbohydr. Polym.*, *74*, 922-929.
[156] Esfandiar, N., Nasernejad, B. & Ebadi, T. (2014). *J. Ind. Eng. Chem.*, *Doi:* 10.1016/j.jiec.2013.12.072.
[157] Cronje, K. J., Chetty, K., Carsky, M., Sahu, J. N. & Meikap, B. C. (2011). *Desalination*, *275*, 276-284.

[158] Guimarães Gusmão, K. A., Alves Gurgel, L. V., Sacramento Melo, T. M. & Gil, L. F. (2012). *Dyes Pigm.*, *92*, 967-974.

[159] Gusmao, K. A., Gurgel, L. V., Melo, T. M., Carvalho Cde, F. & Gil, L. F. (2014). *J. Environ. Manage.*, *133*, 332-342.

[160] Pereira, L. G., Chagas, M. F., Dias, M. O. S., Cavalett, O. & Bonomi, A. (2014). *J. Clean. Prod.*, Doi: 10.1016/j.jclepro.2014.01.059.

[161] Mariano, A. P., Dias, M. O., Junqueira, T. L., Cunha, M. P., Bonomi, A. & Filho, R. M. (2013). *Bioresour. Technol.*, *135*, 316-323.

[162] Faria, K. C., Gurgel, R. F. & Holanda, J. N. (2012). *J. Environ. Manage.*, *101*, 7-12.

[163] Teixeira, S. R., Magalhaes, R. S., Arenales, A., Souza, A. E., Romero, M. & Rincon, J. M. (2014). *J. Environ. Manage.*, *134*, 15-19.

[164] Alves, J. O., Soares Tenório, J. A., Zhuo, C. & Levendis, Y. A. (2012). *J. Mater. Res. Technol.*, *1*, 31-34.

[165] Fujimoto, H., Nishihara, H., Kyotani, T. & Hisa, M. (2014). *Carbon*, *68*, 814-817.

[166] Li, J. Y. & Jiang, Y. X. (2013). *Adv. Mater. Res.*, *634-638*, 192-197.

[167] Daniel, J. L. P., Capelesso, A., Cabezas-Garcia, E. H., Zopollatto, M., Santos, M. C., Huhtanen, P. & Nussio, L. G. (2014). *Grass Forage Sci.*, *69*, 176-181.

[168] Archimède, H., Martin, C., Periacarpin, F., Rochette, Y., Etienne, T. S., Anais, C. & Doreau, M. (2014). *Anim. Feed Sci. Tech.*, *190*, 30-37.

[169] Martin, C., Morgavi, D. P. & Doreau, M. (2010). *Animal*, *4*, 351-365.

[170] Archimède, H., Eugène, M., Marie Magdeleine, C., Boval, M., Martin, C., Morgavi, D. P., Lecomte, P. & Doreau, M. (2011). *Anim. Feed Sci. Technol.*, 166-167, 59-64.

[171] Preston, T. R. (1977). *Trop. Anim. Prod.*, *2*, 125-142.

[172] Hunter, R. A., Mc Intyre, B. L. & Mc LLroy, R. J. (1970). *J. Sci. Food. Agric.*, *21*, 400-405.

[173] Feng, S., Luo, Z., Zhang, Y., Zhong, Z. & Lu, B. *Food Chem.*, 2014, *151*, 452-458.

[174] Duarte-Almeida, J. M., Negri, G., Salatino, A., de Carvalho, J. E. & Lajolo, F. M. (2007). *Phytochemistry*, *68*, 1165-1171.

[175] Duarte-Almeida, J. M., Salatino, A., Genovese, M. I. & Lajolo, F. M. (2011). *Food Chem.*, *125*, 660-664.

[176] Takara, K., Otsuka, K., Wada, K., Iwasaki, H. & Yamashita, M. (2007). *Biosci. Biotechnol. Biochem.*, *71*, 183-191.

[177] Guan, Y., Tang, Q., Fu, X., Yu, S., Wu, S. & Chen, M. (2014). *Food Chem.*, *152*, 552-557.

[178] Baysal, T., Ersus, S. & Starmans, D. A. J. (2000). *J. Agric. Food Chem.* 2000, *48*, 5507-5511.

[179] Mandelli, F., Brenelli, L. B., Almeida, R. F., Goldbeck, R., Wolf, L. D., Hoffmam, Z. B., Ruller, R., Rocha, G. J. M., Mercadante, A. Z. & Squina, F. M. (). *Ind. Crop Prod.*, *52*, 770-775.

[180] Wang, J. & Wan, W. (2009). *Int. J. Hydrogen Energy*, *34*, 799-881.

[181] Hawkes, F. R., Dinsdale, R., Hawkes, D. L. & Huss, I. (2002). *Int. J. Hydrogen Energy*, *27*, 1339-1347.

[182] Abdalla, Z. E. A., Li, B. & Tufail, A. (2009). *Colloids Surf. A*, *341*, 86-92.

[183] Abd-Alla, M. H., Morsy, F. M. & El-Enany, W. (2011). *Int. J. Hydrogen Energy*, *36*, 13518-13527.

[184] Abd-Alla, M. H. & El-Enany, E. (2012). *Biomass Bioenergy*, *42*, 172-178.

[185] Morsy, F. M. (2011). *Int. J. Hydrogen Energy*, *36*, 14381-14390.

[186] Das, D. & Veziroglu, T. N. (2001). *Int. J. Hydrogen Energy*, *26*, 13-28.

[187] Das, D. & Veziroglu, T. N. (2008). *Int. J. Hydrogen Energy*, *33*, 6046-6057.

[188] Li, T., Yan, Y. & He, J. (2014). *Bioresour. Technol.*, *155*, 220-228.

[189] Nagel, N. & Lemke, P. (1990). *Appl. Biochem. Biotechnol.*, *24*, 355-361.

[190] Koutb, M. & Morsy, F. M. (2011). *Biomass Bioenergy*, *35*, 3182-3187.

[191] Gao, D., Zeng, J., Zheng, Y., Yu, X. & Chen, S. (2013). *Bioresour. Technol.*, *133*, 315-32

[192] Leiva-Candia, D. E., Pinzi, S., Redel-Macías, M. D., Koutinas, A., Webb, C. & Dorado, M. P. (2014). *Fuel*, *123*, 33-42.

[193] Beopoulos, A., Cescut, J., Haddouche, R., Uribelarrea, J. L., Molina-Jouve, C. & Nicaud, J. M. (2009). *Prog. Lipid. Res.*, *48*, 375-387.

In: Sugarcane
Editor: Eleanore Webb
ISBN: 978-1-61122-356-9

Chapter 12

APPLICATION OF VINASSE TO SUGARCANE

Aneeza Soobadar
Mauritius Sugarcane Industry Research Institute,
Réduit, Mauritius

ABSTRACT

Among the disposal means for vinasse, application to agricultural land is believed to represent the most sensible economic option from both the agronomic and environmental point of view. This belief stems from the numerous studies that have been conducted in sugar producing countries to determine the impacts of the vinasse, often at high doses, on soil quality, on the sugarcane plant and on groundwater quality. The vinasse,, is very variable in chemical composition but from analyses on samples collected at regular intervals of three months during 2005 to 2008 in Mauritius contain on average 9.37 g/L of K. Its fertilizer value as found everywhere is therefore mostly as a source of potassium. However at an application rate of for instance 100 m^3/ha, vinasse can in addition represent a significant source of N (average of 122 kg N/ha) and of organic matter (average of 8-15 % dry matter).

Apart from K, organic matter and N, vinasse contains heavy metals (Cu, Zn, Ni, Mn, Pb) but their concentrations are in general negligible and most often the heavy metals are below their detection limits on the atomic absorption spectrophotometer (5 mg/kg for Cu, Zn, Ni, Pb and 10mg/kg for Mn). Analyses of soils have shown that application of vinasse may, on the other hand, initially lower soil pH, e.g. from 5.9 to 5.4. but the soil pH will invariably return to its original value a few months afterwards. At high rates of the order of 100 m^3/ha, vinasse will in addition raise the electrical conductivity of the soil, but in spite of this increase, the electrical conductivity will remain below the threshold value of 1700 μS/cm accepted for sugarcane. Despite its high K content, analyses of soils have further showed that after its application, even at 100 m^3/ha, vinasse will have no adverse bearing on the exchangeable Ca status of the soils.

Field trials have often demonstrated that vinasse gives a higher cane yield than with NPK fertilizers alone. Additionally, because of its low heavy metal concentrations, vinasse would not increase the heavy metal concentration in the sugarcane plant. Measurements of the effects on groundwater quality of applying vinasse to soil at high rates in different soil types and rainfall regimes moreover showed that the vinasse would not enhance the leaching loss of N in the form of nitrate. The heavy metals (Cu, Ni and Zn) known to be mobile, when they were detected in drainage water, would remain well below the drinking water limits proposed by the World Health Organization (1 mg/L for Cu, 5 mg/L for Zn and 0.02 mg/L for Ni). Indeed the studies tend to show that application of high rates of vinasse is environment friendly and will not be to the detriment of the soil quality or of the sugarcane plant. In particular vinasse will not lead to any contamination of groundwater under sugarcane fields.

Keywords: Sugarcane industry, nutrients, soil quality, heavy metals, groundwater

Introduction

Sugarcane, a semi-perennial C4 grass species, is cultivated on 25.4 million hectares in more than 100 countries for a total production of some 1.8 billion tonnes of cane (FAOSTAT, 2011). This harvested area places sugarcane in 12th place among 161 crops grown in 2011. As with all sugarcane production systems, the disposal of large volumes of waste materials during processing can be expensive and environmentally threatening. However if alternative uses can be found for the wastes, disposal costs can be avoided and added economic value can be obtained from the conversion of the wastes into co-products. The main co-products of sugar production are molasses and bagasse, which can be re-processed into value added commodities, notably ethanol from the molasses, and electricity, from the burning of bagasse during cogeneration. Figure 1 illustrates the co-products that are obtained in a biomass flexi-factory.

For centuries, sugarcane has widely been used for the production of sugar and, more recently, for the production of biofuels; either in first-generation processes by fermentation of the sugars contained in the juice or molasses, or in second-generation processes by hydrolyzing the cellulose and hemi-cellulose contained in sugarcane fibres (bagasse) or in the leaf matter (Burnquist, 2013). Brazil has shown the way, producing very large quantities, currently over 25 billion liters (25 million m^3) of ethanol per annum with slightly more than half the sugarcane grown in Brazil being devoted to ethanol production (Gosnell, 2011).

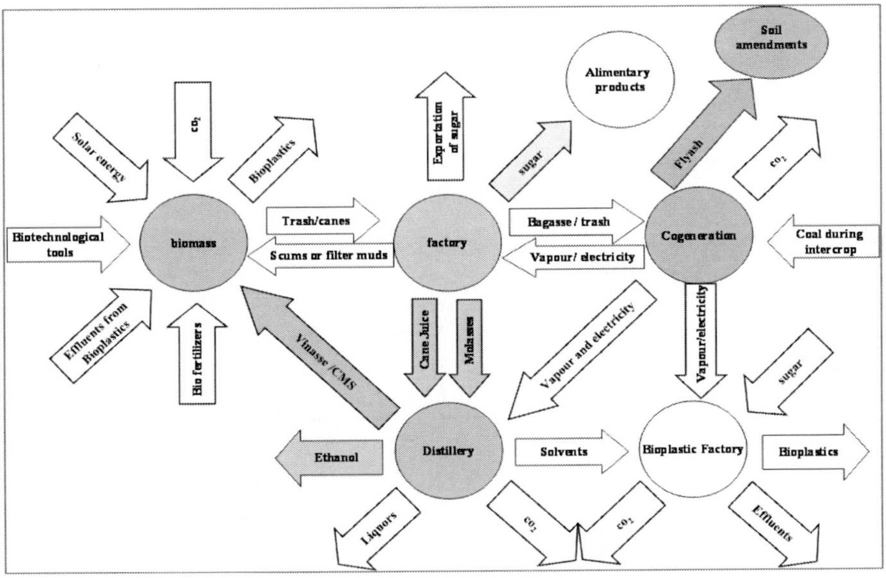

Figure 1. Uses of the co-products of sugarcane within a flexi-factory (Cheeseman, 2004).

The alcohol production generates large quantities of agro-industrial residues, the main one being vinasse, the aqueous effluent of the distillation unit in the sugar-alcohol industry. The quantity of vinasse produced depends on the processing technique employed and also on the alcohol composition, varying between 10-18 litres per litre of alcohol produced (Mello Prado et al., 2013). The vinasse originates from three sugary musts: molasses, mixed musts and juice. Wash water used to clean the fermenters, cooling water blown down and broiler water blown down further contribute to its variability (Mohana et al., 2009). Vinasse has an acid pH around 4.0 to 4.6 and a very high organic content, (5.26%, Tejada, 2010). Its disposal, treatment or further utilisation has therefore to be judiciously established to avoid any subsequent environmental impacts.

Sugar and Alcohol Production

A flow chart representing the major procedures in sugar and alcohol production is shown in Figure 2 below. Sugarcane is first washed to eliminate coarse impurities such as soils and rocks. The juice is obtained by cutting, crushing and saturating the bagasse with water. The raw juice is then filtered and then chemically purified using a clarification process which is designed to remove all soluble and insoluble impurities that will interfere with sucrose extraction. This process consists of the addition of lime, lime and phosphate or lime and sulphur dioxide (sulfitation) all in combination with heat. Lime is added to neutralize the acidity of the juice.

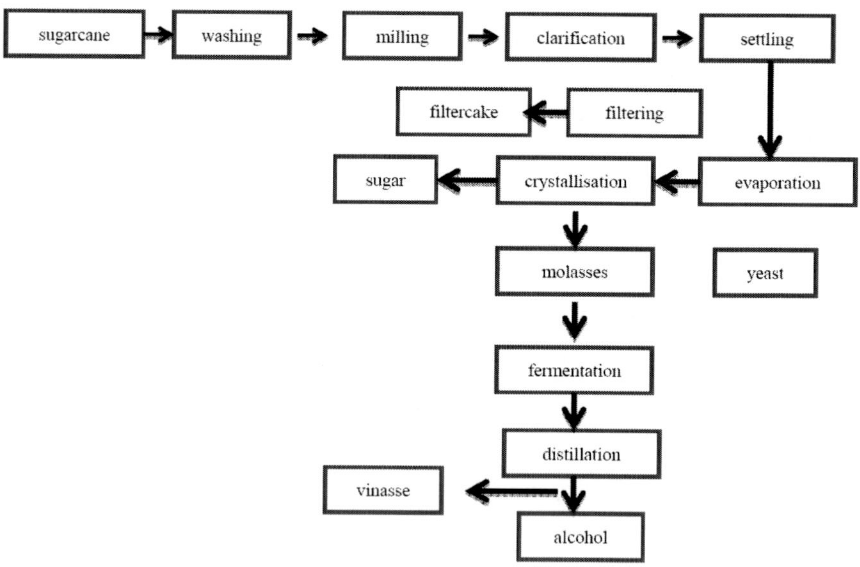

Figure 2. Flow chart illustrating alcohol production (simplified version, Nandy et al., 2002).

The impurities (mud) flocculated during the clarification treatment are separated from the clear juice by sedimentation followed by filtration using rotary drum filters.

Vinasse Composition

Chemically, vinasse composition also varies according to the soil, sugarcane variety, harvesting methods and the industrial process used in the production of ethanol. Table 1 summarizes vinasse composition as stated by various authors. Wash water used to clean the fermenters, cooling water blow down, and boiler water blow down may all be combined with the stillage to contribute to its variability in chemical composition. Its colour, total solid contents and acidity vary according to the type of vinasse processes and treatment. For instance, biological aerobic treatment employing fungi and bacteria can decolorize vinasse (Kumar et al., 1998) whilst physicochemical treatments are effective in removing both colour and COD (Pena et al., 2003). Vinasses contain in general unconverted sugars, non- fermented carbohydrates, dead yeast and a variety of organic compounds (Johnson et al., 2008) and effluents from the distillation of molasses differ from those of sugarcane juice.

Molasses mosto has higher concentrations of organic matter, potassium, calcium, and magnesium. Sugarcane mosto, on the other hand, has considerably lower concentrations of these elements (Robertiello, 1982). In general, vinasse presents a dark color and consists of basically water (93%) and organic solids and minerals (7%). It has high levels of organic matter, but is low in N and P. The main component of vinasse is organic matter in the form of organic acids and cations such as K, Ca and Mg, (Christofoletti et al., 2013).

The concentration of sugars in sugarcane molasses by crystallization and evaporation of the cane juice increases the content of non-fermentable organics which remain in the stillage after fermentation, thereby augmenting the COD and hence increasing the COD/BOD ratio (Wilkie et al., 2000). The organic components of vinasse have been studied by many researchers and the principal low molecular weight components were found to be lactic acid, glycerol, ethanol and acetic acid, (Dowd et al., 1994). Parnaudeau et al., (2008), have observed that the raw material used for fermentation had the greatest effect on the nature and size of the resistant organic pool.

Besides the organic content of vinasse, there are other characteristics which are important from the environmental point of view such as colour and heavy metals. Highly coloured vinasses can have negative impact on the environment if they are released into surface waters where they may disrupt the growth of aquatic flora.

Table 1. Chemical composition of vinasse from sugarcane

Parameter	Silva et al., (2011)	Aguiar et al., (2010)	Ribas et al., (2009)	Khardenavis et al., (2009)	Chang et al., (1990)	Feder et al., (2004)	Resende et al., (2006)	Nandy et al., (2002)	Soobadar (2009) Alcodis	Soobadar (2009) Medine	Goyal et al., (1996)
Total sugars (mg/L)	1800			15750							
Reducing sugars (mg/L)		962		12.5							
Glucose (mg/L)	1400										
Sucrose (mg/L)	400										
Solubility (%)	6.51										
Water content (%)	98.0										
Protein (%)	0.33										
Ash (%)	0.24										
Ether extract (%)	0.04										
Nitrate (g/100 g)	31			0.0000075							
Total N (g/L)		0.7	2.4	1.33	0.003	2.9	0.28	1.7	0.6	1.84	1.2
Total pectin (mg/100g)	91.4										
Tannin (mg/100 g)	80.47										
COD (g/L)	57.5	42	51	124	7.4	103		92			80
BOD (g/L)	1.9	11.3			3			55			50
DO (mg/L)	1.0										
Total solids (g/L)	24.56	6			19			2			70
TDS (ppm)		152126									

Parameter	Silva et al., (2011)	Aguiar et al., (2010)	Ribas et al., (2009)	Khardenavis et al., (2009)	Chang et al., (1990)	Feder et al., (2004)	Resende et al., (2006)	Nandy et al., (2002)	Soobadar (2009) Alcodis	Soobadar (2009) Medine	Goyal et al., (1996)
Potassium(g/L)		2.3	6.1			14.4	1.2	9	4.16	14.58	7
Phosphate (as P) g/L		0.2		0.13	0.00017	0.24	0.1	0.35	0.02	0.20	1
Sulphate (g/L)			1.8	0.00055	0.04		0.44	2.2	0.44	1.54	4
Cl (g/L)				3.5		5.6		6.7			7
Ca (g/L)						2.3	0.33	75	0.74	1.85	
Mg (g/L)						0.0012	0.125		0.65	1.69	
Na (g/L)									0.88	5.85	
Pb (ppm)									12.05	8.98	
Cu (ppm)									14.62	1.21	
Ni (ppm)									4.43	3.56	
Zn (ppm)									10.64	2.49	
Mn (ppm)									61.17	10.22	
Fe (ppt)									1.1	0.04	
pH	7.27	3.95	4.6	4.3					4.26	4.75	

Indeed phenolics such as tannic and humic acids from the feedstock (Sierra-Alvarez et al., 1994), melanoidins from Maillard reaction of sugars with proteins (Chem and Chou, 1993), caramels from overheated sugars (Rivard and Grohmann, 1991) and furfural from acid hydrolysis can all contribute to the colour intensity of vinasse.

The dissolved organic matter fraction of vinasse has been extensively studied by Zarruk et al., (2007) who fractionated the vinasse through dialysis into four molecular sizes namely MW<500, 500<MW<3500, 3500<MW<12000-14000, and MW>12000-14000 Da. They showed that vinasse in fact contains around 70% of the total organic carbon infractions with low molecular weights (MW<3500 Da).The latter has the highest potential for metal binding and thus mobilization of heavy metals down the soil profile towards ground water.

Heavy metals have also been detected in vinasse but as expected their concentrations are low and not significantly different from sugar beet molasses which are used as raw materials in some sugar producing countries, for ethanol fermentation (Diaz et al., 2002). Wilkie et al., (2000) have reported detectable levels of chromium, copper, nickel and zinc in vinasse.

Whilst some heavy metals may be introduced from the feedstock and chemical used during ethanol production, corrosion of piping, tanks and heat exchangers is expected to contribute to the heavy metals in the effluent. Indeed though processing equipment used in acid hydrolysis is often made of corrosion-resistant alloys to withstand the high temperature and acidic conditions of hydrolysis heavy metals contained in these alloys may nevertheless leach into the feedstock during hydrolysis, resulting in their detectable levels in the vinasse.

Vinasse Disposal

The topic of vinasse disposal has been comprehensively covered in the literature (e.g Verma et al., 2011). Some of the disposal routes discussed are mentioned briefly below:

- Vinasse has been disposed through irrigation, which represents one of the simplest methods of disposal provided that sufficient land is available.
- Evaporation and concentration of vinasse to reduce its volume by almost 50% and to facilitate its transport. The resulting concentrated

molasses stillage can either be supplemented with N and P fertilizers and applied to sugarcane fields or it can be used as an animal feed. Additionally since the water content is significantly reduced after evaporation, the dried substrate can be incinerated (Gate, 2000). The ashes can be utilised as fertiliser or, with the addition of chemicals, for the recuperation of K_2SO_4 and $CaSO_4$. The adverse environmental effects of disposal of untreated stillage are thus avoided

- Spray drying of vinasse to produce ethyl concentrate feed
- Vinasse incineration with zero liquid discharge is one of the methods of disposal that is gaining worldwide attention. This new technique which is being implemented in India requires that ethanol plants adopt new technologies for effluent discharge (ISSCT, 2009).
- Composting of vinasse with other wastes is also commonly practised in many sugar producing countries. Diaz et al., (2002) have shown that composting of vinasse with cotton gin wastes can in fact enhance the level of organic matter in soils of low fertility.

Current Vinasse Treatments

The methods currently available for the physico-chemical and biological treatments of vinasse are highly capital intensive and may exceed the cost of ethanol itself. Moreover these treatments do not reduce the BOD to the permissible limits often making it necessary to dilute the treated vinasse before discharge.

Physico-chemical Treatments Current physico-chemical treatments adopted are based on the following processes:

- Ion exchange
- Coagulation-flocculation
- Carbon adsorption

The three methods can only be used as pre-treatments as they do not lead to the desired reduction in BOD. The pre-treated vinasse can then be subjected to biological treatments below.

Biological Treatments

The biological treatments existing make use of the following processes

- Anaerobic digestion
- Anaerobic lagooning
- Ammonification process
- Activated sludge method
- Tricking filters
- Facultative lagoons
- Aerated lagoons
- Bio-methanation
- Bio-composting

The biological treatments can result in 80-95% BOD removal, 65-70% COD removal and significant energy recovery in the form of biogas. However the BOD of biomethanated effluent is still high.

Emerging treatments:

New treatments that are either more cost effective or give more value added to the waste are emerging, for instance :

- Biomass production
- Oxidation
- Electrochemical oxidation
- Evaporation, incineration, combustion
- Membrane technologies- reverse osmosis, electrodialysis, nanofiltration, VSEP membrane technology

These methods for treating vinasse have both their advantages and inconveniences. The treated effluent can be used for irrigation, but most of the time represents another major concern for its safe disposal in the environment. In this context, agricultural recycling of raw vinasse is still the best option to safely dispose of this liquid waste, provided that soil and groundwater quality are not affected beyond threshold limits in terms of pollutants.

IMPACT ON SOIL QUALITY

Soil pH

Soobadar and Ng Kee Kwong, (2012), have shown that in spite of the high acidity of vinasse, only a slight decrease in soil pH could be recorded one month after the application of 100m^3 /ha vinasse as shown in Figure 3. This drop in soil pH was only transitory as in effect a rise in soil pH from 5.00 to 5.47 was observed 12 months after the vinasse application.

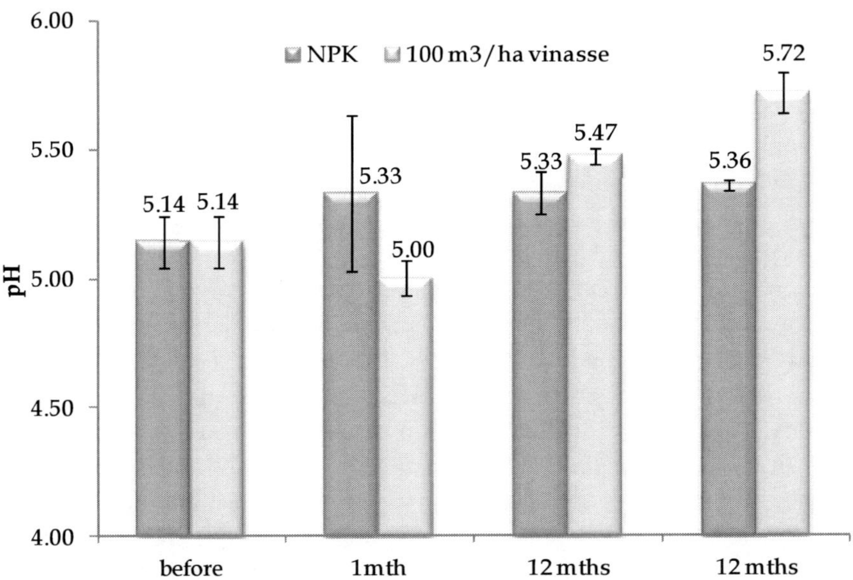

Figure 3. Effect of 100 m^3 /havinasse on pH of soils in Mauritius, mean of 4 sites, (Soobadar and Ng Kee Kwong, 2012).

This rise in soil pH is due to the oxidation and thus elimination of organic acids that contributed to the acidity of the vinasse. Similar observations have been made elsewhere, e.g. by Tejada (2010) who reported a temporary decrease in soil pH following application of vinasse diluted at 10%. Rodella et al. (1983) had previously observed in laboratory experiments that vinasse, apart from increasing the soil concentration of base ions, caused a chelation of the liberated Al ions by the organic compounds present in its soluble fraction thereby leading to an increase in soil pH. Benke (1998) has also noted a rise in soil pH following the application of $150m^3$/ha of vinasse in the oxisols of Brazil.

Jiang et al., (2011), have shown that continuous soil application of vinasse at 75 T/ha for two to three years did not alter significantly the soil pH as compared to the control. Da Silva et al., (2011) have shown that soil acidification occurs initially in the days following concentrated vinasse application to supply an equivalent amount of 270 kg K_2O/ha but soil pH gradually increases with time. They attributed this to the result of complexation reactions that occur between Al^{3+} and colloids of vinasse, reduction reactions that consume hydrogen ions and the increase in exchangeable bases. Yan et al., (1996) have also demonstrated that the effects of vinasse on soil pH are only temporary, returning to their original values after a certain period of time due to H^+ ionization of carboxylic, phenolic acids and alcohols from organic matter present in the vinasse. Bueno et al., (2009), have demonstrated little variation in soil pH after application of vinasse at 160 T/ha in two different soil types, luvisol and vertisol. They have attributed this observation to the acid neutralizing capacity of the soils due to the calcium carbonate and clay contents. Sayed and Elazim, (2002), have reported that fertigation with vinasse had only a slight effect on soil pH and did not affect cane growth and cane yields. The pH of amended soils declined slightly during the study and stayed below the level of 5.0 (Table 2). The literature is therefore full of studies showing that effects of vinasse on soil pH are only temporary, the pH after declining will invariably return to its original value after a certain period of time and is of little consequence to the growth of the sugarcane. For instance, some studies have revealed an initial decline in soil pH following vinasse application and have connected it with the decomposition of incorporated organic matter (Bronick and Lal, 2005; Giller et al., 2001). Products of mineralization and humification such as CO_2 and organic acids are known to intensify leaching of base cations while dissociation of organic compounds may release acidity into the active form (Bertand et al., 2007). Furthermore, ammonium-forming organic materials are potential acid formers in soil, since nitrification generates acidity.

Soil Electrical Conductivity

Vinasse is reported to increase soil salinity under dry conditions where there was practically no leaching of soluble salts (Tejada & Gonzalez 2005). Soobadar and Ng Kee Kwong (2012) indeed noted a significant increase in soil electrical conductivity one month after vinasse application, but this increase was only temporary because 12 months later, the electrical conductivity had reverted to more or less its initial value (Figure 4). In spite of that initial increase it must be noted that the soil electrical conductivity remained well below the threshold value of 1700 µS/cm as proposed for sugarcane (Rhoades & Loveday 1990). From a review of the literature it can be inferred that the high rainfall generally encountered in tropical areas would probably rapidly leach any soluble salts added by the vinasse and return soil salinity to its initial state some12 months later. Madejon et al. (2001) indeed reported a rapid initial rise and a subsequent decrease of soil electrical conductivity with the application of 100 m^3 /ha of vinasse. Bueno et al., (2009), have also observed an increase in soil salinity following fertigation with vinasse. The spatial and temporal variability of salinity are affected by the complex movement of electrons through soil.

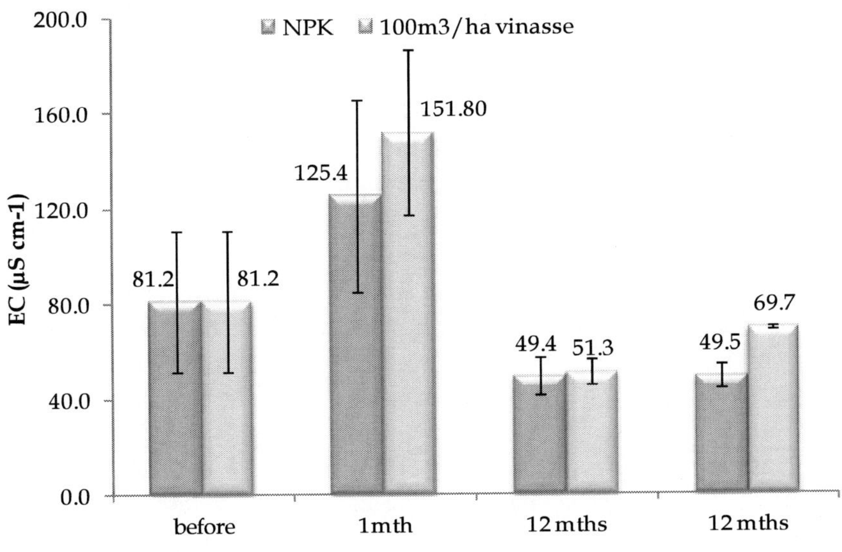

Figure 4. Effect of 100 m^3 /ha vinasse on salinity of soils in Mauritius mean of 4 sites, (Soobadar and Ng Kee Kwong, 2012).

Electrons may travel through the water in soil macropores, along the surfaces of soil minerals (i.e., via exchangeable ions), or through alternating layers of particles and solution (Rhoades et al., 1989).Apart from rainfall, soil properties such as, bulk density, structure, clay and organic matter may also influence soil salinity by vinasse. Hati et al., (2007) demonstrated increased salinity of soils with increasing vinasse application rates. The effluent treated plots showed significantly higher electrical conductivity than in the control (0.47 dS /m) and in the NPK + FYM treatment (0.58 dS/ m). Bueno et al., (2009) also noted an increase in soil salinity following fertigation with vinasse. Thus in certain soils even if the application of vinasse will impart agronomic benefits, the level of application must be within the acceptable limits to avoid development of salinity in the long run.

Soil Organic C

Modern farming practices currently adopted such as intensive cropping, tillage and removal of crop residues, contribute to depletion of soil organic matter in arable land across the world. The addition of organic matter to soils has recently gained further importance as a consequence of worldwide environmental concerns with the contribution of agricultural ecosystems to atmospheric concentrations of greenhouse gasses as well with the potential of arable soils to sequester C. The sink capacity of soil organic matter (SOM) for atmospheric CO_2 is significantly enhanced when degraded soils and impoverished ecosystems such as those in arid and semiarid areas are restored, (Lal, 2004). Bustamante et al., (2010), have shown that winery distillery wastes when applied to sandy-loam and clay-loam soils enhanced the labile organic C pool and soil respiration considerably.

Organic amendments are known to increase the organic matter content of soils both directly by enhancing plant growth and indirectly thereafter through a larger volume of crop residues incorporated (Skowronsaka 2008). Soobadar and Ng Kee Kwong, (2012) have shown that despite the fact that the vinasse used was rich in organic carbon (as much as 12.63 g/l) the increase in soil organic carbon was not as large as expected (from 25.8 to 29.7 mg/ kg soil) after an application of 100m^3 /ha of vinasse, as shown in Figure 5.

This implied that either the major part of the organic carbon added through the vinasse could have been mineralized over the 12 month period or that the vinasse, by increasing soil microbial activity, had stimulated decomposition of the soil organic carbon. As a result, in both instances, soil organic carbon would not be increased very substantially by the application of vinasse. Other authors, for instance Benke et al. (1999) made similar observations and explained that the lower than expected increase in soil organic carbon was due to the leaching of the dissolved organic carbon in the vinasse down the soil profile because of a poor adsorption capacity of the soil for the dissolved organic carbon compounds.

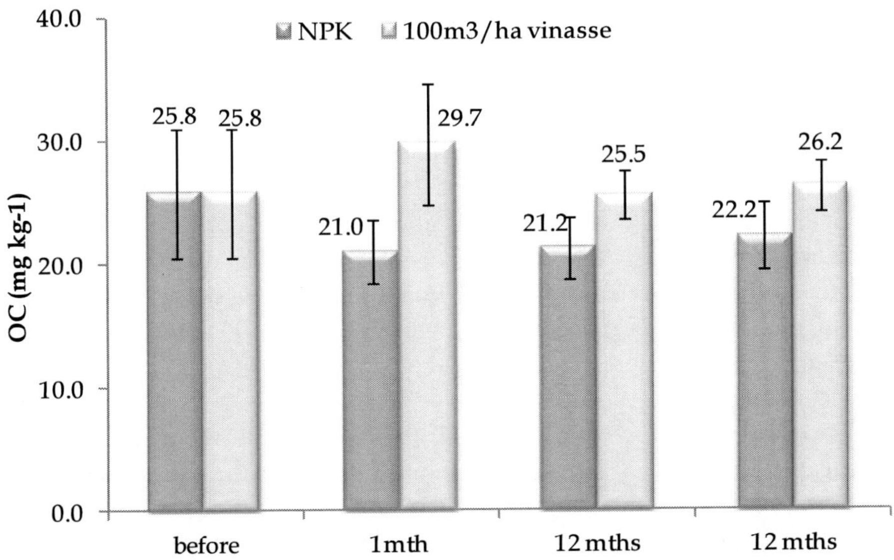

Figure 5. Effect of 100 m^3/ha vinasse on organic C content in soils in Mauritius, mean of 4 sites, Soobadar and Ng Kee Kwong, 2012).

On the other hand, there are just as many reports showing a significant beneficial contribution of vinasse to the organic matter status of soils. Biswas et al., (2009), reported that soils treated with spent wash had improved levels of macro-aggregate-associated carbon and soil organic carbon.

Kaushik et al., (2005), have concluded that long-term irrigation with post methanated effluent in sodic soil improved soil organic carbon, nitrogen, soil dehydrogenase and invertase activities, potassium and phosphorus in the soil, all of which are favorable for plant growth. Tejada and Gonzalez, (2005), have shown that when beet vinasse is applied at low rates (3 and 6 T/ha) under dry land conditions, the organic C of the soil increased, but at higher rates (20 and 40 T/ha) soil microbial biomass and soil structure was negatively affected. Madejon et al., (2001), have reported significant increases in soil organic matter content following repeated applications of composted vinasse on a calcareous loamy sand soil.

From the existing report in the literature it is therefore evident that vinasse contributes towards an increase in soil organic matter status though in many instances, the increases may not be as large as anticipated.

Soil Exchangeable Bases

Though vinasse contains appreciable concentrations of exchangeable bases as shown in Table 1, a notable increase in exchangeable Ca, Mg and Na may not observed following an application of vinasse as demonstrated by Soobadar and Ng Kee Kwong (2012) with 100 m^3/ha of vinasse. But as expected and as illustrated in Table 2 with an of application of 100m^3 /ha of vinasse, the latter being rich in K (average of 937 kg/ 100m^3) would increase soil exchangeable K more significantly than mineral fertilizers immediately or soon(1 month) after application.

However with time e.g.12 months after application, there would be no significant difference in soil exchangeable K between the vinasse treatment and the one receiving mineral fertilizer. The absence of any beneficial increase in soil exchangeable Ca, Mg and Na is understandable and may be explained by the predominance in the vinasse of K which at high concentrations would displace the other cations into soil solution and consequently rendering them more prone to leaching.

As stated by Mazza et al. (1986), the presence of sulphate anions in vinasse would also help to enhance the movement of the base cations down the soil profile. As for the absence of any noted increase in soil exchangeable K 12 months after application of the vinasse, it is often due to the fact that in soils especially those with low cation exchange capacity the large amount of K supplied by the vinasse could not be retained by the soils and had been mostly lost by leaching. This implies that K in vinasse would often exhibit little residual value and to be an effective K source to sugarcane, vinasse would have to be applied at low rates yearly to the sugarcane crop. Indeed Perez et al., (2013) used between 10 and 30 m^3 of vinasse per year and showed that continuous applications of such low doses of vinasse over six years caused a linear increase of exchangeable K in the soil, especially in the first 75 cm with the highest concentration of K found within the first 25 cm from the surface..

Table 2. Effect of 100 m^3 /haof vinasse on soil exchangeable bases ($cmol^+ kg^{-1}$) in Mauritius,mean of 4 sites ± SE, (Soobadar and Ng Kee Kwong, 2012)

Exchangeable bases	1 month after 1st application		12 months after 1st application		12 months after 2nd application	
	Mineral treatment NK	100 m^3/ha Vinasse	Mineral treatment NK	100 m^3/ha Vinasse	Mineral treatment NK	100 m^3/ha Vinasse
Ca	3.27± 0.99	3.11± 0.98	3.18± 1.15	3.34± 0.62	2.02 ± 0.50	2.53 ± 0.55
Mg	1.87± 0.64	2.04± 0.76	1.61± 0.57	1.61 ± 0.51	1.29 ± 0.88	1.57 ± 0.46
K	0.84 ±0.14	1.00 ± 0.25	0.43 ± 0.11	0.51 ± 0.10	0.48 ± 0.15	0.43 ± 0.08
Na	0.38 ± 0.07	0.39 ± 0.07	0.18 ±0.04	0.24 ± 0.06	0.15± 0.04	0.20 ± 0.05

Heavy Metal Content of Soil

One of the most important factors invariably considered before recommending large scale application of organic wastes to sugarcane land is their potential effect on heavy metal accumulation in the soil. In this context, Soobadar and Ng Kee Kwong (2012) showed that total heavy metal concentration in soil would not be affected by high doses of vinasse.

Studies conducted, for instance by Lee & Lee, (2008), have led to the same conclusion which should not be surprising since heavy metal concentrations in the vinasse are in general so low (Table 1) that they should not have had any notable bearing on the heavy metal level in the soils.

Impact on Sugarcane Yield

The effects of vinasse on sugarcane productivity are variable depending on the dosage and methods of its application in the fields. Soobadar and Ng Kee Kwong, (2012), have shown that an application rate of 100 m^3 /ha vinasse gave a better cane yield (84.9 TCH), than mineral fertilizers (77.4 TCH) supplying the same amount of nitrogen. The same observation was found with sugar yield implying that there had not been any difference in sucrose content of sugarcane plant caused by the increasing rates of vinasse. Lee and Lee (2008) have also observed significant increase in cane yield, up to 102%, on soil irrigated with vinasse (diluted with mill's residue water in the proportion of 1:10) compared to soil without vinasse. Yang et al. (2006) reported that application of vinasse improved cane yield and sucrose percentage in cane when compared with mineral fertilizer. De Resende et al., (2006), reported increased cane and sugar yields by 12-13% when vinasse was applied at a rate of 80 m^3 /ha over a 16- year period.

Rodriguez, (2000), has shown increases in plant cane yield when vinasse was applied at 50m^3/ha, and in the first and second ratoons when vinasse was applied at 100m^3/ha. Mo et al., (2008), have also shown that application of vinasse at 45 and 75 T/ ha increased cane yield compared to routine fertilization treatment. Li et al., (2007), have tested a novel method of vinasse application which comprises pre-emergence application of vinasse in furrows followed by covering with plastic film and showed that vinasse applied at 105 T/ha did improve sugarcane yield by 14.92%. Perez et al., (2013), have shown that increasing the doses of vinasse (highest dose of 120 m^3 /ha) led to consistent increases in sugarcane yield. In Brazil, long term application of vinasse, at 150 m^3/ha/yr in sugarcane, has confirmed positive effects on sugarcane productivity, (Prado et al., 2013).

Table 3. Expected concentrations of heavy metals in soils after application of vinasse at two different rates at sites in Mauritius (Soobadar and Ng Kee Kwong, 2012)

	Cu	Ni	Zn	Mn	Pb	Hg
	mg/kg					
Belle Rive						
Before application	*52.76*	*65.54*	*61.01*	*167.6*	*49.44*	*0.33*
50 m^3/ha vinasse	53.00	65.61	61.19	168.62	49.64	0.33
100 m^3/ha vinasse	53.25	65.69	61.36	169.64	49.84	0.33
Union Park						
Before application	*106.06*	*312.84*	*117.12*	*793.7*	*54.91*	*0.41*
50 m^3/ha vinasse	106.30	312.91	117.30	794.72	55.11	0.41
100 m^3/ha vinasse	106.55	312.99	117.47	795.74	55.31	0.41
Pamplemousses						
Before application	*118.15*	*405.2*	*188.5*	*2555.90*	*53.75*	*0.08*
50 m^3/ha vinasse	118.39	405.27	188.68	2556.92	53.95	0.08
100 m^3/ha vinasse	118.64	405.35	188.85	2557.94	54.15	0.08
Médine						
Before application	*68.24*	*138.08*	*150.22*	*1335.07*	*35.43*	*0.25*
50 m^3/ha vinasse	68.48	138.15	150.40	1336.09	35.63	0.25
100 m^3/ha vinasse	68.73	138.23	150.57	1337.11	35.83	0.25

On the other hand, there is one study by Tejada & Gonzalez, (2005) which reported that high doses of vinasse (40 T/ha) can induce a decline in crop yield by creating anaerobic conditions in the soil. Apart from that report of Tejada & Gonzalez (2005) there exists a general consensus in the literature that there is no deleterious effect of vinasse on sugarcane production and therefore there is no valid reason why vinasse cannot be used as a source of K for sugarcane and that rates of up to 100-120 m^3/ha can in fact be safely applied to sugarcane.

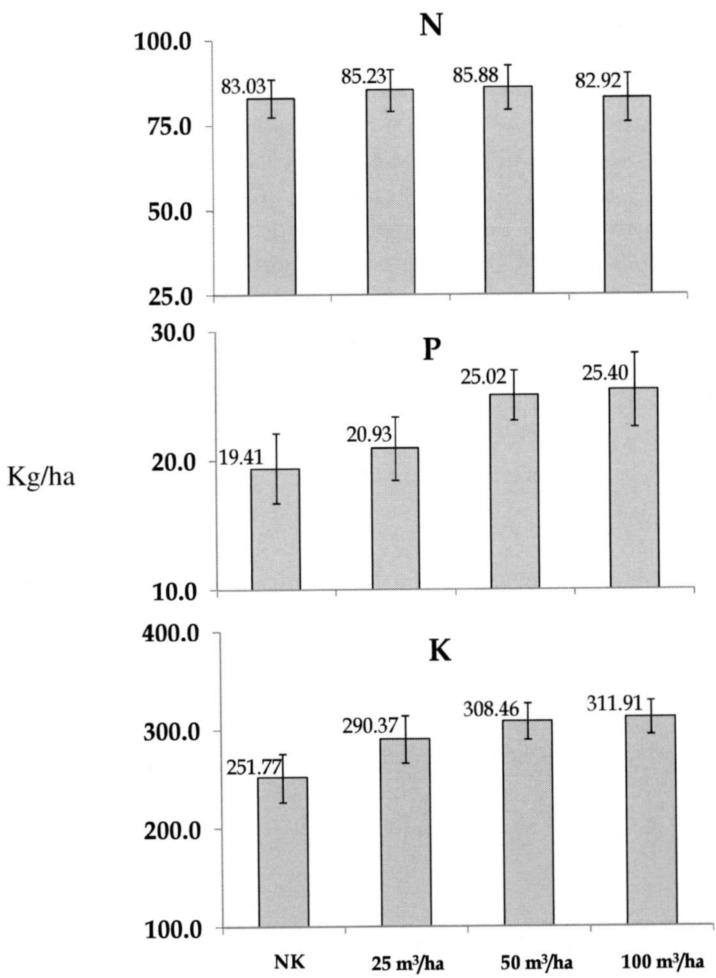

Figure 6. NPK uptake by sugarcane (mean of plant cane and first ratoon) in Mauritius after application of mineral NK fertilizers and vinasse at 25, 50 and 100 m^3 /ha. (Soobadar and Ng Kee Kwong, 2012).

Impact on NPK Uptake by Sugarcane

As summarized by Pautremat et al. (2003), a major fraction of N present in vinasse is in organic form and would hence be less bioavailable to the sugarcane than the N in mineral fertilizers.

Yet Soobadar and Ng Kee Kwong, (2012), demonstrated that there was no significant difference in the uptake of N by the sugarcane plant when vinasse at 100 m^3 /ha was compared with mineral fertilizers (Figure 6). The lack of any difference in N uptake was not surprising as the amount of N applied was the same.

The results also showed that vinasse would not have any adverse effects on N uptake in spite that part of the N present being initially in an organic state. As further shown in Figure 6, vinasse tended to enhance P uptake by sugarcane particularly when applied at high rates of 50 to 100 m^3 /ha. This enhanced P nutrition by in the presence of high rates of vinasse is also not surprising as some 44 kg P_2O_5 /ha had been supplied by the vinasse in comparison to the mineral fertilizer treatment where no P was applied since soil testing, had prior to planting, indicated adequate P in the soils for sugarcane.

The impact of vinasse application on NPK uptake has also been studied in China using 75 T /ha vinasse (Su et al., 2011). It was observed that the vinasse treatment decreased N content and increased P and K contents in sugarcane plants compared to controls.

The conclusion to be drawn regarding uptake of nutrients by sugarcane when vinasse had been discharged in the fields is that the absorption of the nutrients are more related to their amount or concentration than to their chemical form in the vinasse.

Effect of Vinasse on Heavy Metal Uptake by Sugarcane

Since the heavy metal concentrations in vinasse are low and have no significant bearing on heavy metal concentration in soil, no difference in heavy metal uptake should be observed between sugarcane receiving vinasse and that to which mineral fertilizer was applied. As shown by Soobadar & Ng Kee Kwong (2012), 100 m^3/ha of vinasse contributed only 1.46 kg Cu /ha, 0.44 kg Ni/ ha, 1.06 kg Zn/ ha, 6.123 kg Mn/ ha and 1.21 Pb kg/ ha. These were negligible amounts and did not have any impact on the concentrations of the heavy metals in the aboveground sugarcane. Lahlah, (2009) has however drawn attention to the fact that on account of its high levels of organic matter, vinasse when spread on soils may lead to anoxia, to alterations of mineral forms (i.e. complexation and solubilization) and to metal mobilization, all leading to potential heavy metal uptake by crops.

IMPACT ON GROUNDWATER QUALITY

Electrical Conductivity

Vinasse even when applied at 100m^3 /ha would not increase the electrical conductivity of groundwater as much as mineral NPK fertilizers (Soobadar & Ng Kee Kwong, 2013). Any increase in the electrical conductivity that vinasse can cause will not undermine groundwater quality up to the point where its consumption will pose a threat to human health as the WHO drinking water guideline for electrical conductivity is 500 μS/cm, a value which was not exceeded as noted by Soobadar and Ng Kee Kwong (2013) in their study with vinasse at100m^3 /ha.

Impact on Nitrate-N

A concentration of nitrate-N in drinking water exceeding 10 mg N-NO^-_3/L is not recommended by the World Health Organization (WHO) because of the possible carcinogenic effect of nitrate on humans. In humid tropical countries where the soils contain significant proportions of kandites or oxides of Al and Fe, adsorption of nitrate on the surfaces of the soils has been postulated to be an effective mechanism minimizing the leaching of N (Mulla and Strock, 2008).

In addition, biological immobilization by both plants and microbes has been reported to further minimize N leaching (Myrold and Bottomley, 2008). An intense microbial activity induced by the addition of labile carbon in the vinasse would further contribute towards an even more efficient mitigation of leaching of nitrate. Orlando et al. (1996) confirmed a decrease in amount of N leached when vinasse was applied at 600 m^3/ha However as illustrated in Figure 7, though the cumulative amount of N leached may be decreased, concentrations higher than 10 mg N-NO_3^-/L may on occasions be observed when vinasse had been applied e.g. at 100 m^3/ha.

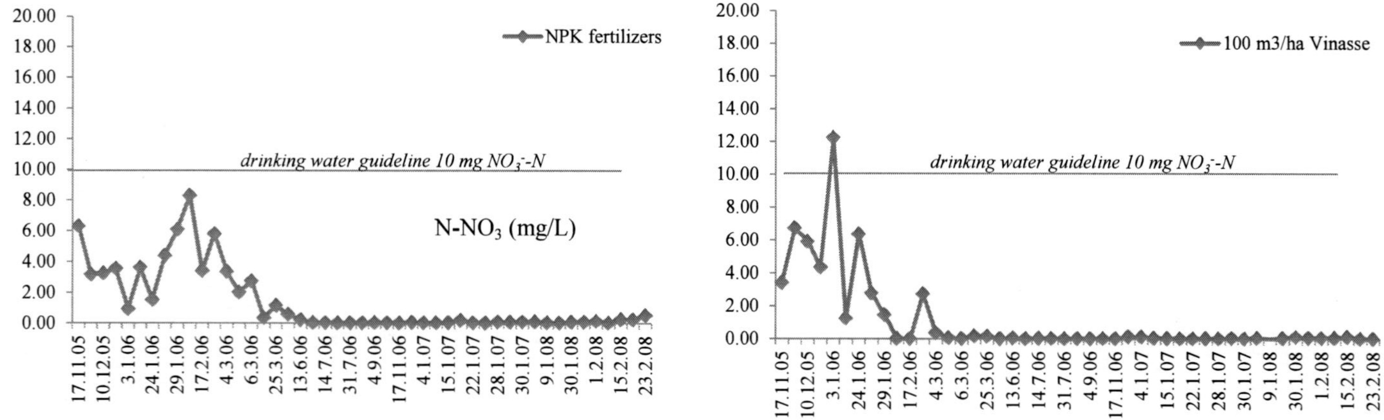

Sampling dates of leachates.

Figure 7. Impact of 100 m^3/ha of vinasse on nitrate-N concentrations of drainage water collected at 1m depth from lysimeters (Soobadar & Ng Kee Kwong, 2013).

The high concentration of the nitrate found in the drainage water with vinasse support the view that there could be a re-organization of the N in the vinasse when applied to the soil resulting in the organic N it contained to be partially or fully mineralized.

Heavy Metals

As the concentrations of heavy metals Cd, Hg, Pb, Ni, Zn and Cu in vinasse are low, their concentrations in drainage water would be low. As shown by Soobadar and Ng Kee Kwong (2013) the concentrations of the heavy metals in water draining at one metre depth remained mostly below detection limits of the measuring atomic absorption spectrometer even when vinasse was applied at 100 m^3/ha.

Whilst trace amounts of Zn (190 μg/L), Cu (3.67 μg/L) and Ni (17.59 μg/L) were detected these concentrations were not significantly different than those in the drainage water coming plots receiving only mineral fertilizers and they were well below the drinking water guidelines of 5000 μg Zn/L, 1000 μg Cu/L and 20 μg Ni/L.

Conclusion

Vinasse which is produced in large quantities by ethanol agro industries in sugar producing countries has great potential for agricultural use. It is increasingly being used in fertigation in agricultural areas, substituting for mineral K fertilizers and supplying water, organic matter and other mineral nutrient elements to sugarcane and other crops This review has provided evidence that vinasse, far from being a nuisance and a waste, is a valuable resource to reduce production costs and enhance sugarcane productivity.

A review of the various studies carried out in the world shows that vinasse can increase sugarcane yield significantly as compared to mineral NPK fertilizers, and that its application at high rates will not affect soil properties, the quality of the sugarcane and that of groundwater, thus dispelling doubts that the sustainability of the production capacity of soils will be negatively affected by disposal of vinasse in soils.

REFERENCES

[1] Aguiar MM, Ferreira LFR and Monteiro RTR. (2010). Use of vinasse and sugarcane bagasse for the production of enzymes by lignocellulolytic fungi. *Brazilian archives of biology and technology*, 53 (5): 1245-1254.

[2] Benke, MB. (1999). Characterization and interaction of sugarcane industry residues with soil, kaolinite and Fe-oxides. PhD Thesis, University of Saskatchewan, Saskatoon, Canada.

[3] Bertrand I, Delfosse O and Mary B (2007) "Carbon and nitrogen mineralization in acidic, limed and calcareous agricultural soils: apparent and actual effects", Soil Biology & Biochemistry, 39, 276-288.

[4] Bioenergy for Sustainable Development and Global Competitiveness, Available at: www.carensa.net.

[5] Biswas AK, Mohanty M, Hati KM and Misra AK. (2009). Distillery effluents effect on soil organic carbon and aggregate stability of a Vertisol in India

[6] Bronick CJ and Lal R (2005) "Soil structure and management: a review", Geoderma, 124, 3-22.

[7] Bueno PC, Rubi JAM, Gimenez RG and Ballesta RJ. (2009). Impacts caused by the addition of wine vinasse on some chemical and mineralogical properties of a Luvisol and a Vertisol in La Mancha Central Spain. *Journal of Soil Sediments*, 8pp.

[8] Burnquist WL. (2013). Sugarcane research and development: a view from the private sector. *Proceedings of International Society of Sugarcane Technologists,* Vol. 28, 2013.

[9] Bustamante MA, Said-Pullicino D, Paredes C, Cecilia JA and Moral R. (2010). Influences of winery-distillery waste compost stability and soil type on C dynamics in amended soil. *Waste Management*, 30: 1966-1975.

[10] Chang L. J., P.Y. Yang and S.A. Whalen (1990). Management of sugarcane mill wastewater in Hawaii. *Water Science Technology* 22(9), 131-140.

[11] Cheeseman O D. (2004). Environmental impacts of sugar production. Crownwell Press, UK, 151-172.

[12] Chem JCP and Chou CC. (1993). Cane sugar handbook: a manual for cane sugar manufacturers and their chemists. 12th Ed, New York, Wiley, 1993.

[13] Christofoletti CA, Escher JP, Correia JE, Marinho JFU and Fontanetti CS. (2013). Sugarcane vinasse: environmental implications of its use. *Waste Management*, 33: 2752-2761.

[14] Da Silva A, Rosetto R, Bombecini J, Piemonte M and Muraoka T. (2011). Nitrogen mineralization from vinasse. In 4th IAPSIT proceedings 'Balancing sugar and energy production in developing countries: sustainable technologies and marketing strategies. India 21-25 November 2011, 267-274.

[15] De Resende, S A, Xavier R P, Oliveira de OC, Urquiaga S, Alves JR and Boddey RM. (2006). Long term effects of preharvest burning and nitrogen and vinasse applications on yield of sugarcane and soil carbon and nitrogen stocks on a plantation in Pernambuco, NE, Brazil. Plant and Soil, 281, 339-351.

[16] Diaz MJ, Madejon E, Lopez F, Lopez R and Cabrera F. (2002). Composting of vinasse and cotton gin waste by using two different systems, Resources, *Conservation and Recycling*, 34: 235– 248.

[17] Dowd MK, Johansen SL, Cantarella L, Reilly PJ. (1994). Low molecular weight organic composition of ethanol stillage from sugarcane molasses, citrus wastes and sweet whey. *Journal of Agricultural Food Chemistry*, 42: 283-288.

[18] Feder F, Saint Macary H, Gosme M. (2004). Recyclage de déchets par la canne à sucre: essais avec des vinasses de distillerie. Actes du séminaire, 1-2 Juillet 2004, (Recycling of waste sugarcane : trials with distillery slops . Proceedings of the Seminar , 1-2 July 2004) Montpellier, France, 69pp.

[19] Gate (2000). Technical information W4e, gtz, gate Information Service / gtz, PO Box 5180, 65726. Eschborn, Germany, http://www.gtz.de/gate/gateid.afp.

[20] Giller KE, Cadisch G and Palm C (2001) "The North-South divide! Organic wastes, or resources for nutrient management" in *Book of abstracts: 11-th Nitrogen Workshop, 9 – 12. 09. 2001, Reims, France*, 11-12.

[21] Gosnell J. (2011). Biomass management: ethanol and electricity generation. In Ed. Ian Meyer 'Good management practices manual for the cane sugar industry'. Publisher: PGBI house, South Africa, 447-475.

[22] Goyal S K., R. Seth and B.K. Handa (1996). Diphasic fixed-film biomethanation of distillery spentwash, *Bioresource Technology* 56, 239-244.

[23] Hati KM, Biswas AK, Bandyopadhyay KK and Misra AK. (2007). Soil properties and crop yields on a vertisol in India with application of distillery effluent. *Soil and Tillage Research*, 92: 60-68.

[24] ISSCT (2009). Co-Products workshop, 15-10 March 2009, Coimbatore, India, Available at: http://www.issct.org/past-workshops/co-report09.html.

[25] Jiang ZP, Rui Li Y, Wei GP, Liao Q, Su TM, Meng YC, Zhang HY and Lu CY. (2011). Effect of long term vinasse application on physicochemical properties of sugarcane field soils. In 4th IAPSIT proceedings 'Balancing sugar and energy production in developing countries: sustainable technologies and marketing strategies. India 21-25 November 2011, 287-291.

[26] Kaushik A, Nisha R, Jagjeeta K and Kaushik CP. (2005). Impact of long and short term irrigation of a sodic soil with distillery effluent in combination with bioamendments. *Bioresource Technology* 96 : 1860–1866.

[27] Khardenavis A.A., A.N. Vaidya, M. S. Kumar and T. Chakrabarti (2009). Utilization of molasses spent wash for production of bioplastics by waste activated sludge Waste Management, *J Ind Microbiol Biotechnology,* 31, 2558-2567.

[28] Knoth de Zarruk K, Scholer G, Dudal Y. (2007). Fluorescence fingerprints and Cu2+-complexing ability of individual molecular size fractions in soil- and waste-borne DOM. *Chemosphere* 69, 540-548.

[29] Kumar V, Wati L, Nigam P, banat IM, Yadav BS, Singh D and Marchant R. (1998). Declorisation and biodegradation of anaerobically digested sugarcane molasses spent wash effluent from biomethanation plants by white rot fungi. *Proceedings of Biochemistry*, 33: 83-85.

[30] Lal, R. (2004). Soil carbon sequestration to mitigate climate change. *Geoderma* 123, 1–22.

[31] Lahlah J, Renault P, Cazevieille P, Buzet A, Hazemann J-L and Womes M. (2009). Geochemical Resilience of a Ferralsol Subjected to Anoxia and Organic Matter Amendment. *Soil Science Society of America Journal* 73:1958; doi:10.2136/sssaj2008.0302.

[32] Lee T S G and Lee L L. (2008). Bio recycling of distillery effluent to improve sugarcane productivity. In Li Y R, Nasr M I, Solomon S, Rao G P (Eds.) ''Meeting the challenges of sugar crops and integrated industries in developing countries'', Proceedings of the International Conference, Al Arish, Egypt, 11-14th September 2008: 636- 639.

[33] Madejon, E, Lopez R, Murillo JM and Cabrera F. (2001). Agricultural use of three (sugar-beet) vinasse composts: effects on crops and chemical properties of a Cambian soil in the Quadalquivir river valley. *Agriculture Ecosystem Environment*, 84, 53-65.

[34] Mazza, J.A., Dematê, J.L.I. & Orlando, F.J. (1986). The effects of repeated applications of vinasse on the properties of some soils. *Proceedings of the International Society for Sugarcane Technologists* 19: 108-118.

[35] Mello Prado de R, Caione G and Campos CNS. (2013). Filter cake and vinasse as fertilizers contributing to conservation agriculture. *Applied and Environmental Soil Scienc*e, Vol 2013, 8 pages.

[36] Mo YC, Ye YP, Gui J, Rui Li Y and Yang LT. (2008). Effect of vinasse on photosynthesis, nitrogen metabolism and yield of sugarcane. In Li Y R, Nasr M I, Solomon S, Rao G P (Eds.) ''Meeting the challenges of sugar crops and integrated industries in developing countries'', Proceedings of the International Conference, *Al Arish, Egypt*, 11-14th September 2008: 484-488.

[37] Mulla DJ and Strock JS. (2008). Nitrogen transport processes in soil. In: Schepers J.S. and Raun W.R (Eds.). Nitrogen in Agricultural Systems. *Agronomy Monograph* No 49: 361-400.

[38] Myrold DD and Bottomley PJ. (2008). In: Schepers J.S. and Raun W.R. (eds). Nitrogen in Agricultural Systems. J.S. Schepes and W.R Raven (eds.) *Agronomy Monograph* No 49: 157-172.

[39] Nandy T., S. Shastry and S.N. Kaul (2002). Wastewater management in a cane molasses distillery involving Bioresource recovery, *Journal of Environmental Management* 65, 25-38.

[40] Orlando FJ, Bittencourt V and Alves MC (1996). Vinasse application in a Brazilian sandy soil and nitrogen water table pollution. *Sociedade dos Técnicos Açucareiros do Brasil* (STAB), 13(6):9–13.

[41] Parnaudeau V, Condom N, Oliver R, Cazevieille amd Recous S. (2008). Vinasse organic matter quality and mineralization potential, as influenced by raw material, fermentation and concentration processes. *Bioresource Technology*, 99: 1553-1562.

[42] Pautremat N, Renault P, Cazevieille P and Marlet, S. (2003). 'contribution of glutids, proteids and lipids to soil geochemical evolution following vinasse application: Impact on metal mobilization. PhD Thesis. University of Montpellier, France.

[43] Pena M, Coca M, Gonzalez G, Rioja R and Garcia MT. (2003). Chemical oxidation of wastewater from molasses fermentation with ozone. *Chemosphere,* 51: 893-900.

[44] Perez O, Acan J, Ufer C and Alfaro MA. (2013). Nitrogen response in sugarcane and changes in chemical and biological properties in an andisol due to vinasse application in Guatemala. *Proceedings of International Society of Sugarcane Technologists,* Vol. 28, 10pp.

[45] Rhoades JD, Manteghi NA, Shouse PJ and Alves WJ (1989) Soil electrical conductivity and soil salinity: new formulations and calibrations. *Soil Science Society of America Journal* 53:433–439.

[46] Rhoades J D, Loveday J. (1990). Salinity in irrigated agriculture. *In*: ''Irrigation of Agricultural Crops'' Stewart B A, Nielsen D R. (Eds.) Agronomy 30, ASA, CSSA and SSSA, USA, Madison WI, pp 1089-1142.

[47] Rivard CJ and Grohmann K. (1991). Degradation of furfural (2-furaldehyde) to methane and carbon dioxide by an anaerobic consortium. Applied *Biochemistry and Biotechnology*, 28/29: 285-295.

[48] Robertiello A. (1982). Upgrading of agricultural and agro industrial wastes: the treatment of distillery effluents (vinasses) in Italy. *Agricultural Wastes* 4, 387– 395.

[49] Rodella AA, Zambello JE. and Orlando FJ. (1983). Effects of vinasse added to soil on pH and exchangeable aluminium content. *Proceedings of the International Society of Sugarcane Technologists,* 18, 189-214.

[50] Rodriguez JGO. (2000). Effects of vinasse on sugarcane productivity. Rev. Fac. Agron. (LUZ), 17: 318-326.

[51] Rui Li Y, Zhu QZ, Wang WZ and Solomon S. (2007). Pre-emergence application of vinasse on sugarcane growth and sugar productivity in China. *Sugar Technology*, 9 (2,3): 160-165.

[52] Sayed A and Elazim YA. (2002). Agronomic evaluation of fertilizing efficiency of vinasse. 17th WCSS, 14-21 august 2002, Thailand. 6pp.

[53] Sierra –Alvarez R, Feld JA, Kortekass S, Lettinga G. (1994). Overview of the anaerobic toxicity caused by organic forest industry wastewater pollutants. *Water Science Technology*, 29, 5/6 :353-363.

[54] Silva A.S., H. Inoue., T. Endo., S. Yano., and E. P. S. Bon (2010). Milling pretreatment of sugarcane

[55] bagasse and straw for enzymatic hydrolysis and ethanol fermentation. *Bioresource Technology* 101 (19), 7402-7409.

[56] Skowronska M. (2008). Beet vinasse effects on selected soil properties. 4th European Bioremediation conference, 5pp.

[57] Soobadar A., (2009). Agronomic and environmental impacts of the application of vinasse and coal ash on sugar cane fields in Mauritius., Phd. Thesis, University of Avignon, France.

[58] Soobadar A and Ng Kee Kwong KF. (2012). Impact of high rates of vinasse on some pertinent characteristics and on sugarcane yield in Mauritius. *Journal of Sustainable Agriculture*, 36: 36-53.

[59] Soobadar A and Ng Kee Kwong KF. (2013). Impact of fertilization of sugarcane with high rates of vinasse on groundwater quality in Mauritius. *Proceedings of the International Society of Sugarcane Technologists,* 28, 11pp.

[60] Su TM, Rui Li Y, Wei GP, Jiang ZP, Liao Q, Zhu SB. (2011). Macronutrients absorption and surface runoff losses under different fertilizing treatments in sugarcane fields. *Journal of Soil Science & Development,* 9, 1: 1-7.

[61] Tejada M and Gonzalez JL. (2005). Beet vinasse applied to wheat under dryland conditions affects soil properties and yield. *European Journal of Agronomy,* 23: 336-347.

[62] Tejada, M. (2010). Use of vinasse for soil reclamation and its impact on elemental loads in vertisol soil and groundwater. *Proceedings of the International Society of Sugarcane Technologis*ts, 27, 7 pp.

[63] Verma AK, Raghukumar C and Naik CG (2011). A novel hybrid technology for remediation of molasses-based raw effluents, *Bioresource Technology,* 102, 2411–2418.

[64] Wilkie AC, Riedesel K J and Owens J M. (2000). Stillage characterization and anaerobic treatment of ethanol stillage from conventional and cellulosic feedstocks. *Biomass and Bioenergy*, 19, 63-102.

[65] Yan F, Schubert S and Mengel K. (1996). Soil pH changes during legume growth and application of plant material. *Biology and Fertility of Soils*, 23: 236-242.

[66] Yang LT and Rui LI Y. (2008). Future research priorities of Chinese sugar industry. In IAPSIT proceedings ‘Meeting the challenges of sugar crops and integrated industries in developing countries’. Egypt, 11-14 September 2008, 764-770.

[67] Yang RL, Zhu QZ, Wang WZ and Solomon S. (2006). Multi location trials to evaluate a novel system of vinasse application in sugarcane field and its impact on cane growth and sugar productivity. *Proceedings International Symposium on Technologies to Improve Sugar Productivity in Developing Countries*. Guilin PR, China, pp. 131-139.

INDEX

B

C

D

E

F

G

H

I

J

K

L

M

N

O

P

Q

R

S

T

U

V

W

X

Y

Z